Computational Molecular Science and Engineering Forum 2014

Core Programming Area at the 2014 AIChE Annual Meeting

Atlanta, Georgia, USA
16-21 November 2014

ISBN: 978-1-5108-1254-3

Printed from e-media with permission by:

Curran Associates, Inc.
57 Morehouse Lane
Red Hook, NY 12571

Some format issues inherent in the e-media version may also appear in this print version.

Printed by Curran Associates, Inc. (2015)

For permission requests, please contact AIChE
at the address below.

AIChE
120 Wall Street, FL 23
New York, NY 10005-4020

Phone: (800) 242-4363
Fax: (203) 775-5177

www.aiche.org

Additional copies of this publication are available from:

Curran Associates, Inc.
57 Morehouse Lane
Red Hook, NY 12571 USA
Phone: 845-758-0400
Fax: 845-758-2634
Email: curran@proceedings.com
Web: www.proceedings.com

TABLE OF CONTENTS

380074 Effect of Deviation from Ideality of Different Pore Geometries for Zero Order Electrochemical Reactions

Abstract:

The study of realistic catalytic pore domains in electrochemical systems is vital in determining the region of parameters that are acceptable for describing, accurately, chemical processes taking place within modelling domains. Models are useful in predictions of system behaviors and assessing experimental results relevant to electrochemical applications such as batteries, fuel cells. A "deviation" from ideal geometry of porous domains for both rectangular and cylindrical domains are analyzed and compared with their ideal geometries. These systems are studied by employing the microscopic scale species mass balancing conservation equation with diffusion and zero order electrochemical reaction present. The equation is scaled up to macroscopic level by employing area averaging-based approach to obtain a system level information. The results obtained from the macroscopic level identified potential deviations in the concentration profile. This study shows the impact of irregular geometry (which is more realistic in analyzing porous systems) on using the often assumed ideal geometry and quantifies useful guidelines for modelling these complex systems.

385166 Design of a Novel Polymeric Acid Catalyst for Biomass Hydrolysis

Abstract:

A novel polymeric solid acid catalyst consisting of dual functional chains for biomass hydrolysis was successfully synthesized. A poly (styrene sulfonic acid) (PSSA) polymer chain is immobilized on a substrate and used to catalyze biomass hydrolysis. A neighboring poly (vinyl imidazolium chloride) ionic liquid (PIL) polymer chain is grafted to help solubilize lignocellulosic biomass and enhance the catalytic activity. The immobilized polymeric catalyst demonstrates over 97% total reducing sugar (TRS) yield in ionic liquid. This solid acid catalyst is highly efficient, stable, reusable and environmentally friendly. In order to understand its catalytic activity and further optimize its performance, the acidity of PSSA chains as well as the interactions among the PSSA, PIL polymer chains and cellulose were investigated using combined quantum mechanical (QM) and classical molecular dynamics (MD) simulations. In addition, the interaction free energies were determined using metadynamics (MTD) simulations. Our results demonstrate that polymer chain length and chain density play a crucial role in the catalytic activities of this polymer catalyst. Additional improvement for catalytic activity by substitution or copolymerization was suggested. This solid acid catalyst can be tuned and potentially be used for many other catalytic conversions.

Modeling the Production of Aromatics from Biomass in Solid Acid Zeolites

Ryan Patet, Stavros Caratzoulas, Dionisios G. Vlachos

Center for Catalysis and Energy Innovation (CCEI), Department of Chemical and Biomolecular Engineering, University of Delaware, 221 Academy Street, Newark, DE 19716

The production of value-added fuels and chemicals from lignocellulosic biomass is an active area of research motivated by a supply of unused biomass material, desire for independence from fossil fuels, and reduction in environmental impact from the production and utilization of these products.[1] Aromatic chemicals such as benzene, toluene, and xylene, which are currently manufactured by the catalytic reforming of naphtha, are produced in high volumes for use, for example, in the polymer industry.[2] Production of these high volume chemicals from biomass would help to meet the aforementioned.

In one specific example, production of p-xylene from biomass-derived 2,5-dimethylfuran (DMF) and ethylene has been achieved with high selectivity to the desired aromatic product using solid acid zeolite catalysts through a two-step reaction process.[3,4] The DMF and ethylene initially undergo a Diels-Alder cycloaddition, followed by a dehydration of the resulting oxanorbornene intermediate to form the final aromatic product.[4] The ability of solid acid zeolites to catalyze this reaction sequence has been shown to be due to the ability of Brønsted acids to protonate the intermediate and lower the reaction barrier of the dehydration step.[4]

The production of p-xylene from DMF and ethylene has been successful, but applying this approach to benzene and toluene from furan and 2-methylfuran, respectively, has resulted in lower yields due to the formation of undesired polymer byproducts.[5] By tailoring the acidic environment of the solid-acid zeolites, selectivity towards the desired aromatic products over the undesired polymer byproducts could be achieved. Being able to model this acidic environment in order to understand the chemistry and make predictions about new catalytic environments becomes difficult computationally, due to the number of atoms needed to define the pore of the system. The size constraints of the system can be managed using the ONIOM method, in which a small model portion of the system is treated using expensive quantum mechanical density functional theory (DFT) calculations, while the remainder of the system is treated with a less-expensive molecular mechanics model.[6,7]

In this study, ONIOM models have been developed which can accurately describe the acidic environment found within the pores of different zeolite structures. Comparison to a large dataset of experimental data shows that our model has predictive ability. With these extensively assessed models, the effect of different zeolite frameworks (e.g. H-Beta vs. H-Y), as well as the nature of the acid sites have been investigated for their effect in the chemistry of the production of aromatics. We will discuss how the reactants, intermediates, and products are stabilized within the pore environments of zeolites and use this understanding toward the design of catalysts capable of improving the selectivity to the desired aromatic products.

1. Climent MJ, Corma A, Iborra S. Converting Carbohydrates to Bulk Chemicals and Fine Chemicals Over Heterogeneous Catalysts. *Green Chem.* 2011;13:520-540.
2. Zukauskas V, Perego C. Aromatics. In: Amadei C, ed. *Encyclopaedia of Hydrocarbons.* Vol II. Rome: Istituto della Enciclopedia Italiana; 2006:591-614.
3. Williams CL, Chang c-C, Do P, et al. Cycloaddition of Biomass-Derived Furans for Catalytic Production of Renewable p-Xylene. *ACS Catal.* 2012;2:935-939.
4. Nikbin N, Do PT, Caratzoulas S, Lobo RF, Dauenhauer PJ, Vlachos DG. A DFT Study of the Acid-Catalyzed Conversion of 2,5-Dimethylfuran and Ethylene to p-Xylene. *J. Catal.* 2013;297:35-43.
5. Wang D, Osmundsen CM, Taarning E, Dumesic JA. Selective Production of Aromatics from Alkylfurans over Solid Acid Catalysts. *ChemCatChem.* 2013;5:1-8.
6. Svensson M, Humbel S, Froese RDJ, Matsubara T, Sieber S, Morokuma K. ONIOM: A Multilayered Integrated MO+MM Method for Geometry Optimizations and Single Point Energy Predictions: A Test for Diels-Alder Reactions and Pt(P(t-Bu)(3))(2)+H-2 Oxidative Additiion. *J. Phys. Chem.* 1996;100(50):19357-19363.
7. Dapprich S, Komaromi I, Byun KS, Morokuma K, Frisch MJ. A New ONIOM Implementation in Gaussian98 - Part 1: The Calculation of Energies, Gradients, Vibrational Frequencies and Electric Field Derivatives. *J. Mol. Struct. (Theochem).* 1999;461:1-21.

Mechanistic Insights into Fructose dehydration to 5-(hyroxymethyl)furfural

Vinit Choudhary, Liu Yang, Georgios Tsilomelekis, Stavros Caratzoulas, Vladimiros Nikolakis, Stanley I. Sandler, and Dionisios G. Vlachos
Catalysis Center for Energy Innovation, Department of Chemical and Biomolecular Engineering, University of Delaware, Newark, DE 19716 USA

Development of improved catalysts to design efficient and economic processes for biomass conversion to fuels and chemicals entail mechanistic insights into the reaction pathways. The objective of this study is to gain fundamental insights into the fructose dehydration to 5-(hydroxymethyl) furfural (HMF). Recent studies showed that closed-ring reaction pathway dominates in fructose dehydration to HMF. However, many different closed-ring pathways have been proposed. Additionally, fundamental questions such as the effect of solvent on the dominant reaction pathway remain unanswered. Here we investigate the fructose dehydration to HMF using isotopic-labeling experiments and NMR spectroscopy along with ab initio calculations with implicit and explicit water. A pronounced kinetic-isotope effect (KIE) is observed for HCl catalyzed fructose-D1 conversion in aqueous media, whereas no KIE is observed in fructose-D5 indicating that hydrogen transfer from C1 to C2 is the rate limiting step, consistent with earlier quantum mechanics/molecular mechanics calculations published by some of the authors. A plausible reaction intermediate in the fructose dehydration has been identified, and deuterium atoms from fructose have been mapped to various products using NMR spectroscopy that provide insights into the dominant reaction pathway. Finally, the effect of solvent on the mechanism will be discussed.

***** Formatting Issue - Best Available Paper/Graphic *****

A CFD Pressure Drop Model for Microfibrous Entrapped Catalyst Filters using Micro Scale Imaging

Qiang Gu and Bruce J. Tatarchuk
Center for Microfibrous Material Manufacturing (CM3), Department of Chemical Engineering, 212 Ross Hall, Auburn University, AL 36849

Abstract

A new structure of matter called microfibrous entrapped catalyst (MFEC) was developed to improve overall performance over conventional monolith reactors. This structure is manufactured by wet-lay process of metal fibers. Catalyst support particles (γ-Al2O3 150-250 μm) were entrapped by the sinter-lock structure of the metal fibers. Pressure drop across the reactor is a crucial factor when designing the reactor. This work presents experimental and simulation method used to evaluate pressure drop across this novel material.

Because of the dimension difference in fiber and filters, pressure drop modeling for filter structures is usually difficult if the details like fiber dimension and entrapped particle shapes are considered. This issue is solved by studying the fiber media in both micro and macro scale scope. Micro scale Computational Fluid Dynamics (CFD) pressure drop model for microfibrous material entrapped catalyst (MFEC) is first established using SEM imaging. The micro scale CFD model is based on theoretical calculation of velocity profile within the fiber material, which shows the plug flow pattern. SEM imaging for top and side view of the fiber media is obtained to sketch the geometric structure of the simulation area. Fiber and particle contribution to the pressure drop are studied separately. Shape factor and fiber compressibility are also considered for model accuracy. Micro scale pressure drop simulation results are then transferred into macro scale pleated filter structures for pressure drop modeling. Two kinds of filter fairing are studied for pressure drop reduction performance. All simulation results are compared with experiment data.

Keywords: Pressure Drop, CFD, SEM, MFEC

1. Methods and Experiments

1.1 Velocity profile in flat microfibrous material

Theoretical calculation of flow in microfibrous media is done by solving equation 4 (Figure 1), which is volume average Navier-Stoke equation for porous media developed by Vafai and Tien

— (4)

1.2 Numerical process

All velocity schemes are discretized using central difference for r and z directions with Euler explicit method. The form for iteration is arranged as

— — — — — —

— — — — — — (8)

*** Formatting Issue - Best Available Paper/Graphic ***

1.3 Flat sheet pressure drop measurement

The quadratic form pressure drop was measured using a 1 inch diameter steel pipe. Then 80 psi of house air was supplied as velocity inlet for microfibrous material.

1.4 SEM imaging for model detail

A sample SEM image of the microfibrous media and its corresponding sketch are shown in Figure 2 and Figure 3. Since only the top few layers are visible and easy to distinguish from one another, sketching of the fiber media structure by SEM was based on these layers. As to the inner layers, the top layers are translated because all layers are randomly dispersed and assumed to be similar in performance.

2. Results and Discussion

2.1 Fiber only

The pressure drop across a blank 8μm nickel fiber (Figure 4) was first simulated. The fiber material used is 1.2% in volume loading. The distance between fibers is determined by SEM imaging method. The simulation applied for four layer of unit cell which is 1.1 mm Experimental tests are based on 4 mm thick fiber sheet and these results will be compared on a pressure drop per millimeter basis.

2.2 Fiber with round particles

Since volume loading of particles in entrapped fiber sheets is much higher than fiber itself, the particles contribute to a large portion of pressure drop to the total pressure drop of the sheet. Typically volume loading of particles ranges from 20% - 30% for metal fiber sheet. In this application, particle volume loading is set at 20%. Particles used are ground and sieved to 150-250 m. In this part, the particles are assumed to be ideal in shape (sphere) and average in size (200 m). In the simulation, the particles are sketched as spheres with surfaces defined as wall conditions. Other simulation criteria are the same as fiber only case. shows SEM imaging of particle-entrapped fiber media, which is sketched in as simulation unit cell.

2.3 Fiber with elliptic particles

Shown by SEM imaging, shape of entrapped particles is not perfectly round. In this simulation, shape factor also known as sphericity has been investigated for contribution to the total pressure drop of the fiber material. Karwa *et al.* has empirically obtained shaped factors using Blake Kozeny equation for microfibrous material (Table 1). In this research, 180-210 μm particles were used and shape factor of 0.8 was chose as simulation criteria. Pressure drop data were shown in Figure 5.

2.4 Fiber compressibility

To improve CFD model accuracy, fiber and particle spacing has been adjusted according to the thickness tested (Figure 6). Due to the volume loading increment of fibers and entrapped particles, the pressure model accuracy has been improved dramatically especially at higher face velocities because of the larger compressions at these velocities (Figure 7).

2.5 Pressure drop for pleated MFEC using micro scale data

In high volumetric tests, the microfibrous media is engineered into pleat structure to reduce pressure drop. These structure help with reduction of amount of air going into each of the pleats. The pressure drop and velocity profile for these complicated structures are simulated using previous flat sheet simulation results. In addition, the pleat structure has brought in pleat tips which increased the total pressure drops. Fairing added to the pleat tip has been tested experimentally and by simulation using FLUENT CFD to explain the benefit of those fairings (Figure 8,9,10).

3. Summary

Pleat structure pressure drop of the microfibrous media was simulated based on the simulation of flat media. Fairings added to reduce pressure drop were investigated, which showed triangle fairings help with gradual compression of the air and the total pressure drop. Optimum pleat number was also obtained based on simulation results.

Reference

Angirasa D., Forced convective heat transfer in metallic fibrous materials, Journal of Heat Transfer, Vol. 124, pp 739-745, 2002

Bird, R., W. Stewart, E. Lightfoot, Transport Phenomena, 2nd ed. New York, John Wiley & Son, Inc., 2001

Caesar, T., Schroth T., The influence of pleat geometry on the pressure drop in deep-pleated cassette filters, Filtration and Separation, Vol. 39(9), pp 48-54, 2002

Cahela, D.R. Tatarchuk, B.J., Permeability of sintered microfibrous composites for heterogeneous catalysis and other chemical processing opportunities, Catalysis Today Vol. 69, pp 33-39, 2001

Chen, D.R., Pui, D.Y.H., Tang Y.M., Filter pleating design for cabin air filtration, Technical paper 960944, Annual index/abstract of SAE technical papers, Society of Automotive engineers, Warredale, PA., 1996

Dhandapani B., Oyama S.T., Gas phase ozone decomposition catalyst, Applied Catalysis B: Environmental Vol. 11, pp 120-166, 1997

Harris D.K., Cahela D.R., Tatarchuk B.J., Wet layup and sintering of metal-containing microfibrous composites for chemical processing opportunities, Composites: Part A Vol. 32, pp 1117-1126, 2001

Hill R.J., Koch D.L., The transition from steady to weakly turbulent flow in a close-packed ordered array of spheres, Journal of Fluid Mechanics, Vol. 465 pp 59-97, 2002

Kalluri R.R., Cahela D.R., Tatarchuk B.J., Comparative heterogeneous contacting efficiency in fixed bed reactor: Opportunities for new microstructured systems, Applied Catalysis B: Environmental Vol. 90, pp 507-515, 2009

Kalluri R. R., Duggirala R. K., Cahela D. R., Roy C. J., and Tatarchuk B. J., CFD analysis of mass transfer enhancement due to inert structures present in fluid flow paths in heterogeneous contacting systems, manuscript submitted to computers and chemical engineering, 2008

Kameya T., Urano K., Catalytic Decomposition of ozone gas by a Pd impregnated MnO2 catalyst, Journal of environmental engineering, pp 286-292, 2002

Karwa A.N., Tatarchuk B.J., Pressure drop and aerosol filtration efficiency of microfibrous entrapped catalyst and sorbent media: Semi-empirical models, Separation and Purification Technology Vol. 86 pp 55-63, 2012

Rivers, D.R., Murphy D.J., Air filter performance under variable air volume conditions, ASHRAE Transactions Vol. 106(2), pp 131-144, 2000

Sothen R.A., Tatarchuk B.J., A semi-empirical pressure drop model: part I- pleated filters, HVAC&R Research, Vol. 14(6), 2008

Vafai, K., Tedn, C. L., Boundary and inertia effect on flow and heat transfer in porous media, International J. Heat Mass Transfer, pp 1183-1190, 1982

Wang X., Kim K., Lee C., and Kim J., Prediction of air filter efficiency and pressure drop in air filtration media using a stochastic simulation, Fiber and Polymers, Vol. 9(1), pp 34-38, 2008

***** Formatting Issue - Best Available Paper/Graphic *****

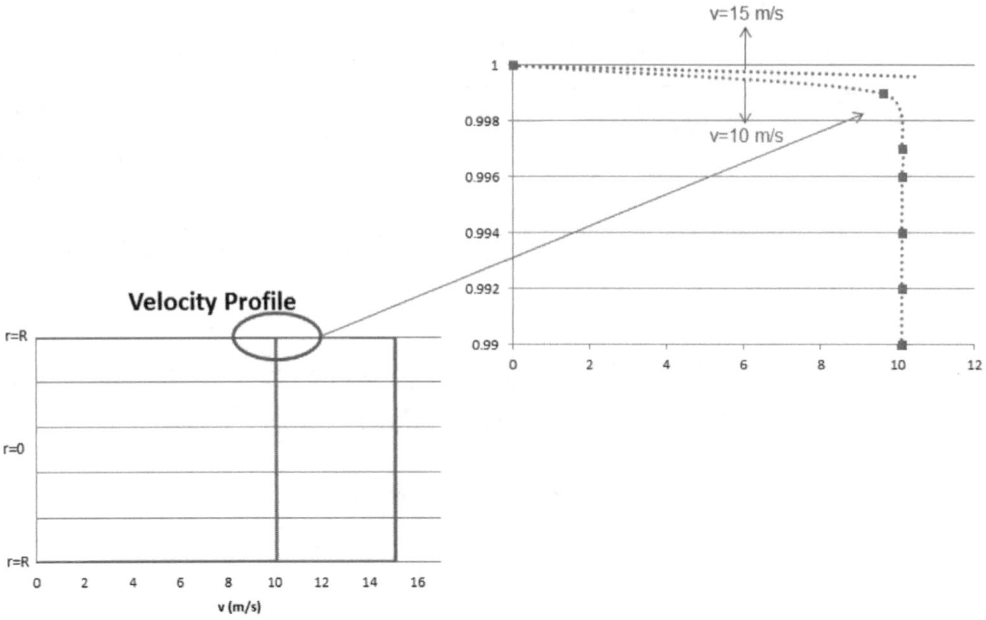

Figure 1 Velocity Profile of microfiber material at two different face velocities with high Reynolds numbers

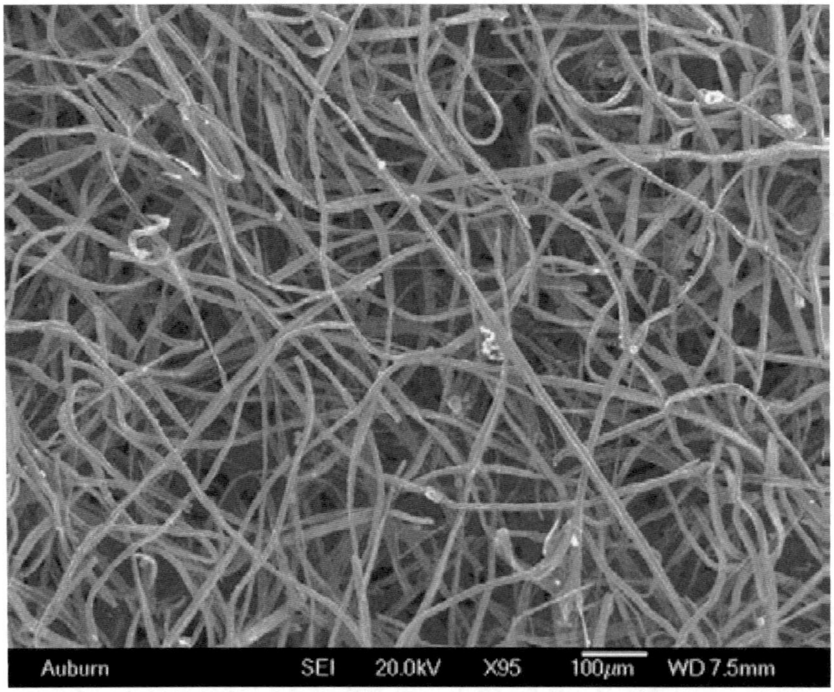

Figure 2 Sample SEM imaging of sintered microfibrous material surface

*** Formatting Issue - Best Available Paper/Graphic ***

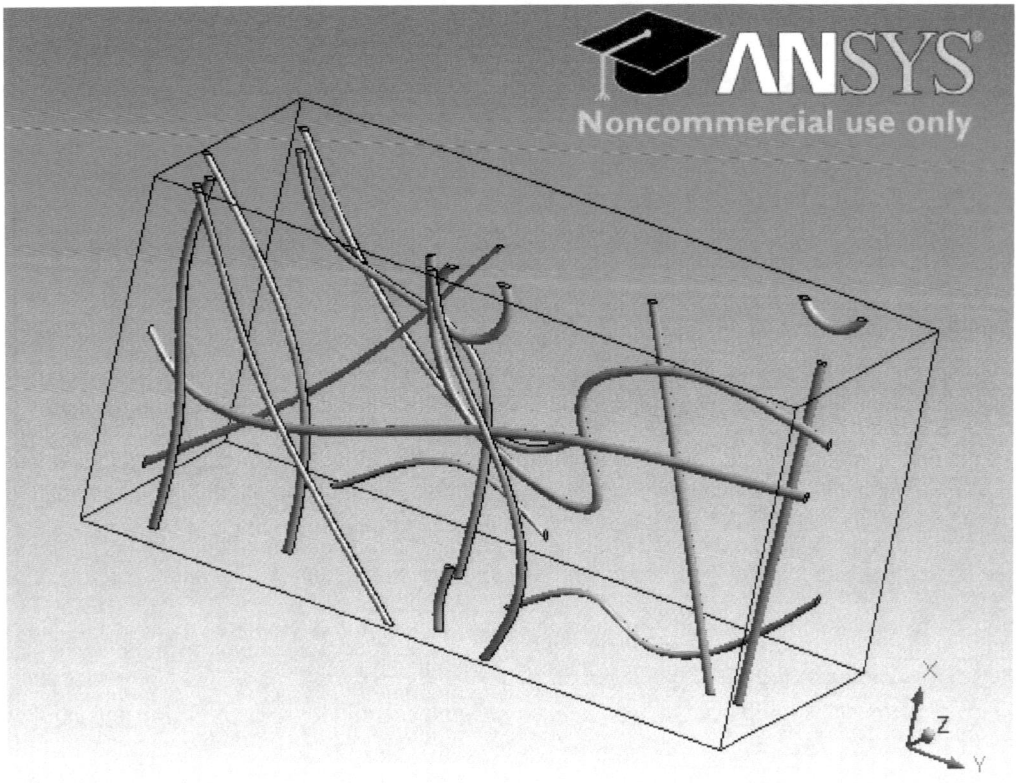

Figure 3 Sketch of selected area of microfibrous material for simulation

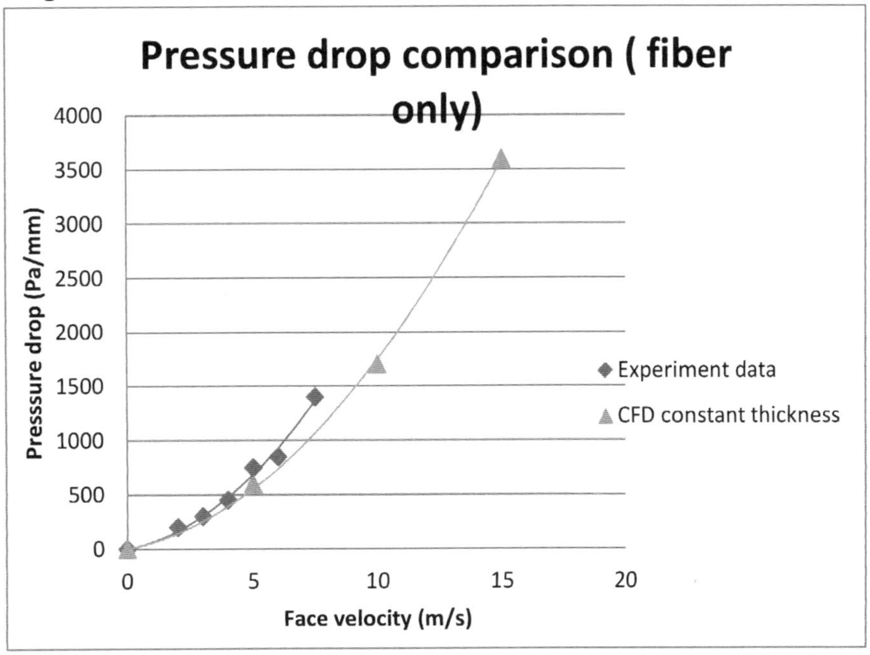

Figure 4 Pressure drop comparison between the model and experiment data

***** Formatting Issue - Best Available Paper/Graphic *****

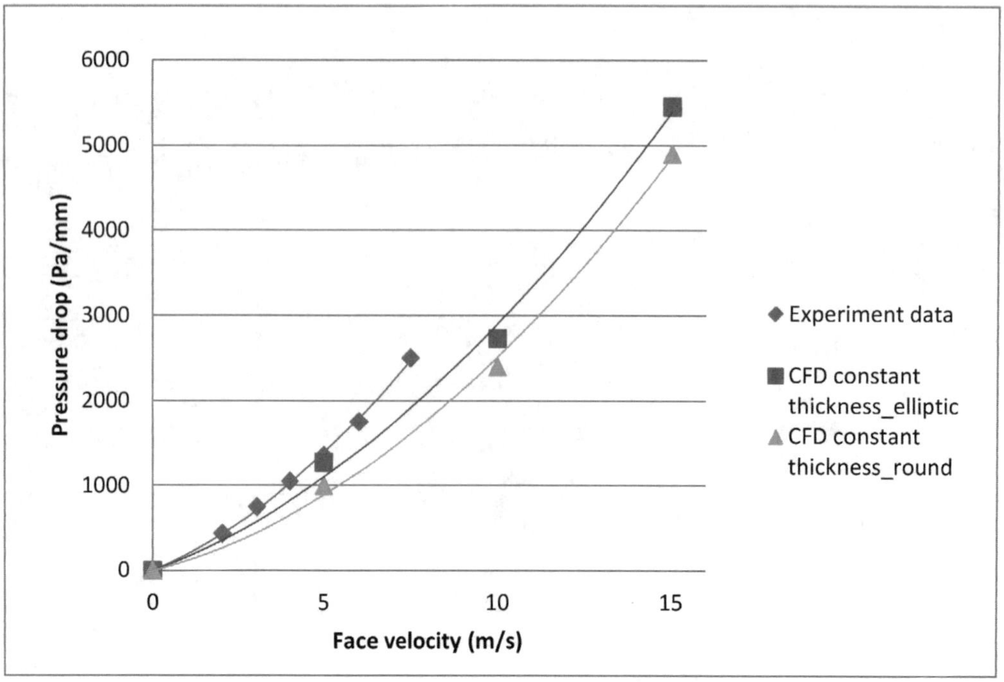

Figure 5 Pressure drop comparison of experiment, round particle and elliptic particle media by CFD simulation

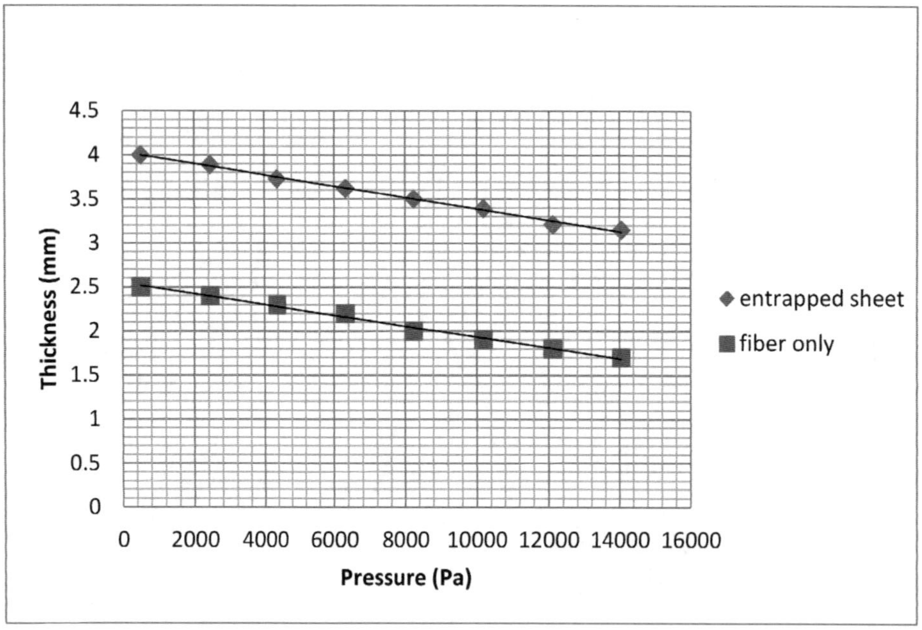

Figure 6 Microfiber media thickness at different pressures; initial volume loading, fiber 1.5%, particles 20%; initial thickness 4 mm

***** Formatting Issue - Best Available Paper/Graphic *****

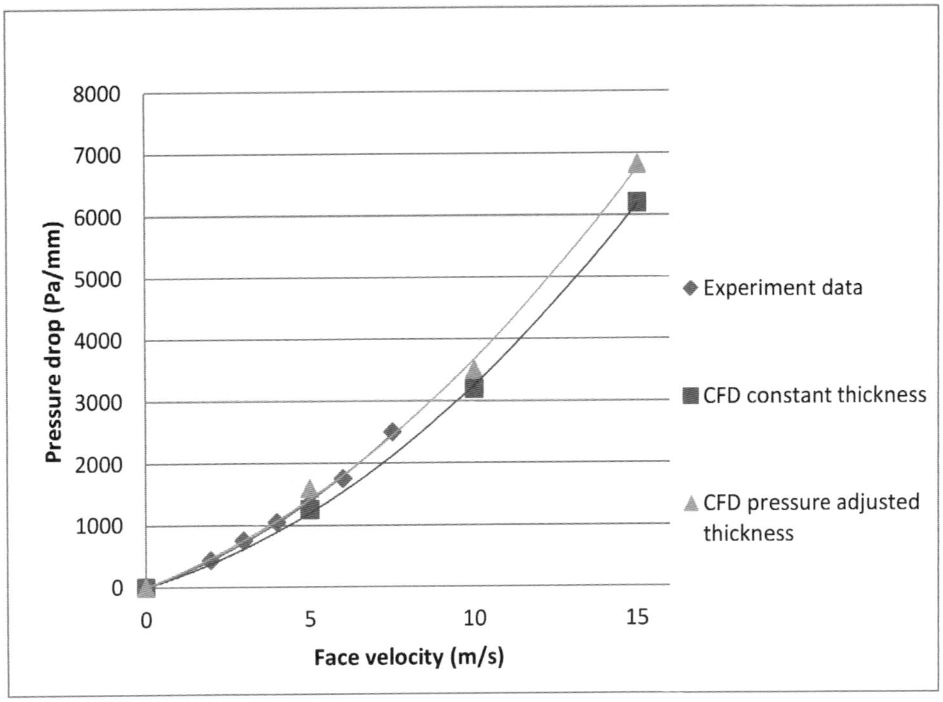

Figure 7 Pressure drop for fiber with elliptic particle case; CFD simulation in both uncompressed and compressed consideration; experiment result collected up to 7.5 m/s

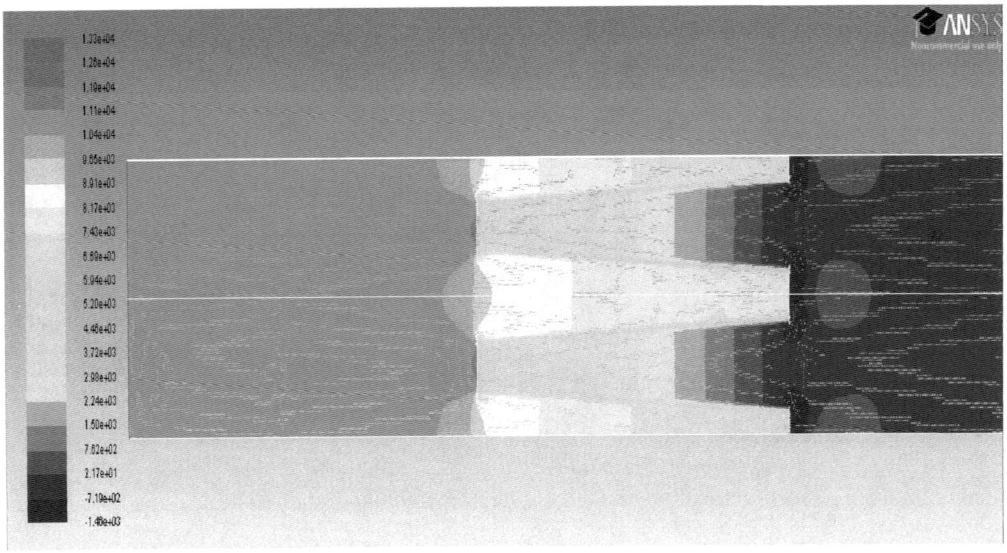

Figure 8 Pressure drop simulation for pleated microfibrous material filter

*** Formatting Issue - Best Available Paper/Graphic ***

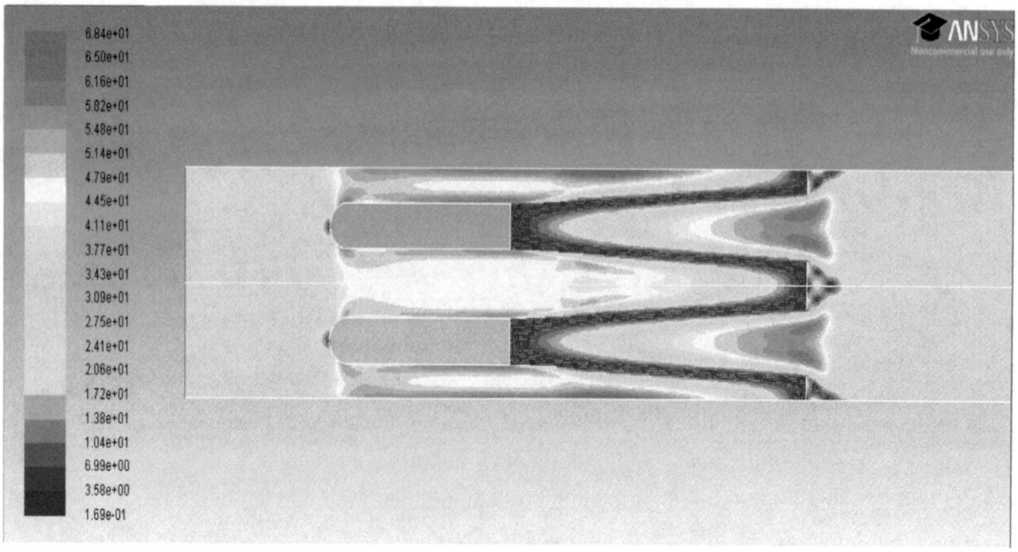

Figure 9 Velocity profile for pleated microfibrous material filter with round-top fairings; eddy flow removed by added fairings

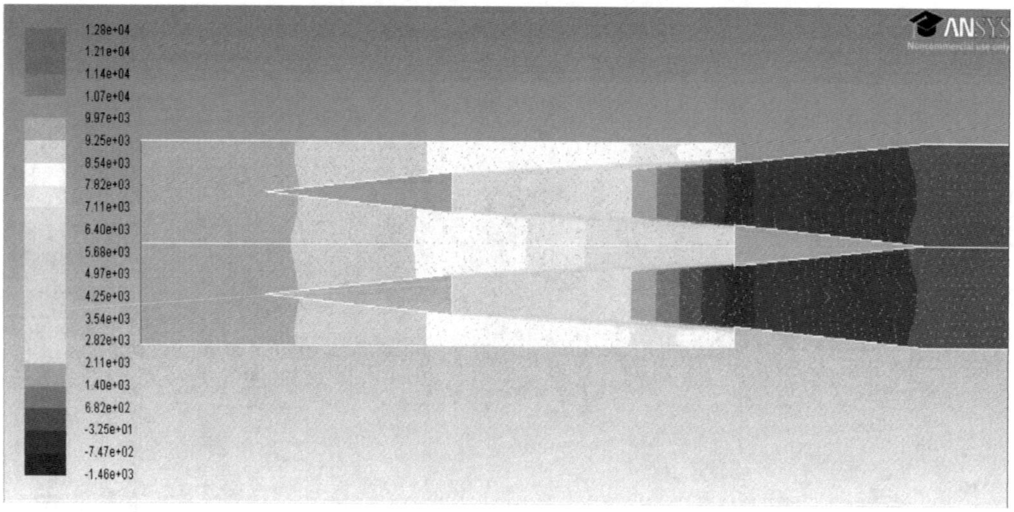

Figure 10 Pressure drop simulation for pleated microfibrous material filters with triangle fairings at both upstream and downstream

Kinetics of Jatropha oil methanolysis

Mario Andrés Noriega Valencia[1], Paulo César Narváez Rincón[1], Christoph Heinz[2], Nevardo Bello Yaya[1], Juan Guillermo Cadavid Estrada[1]

1.Grupo de Procesos Químicos y Bioquímicos, Departamento de Ingeniería Química. Facultad de Ingeniería, Universidad Nacional de Colombia, Bogotá, Colombia.
2 Technische Universität München, Chemistry Department, Lichtenbergstraße 4, D-85748 Garching b. München, Germany.

*pcnarvaezr@unal.edu.co

Abstract

Biodiesel is currently produced by methanolysis of vegetable oils such as rapeseed, palm and soybean oils. However, Jatropha oil is considered one of the most promising sources for producing biodiesel in Asia, Europe and Africa, The knowledge of the kinetics of his methanolysis reaction is extremely important because it enables the design of reactors with suitable dimensions and the operation in optimal conditions. In this work the kinetics of Jatropha oil methanolysis was studied. The effect of temperature, catalyst concentration and methanol to oil molar ratio on Jatropha oil methanolysis were studied in the range from 40 to 60 °C, 0.2 to 1wt% (NaOH) and 3 to 12, respectively. A 100% conversion of Jatropha oil with a 97% yield to fatty acid methyl esters (FAME) were obtained after 60 minutes of reaction at 60°C, with 0.6wt% of NaOH, and methanol to oil molar ratio of 6:1. Catalyst concentration increments between 0.2wt% and 0.6wt% had a positive effect on the reaction rate and on the yield to FAME. However, the 0.2wt% catalyst concentration did not show activity at all reaction temperatures. A second order kinetic model that includes the effect of temperature and catalyst concentration on reaction rate describes adequately Jatropha oil methanolysis, as validated by the Fisher–Snedecor test of unbiased variances. Kinetic model parameters of Jatropha oil methanolysis, Palm oil methanolysis and Palm oil ethanolysis were compared and showed similar values and behavior.

Keywords: Jatropha oil, Palm oil, Methanolysis, Ethanolysis Biodiesel, Kinetic model, Mathematic model

Catalyst-Scale Modeling of Fischer-Tropsch Synthesis from Syngas Using COMSOL Multiphysics

Arvind. Nanduri[1] and Patrick L. Mills[*1]
[1]Texas A&M University-Kingsville, Department of Chemical & Natural Gas Engineering
*Corresponding author: Texas A&M University-Kingsville, Department of Chemical & Natural Gas Engineering, EC 303D, MSC 193, Kingsville, TX-78363-8202, USA. Email: Patrick.Mills@tamuk.edu

Abstract: A 2-D particle model for different catalyst shapes (solid cylinder, hollow cylinder, 4-hole cylinder, modified 4-hole cylinder and 7-hole cylinder) was simulated to compare the particle-level performance for the Fischer-Tropsch Synthesis (FTS). A Fe-based micro-kinetic olefin re-adsorption model developed by Wang *et al.* (2008) was coupled with the Soave-Redlich-Kwong (SRK) equation of state to describe the particle-scale transport-kinetic interactions and phase behavior for the gas-phase FTS [4]. The intra-particle effectiveness factor, liquid-to-vapor ratio, CO conversion and intra-particle volume-averaged concentration of diesel were analyzed at different process conditions to compare the performance of different catalyst particle shapes.

Keywords: catalyst particle shape, Fischer-Tropsch Synthesis, vapor-liquid-equilibrium (VLE), gas phase micro-kinetics.

1. Introduction

The Fischer-Tropsch Synthesis (FTS) is a highly exothermic condensation polymerization reaction of syngas ($CO+H_2$) in the presence of Fe/Co/Ru-based catalysts to produce a wide range of paraffins, olefins and oxygenates, the latter of which is often called *syncrude*. Multi-Tubular Fixed Bed Reactors (MTFBR) and Slurry Bubble Column reactors (SBCR) are widely employed for FTS processes [1]. To understand the reactor-scale catalyst performance, it is important to first analyze particle-scale transport-kinetic interactions.

The FTS reaction network produces hydrocarbons with carbon numbers typically ranging from 1 to 100 so the catalyst pores in this process can be potentially filled with liquid wax (C_{20}+) leading to high diffusional limitations. To model such a reaction network and account for all species, micro-kinetic rate expressions for each individual species must be coupled with the intraparticle solubility of gases in liquid wax and transport equations for the various reaction species in the porous catalyst. This paper describes the initial results obtained from such an effort using COMSOL Multiphysics™ as the numerical engine for solving the model equations.

2. Catalyst Pellet Model

The isothermal 2-D catalyst pellet model used to describe transport-kinetic interactions is presented in this section.

A total number of 20 paraffins (C_1 to C_{20}), 19 olefins (C_2 to C_{20}) and 4 key components (H_2, CO, CO_2, and H_2O) are considered in the reaction network, which leads to 43 second-order nonlinear differential equations for the specie mass balances. The micro-kinetic and thermodynamic expressions used in the model can be found elsewhere [3-5]. Five different particle shapes based on a solid cylinder, sphere, and hollow cylinders having the same effective diffusion length ($L_e = V_p/S_p$) are simulated in this paper [2]. The catalyst shapes are illustrated in Figure 1.

2.1 Specie mass balance equations

The specie mass balance equations for each particle shape are summarized below.

General specie balance

$$\nabla.(-D_{ei}\nabla C_i) = \rho_p \sum_{j=1}^{44}\sum_{i=1}^{43} \alpha_{ij}R_{ij}$$

Specie balance in cylindrical coordinates

$$\frac{1}{r}\frac{\partial}{\partial r}\left(D_{ei} r \frac{\partial C_i}{\partial r}\right) = -\rho_p \sum_{j=1}^{44}\sum_{i=1}^{43} \alpha_{ij}R_{ij}$$

The specie mass balance for all the particle shapes considered in this work are solved in primitive variables where *r* is the radial coordinate. Particle end effects are neglected so specie concentrations are a function of the radial coordinate only.

Numerical solutions showed that the intra-particle temperature gradient was on the order of a few degrees owing to the high effective thermal conductivity of the particle. For this reason, the energy balance was not included in any subsequent simulations.

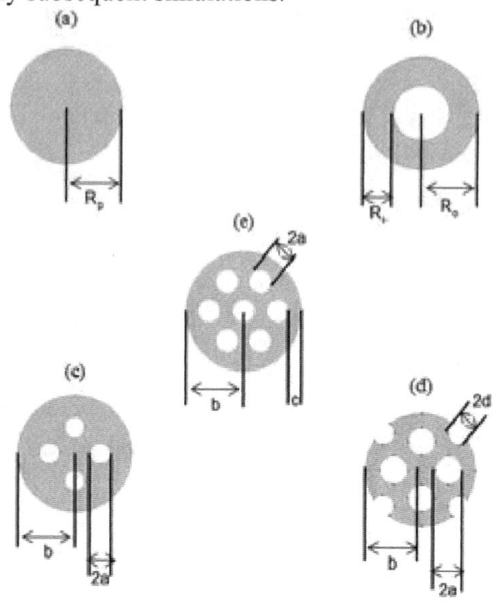

Figure 1. Catalyst particle shapes. (a) cylinder, (b) hollow cylinder (H-Cylinder), (c) 4-hole cylinder (4-H Ring), (d) modified 4-hole cylinder (M-4-H Ring), and (e) 7-hole cylinder (7-H Ring).

2.2 Boundary conditions

The boundary conditions correspond to specified concentrations at the pellet surface (Dirichlet conditions). The boundary conditions are summarized in Table-1.

Table 1. Boundary Conditions.

Cylinder	At $r = R_p$ (surface), $C_i = C_{i,bulk}$
H-Cylinder	At $r = R_o$ (outer surface) and $r = R_i$ (inside surface), $C_i = C_{i,bulk}$
4-H Ring	$C_i = C_{i,bulk}$ on all boundaries exposed to bulk phase
M-4-H Ring	$C_i = C_{i,bulk}$ on all boundaries exposed to bulk phase
7-H Ring	$C_i = C_{i,bulk}$ on all boundaries exposed to bulk phase

2.3 Vapor-Liquid-Equilibrium calculations

The vapor-liquid equilibrium for the FTS reaction mixture is described here using the Soave-Redlich- Kwong equation of state [3]. The flash calculations are based upon the Rachford-Rice objective function:

$$F(\alpha_g) = \sum_i \frac{z_i(K_i - 1)}{\left(1 + \alpha_g(K_i - 1)\right)} = 0$$

In this application, the Rachford-Rice objective function is a polynomial of degree 43 (43 distinct roots). Since the fraction $\alpha_g = V/(V+L)$ is a positive number less than 1, only the roots that make physical sense are considered.

The initial estimate of the distribution coefficient is obtained using Wilson's correlation [3]:

$$K_i^{\text{guess value}} = \frac{P_{ic}}{P}\exp\left(5.37(1 + \omega_i)\left(1 - \frac{T_{ic}}{T}\right)\right)$$

Here, the distribution coefficient is defined as the ratio of the vapor to liquid mixture fugacity coefficients:

$$K_i = \frac{\emptyset_i^{\,V}}{\emptyset_i^{\,L}}$$

3. Modeling Strategy

The 2-D diffusion-reaction system of equations was solved using the transport of diluted species module in COMSOL Multiphysics™ assuming steady-state conditions. The Rachford-Rice objective function (VLE calculations) was solved using the coefficient form of the PDE solver. The catalyst properties and operating conditions used in the models are listed in Tables 2 - 4. The solution for the specie concentration profiles were based upon a convergence factor of 10^{-3}.

To avoid convergence issues, the reaction rates for individual species were multiplied by a perturbation factor of 10^{-4} and this factor was slowly increased to 1. Since the FTS is a relatively fast diffusion-reaction system, numerical instabilities were encountered in the interior region of the particle as the CO and CO_2 concentrations approached zero. The convergence issues were solved by using built-in logical operators, such as CO = if (CO ≤ 0, eps, CO), and CO_2 = if (CO_2 ≤ 0, eps, CO_2) which resulted in positive values for the specie concentrations. Mesh refinement was manually performed until the concentration profiles were relatively constant and satisfied the convergence criterion.

Table 2. Dimensions of different catalyst particle shapes with equal characteristic diffusion length.

Cylinder	L = 3 mm & R = 1 mm
H-Cylinder	L = 3 mm, R_o = 2 mm & R_i = 1 mm
4-H Ring	L = 3 mm, b = 2 mm, a = 0.36 mm & c = 0.54 mm
M-4-H Ring	L = 6 mm, b = 2 mm, a = 0.5 mm, c = 0.5 mm & d = 0.415 mm
7-H Ring	L = 6 mm, b = 3 mm, a = 0.6 mm & c = 0.6 mm

Table 3. Catalyst properties.

Density of pellet, ρ_p	1950 (kg/m^3)
Porosity of pellet, ε	0.51
Tortuosity, τ	2.6

Table 4. Operating conditions.

Temperature, K	493, 523 & 533
Pressure, bar	25 & 30
H_2/CO	2

4. Results and Discussion

4.1 Concentration profiles of key components, fuel gas and diesel

The 2-D isothermal intra-particle concentration profiles of the key components (H_2, CO, CO_2 and H_2O), fuel gas (C1-C2) and diesel (C13-C17) for different catalyst particle shapes are shown in Figure 2 - 7. The volume-averaged concentration of each component is used as the basis for comparison. The average concentration profiles for all operating conditions are shown in Figure 8-19. The profiles show that the CO concentration approaches zero for all shapes due to the presence of significant diffusional limitations. The CO_2 concentration reaches a maximum and then decreases due to the reverse Water-Gas-Shift (WGS) reaction. The reverse WGS reaction produces CO, which is then consumed in the subsequent hydrocarbon producing FT reactions. Figure 12 and 13 show that the reverse WGS reaction does not occur in the M-4-H ring.

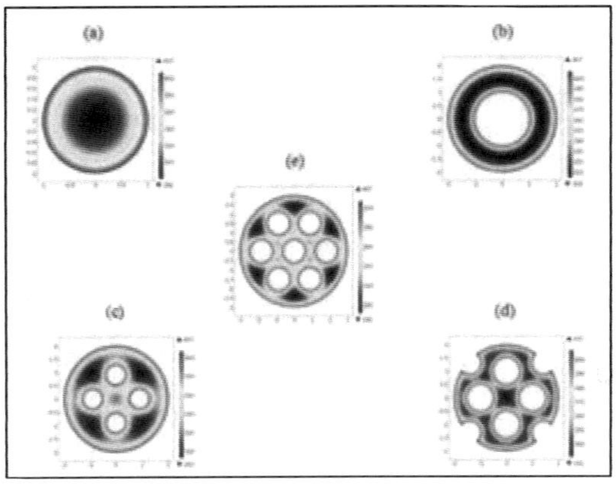

Figure 2. Intra-particle concentration profile of H_2 (mol/m^3) at T = 493 K and P = 25 bar. (a) cylinder, (b) hollow cylinder (H-Cylinder), (c) 4-hole cylinder (4-H Ring), (d) modified 4-hole cylinder (M-4-H Ring), and (e) 7-hole cylinder (7-H Ring).

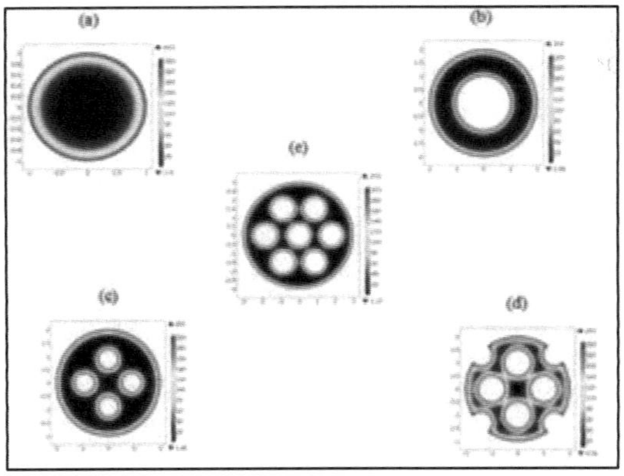

Figure 3. Intra-particle concentration profile of CO (mol/m^3) at T = 493 K and P = 25 bar. (a) cylinder, (b) hollow cylinder (H-Cylinder), (c) 4-hole cylinder (4-H Ring), (d) modified 4-hole cylinder (M-4-H Ring), and (e) 7-hole cylinder (7-H Ring).

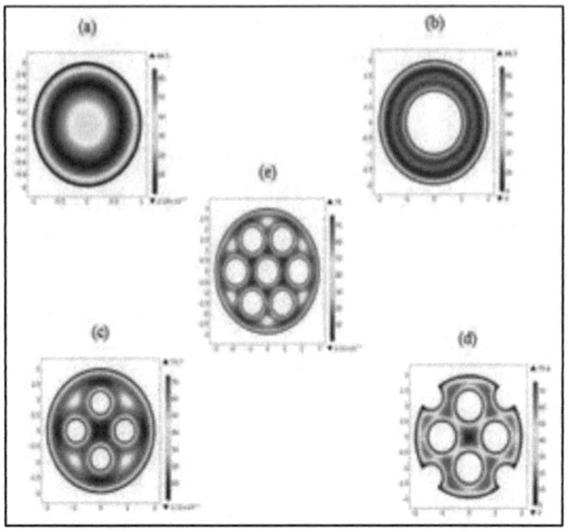

Figure 4. Intra-particle concentration profile of CO_2 (mol/m^3) at T = 493 K and P = 25 bar. (a) cylinder, (b) hollow cylinder (H-Cylinder), (c) 4-hole cylinder (4-H Ring), (d) modified 4-hole cylinder (M-4-H Ring), and (e) 7-hole cylinder (7-H Ring).

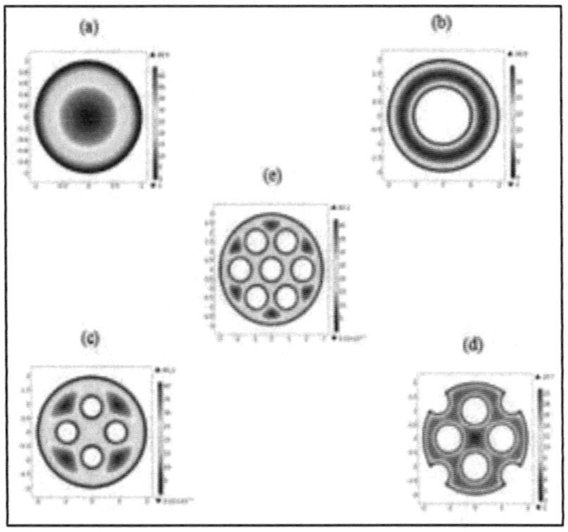

Figure 5. Intra-particle concentration profile of H_2O (mol/m^3) at T = 493 K and P = 25 bar. (a) cylinder, (b) hollow cylinder (H-Cylinder), (c) 4-hole cylinder (4-H Ring), (d) modified 4-hole cylinder (M-4-H Ring), and (e) 7-hole cylinder (7-H Ring).

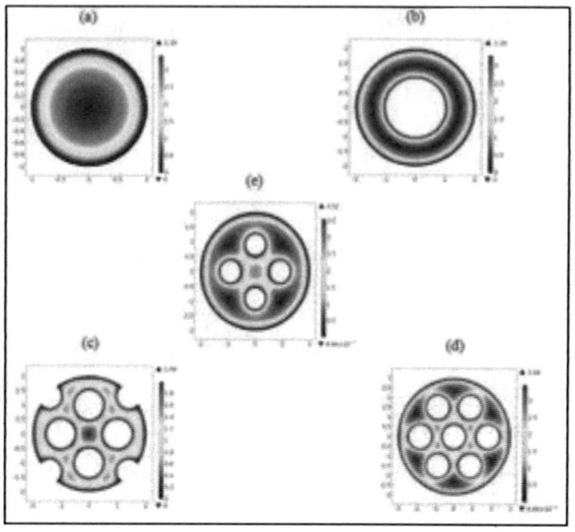

Figure 6. Intra-particle concentration profile of fuel gas (mol/m^3) at T = 493 K and P = 25 bar. (a) cylinder, (b) hollow cylinder (H-Cylinder), (c) 4-hole cylinder (4-H Ring), (d) modified 4-hole cylinder (M-4-H Ring), and (e) 7-hole cylinder (7-H Ring).

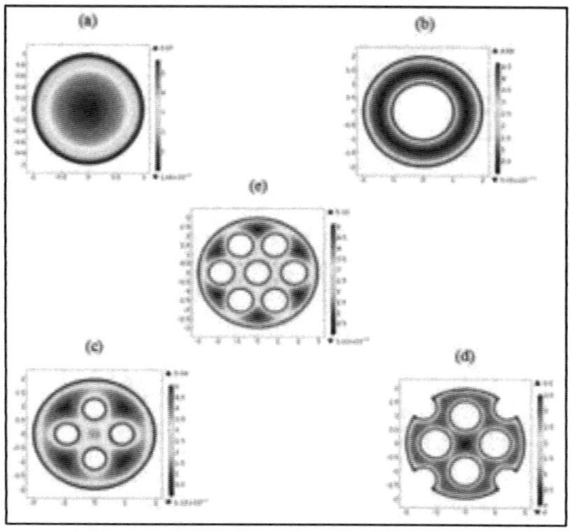

Figure 7. Intra-particle concentration profile of diesel (mol/m^3) at T = 493 K and P = 25 bar. (a) cylinder, (b) hollow cylinder (H-Cylinder), (c) 4-hole cylinder (4-H Ring), (d) modified 4-hole cylinder (M-4-H Ring), and (e) 7-hole cylinder (7-H Ring).

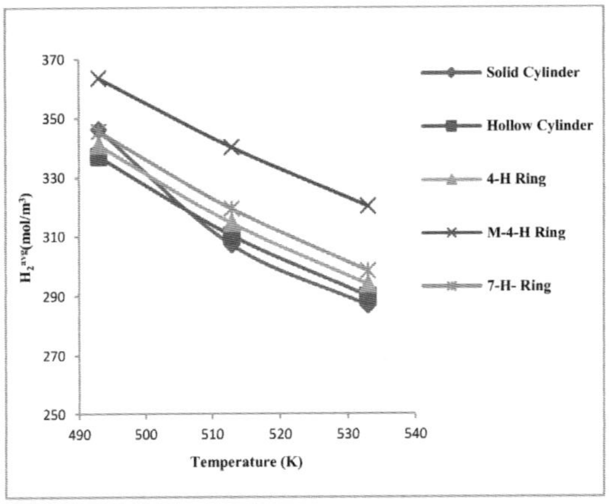

Figure 8. Average concentration of H_2 (mol/m^3) for T = 493 K, 513 K and 533 K and P = 25 bar.

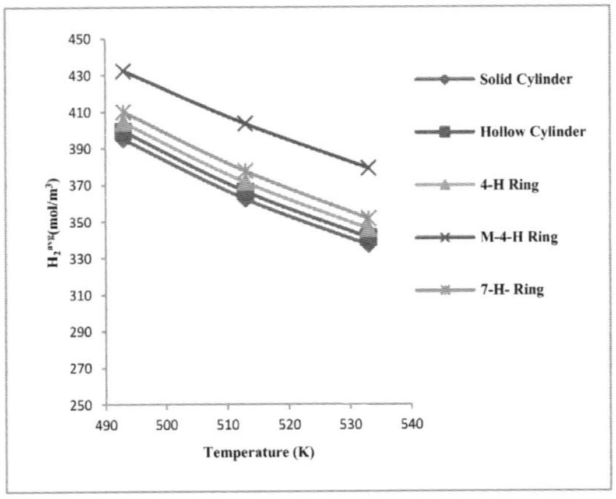

Figure 9. Average concentration of H_2 (mol/m^3) for T = 493 K, 513 K and 533 K and P = 30 bar.

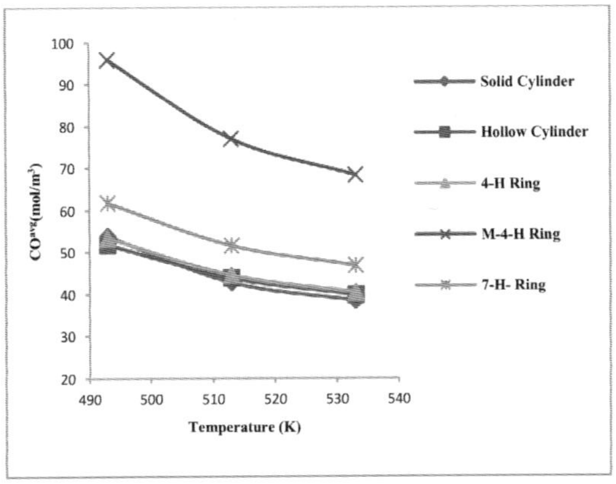

Figure 10. Average concentration of CO (mol/m^3) for T = 493 K, 513 K and 533 K and P = 25 bar.

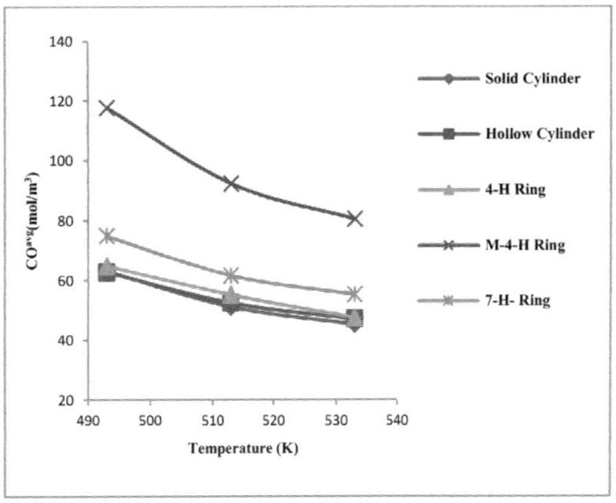

Figure 11. Average concentration of CO (mol/m^3) for T = 493 K, 513 K and 533 K and P = 30 bar.

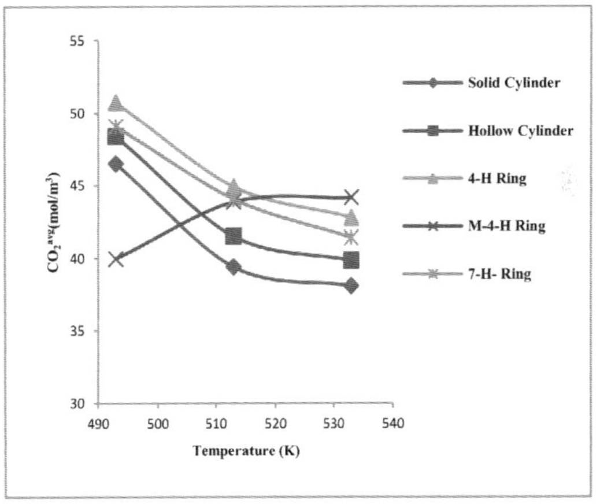

Figure 12. Average concentration of CO_2 (mol/m^3) for T = 493 K, 513 K and 533 K and P = 25 bar.

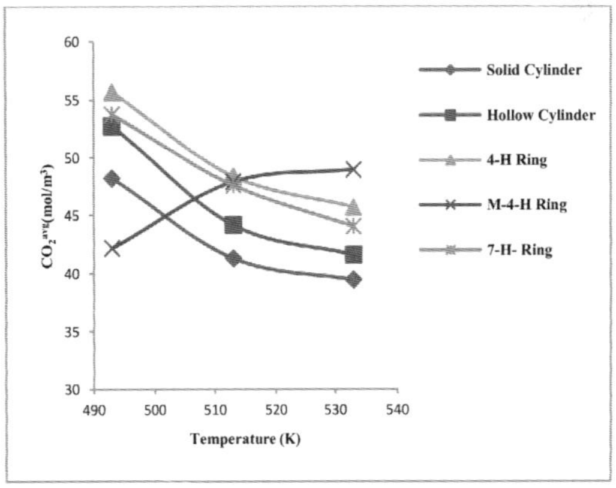

Figure 13. Average concentration of CO_2 (mol/m^3) for T = 493 K, 513 K and 533 K and P = 30 bar.

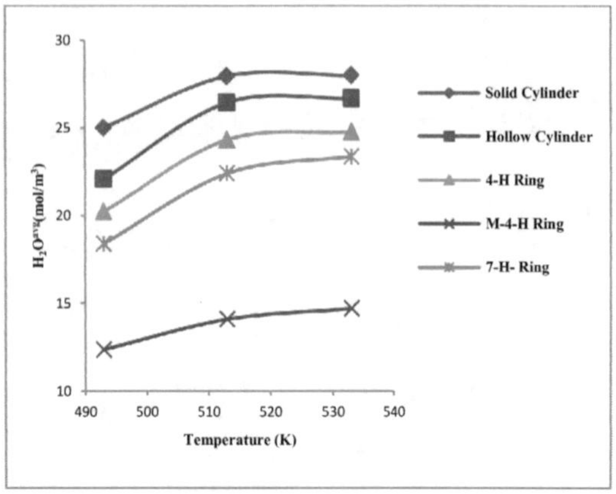

Figure 14. Average concentration of H_2O (mol/m^3) for T = 493 K, 513 K and 533 K and P = 25 bar.

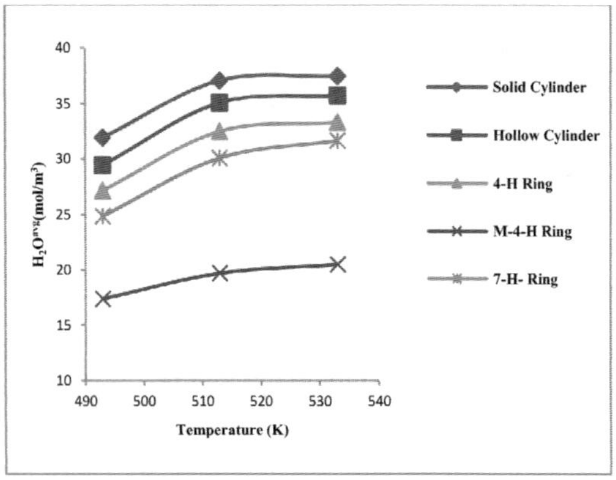

Figure 15. Average concentration of H_2O (mol/m^3) for T = 493 K, 513 K and 533 K and P = 30 bar.

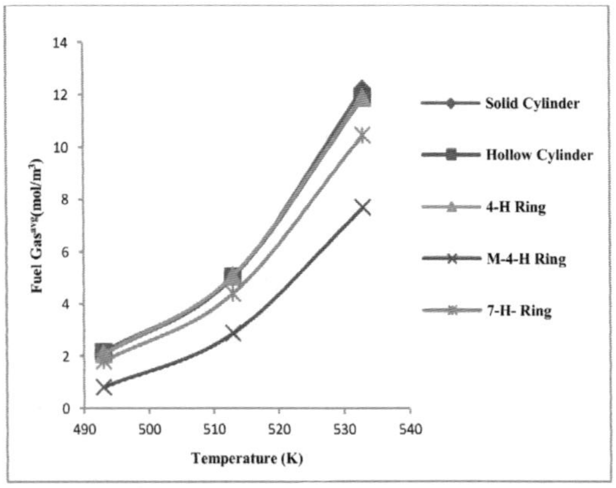

Figure 16. Average concentration of fuel gas (mol/m^3) for T = 493 K, 513 K and 533 K and P = 25 bar.

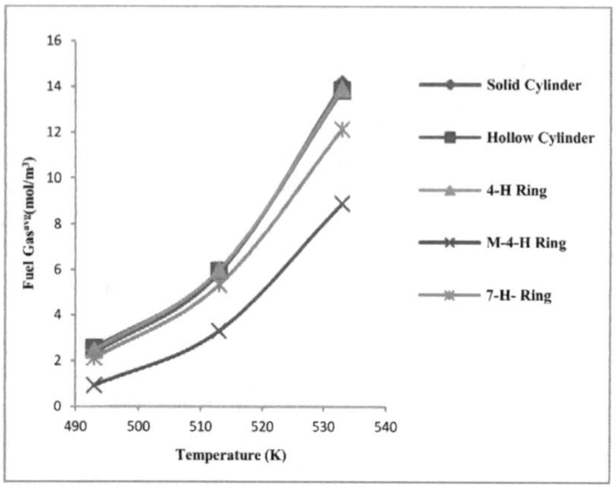

Figure 17. Average concentration of fuel gas (mol/m^3) for T = 493 K, 513 K and533 K and P = 30 bar.

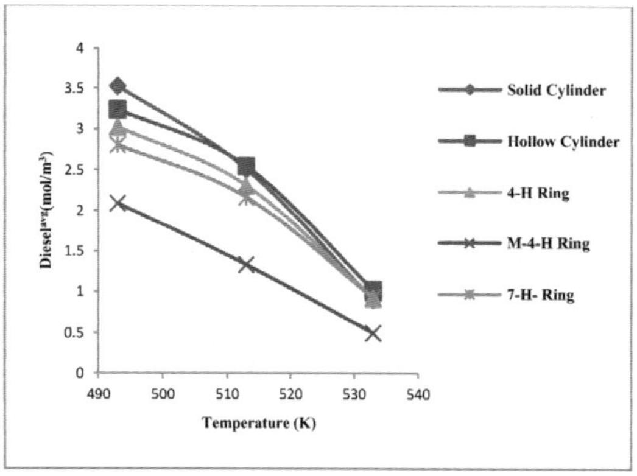

Figure 18. Average concentration of diesel (mol/m^3) for T = of 493 K, 513 K and 533 K and P = 25 bar.

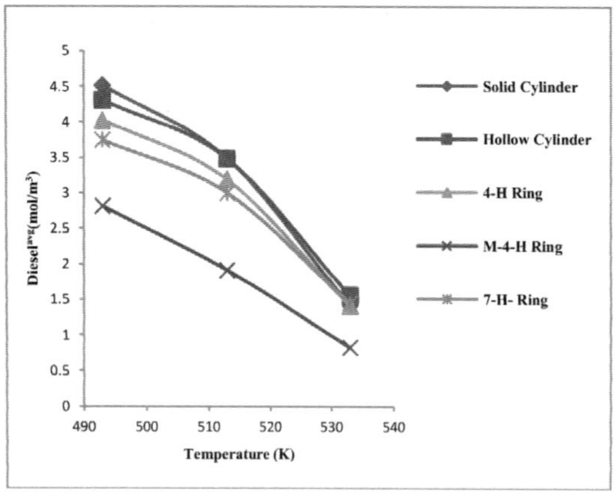

Figure 19. Average concentration of diesel (mol/m^3) for T = 493 K, 513 K and 533 K and P = 30 bar.

4.2 Overall effectiveness factor

The overall effectiveness factor (η) for different particle shapes are shown in Figure 20 - 21. The effectiveness factor has a decreasing trend with temperature and as the center of the particle is approached. It can be observed that the M-4-H ring has the highest effectiveness factor, but the reactor-scale performance and diesel range selectivity for all catalyst particle shapes has to be analyzed in more detail to identify the preferred catalyst shape. Other factors must also be considered, such as mechanical strength of the catalyst as well as pressure drop.

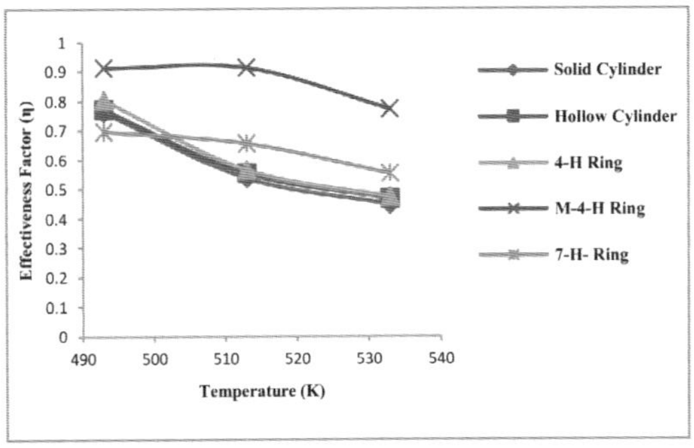

Figure 20. Effectiveness factor (η) for T = 493 K, 513 K and 533 K and P = 25 bar.

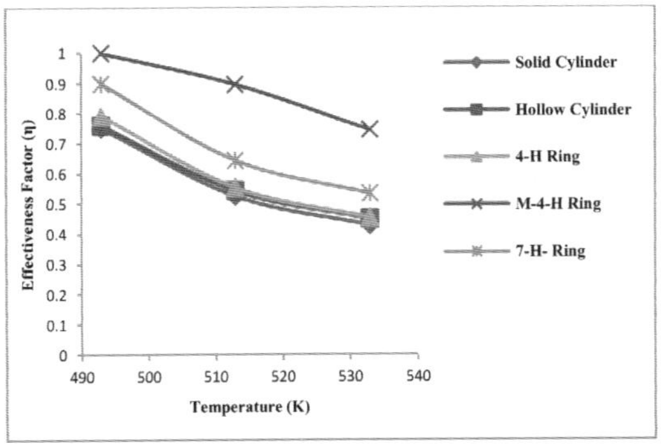

Figure 21. Effectiveness factor (η) for T = of 493 K, 513 K and 533 K and P = 30 bar.

4.3 Intra-particle and average liquid-to-vapor ratio

The intra-particle and volume average liquid-to-vapor molar ratio (L/V) for different particle shapes is shown in Figures 22 - 24. These results show that the M-4-H ring and 7-H ring particles have the smallest liquid fraction when compared to the solid cylinder, the H-cylinder, and the 4-H ring. This result is also supported by comparing concentration profiles of key components (see Figures 2 - 7) where the later particle shapes exhibit the most significant diffusional limitation. These results also show the importance of understanding the phase behavior of FT products in the catalyst pores since the reaction kinetics for liquid phase and gas phase would differ. However, this lies outside the scope of the current work.

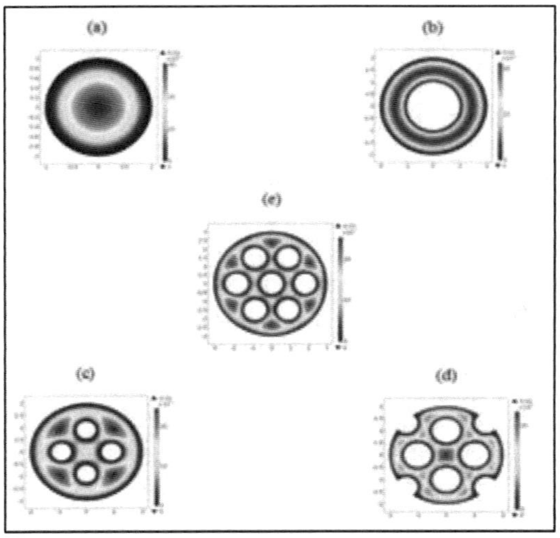

Figure 22. Intra-particle liquid-to-vapor ratio at T = 493 K and P = 25 bar. (a) cylinder, (b) hollow cylinder (H-Cylinder), (c) 4-hole cylinder (4-H Ring), (d) modified 4-hole cylinder (M-4-H Ring), and (e) 7-hole cylinder (7-H Ring).

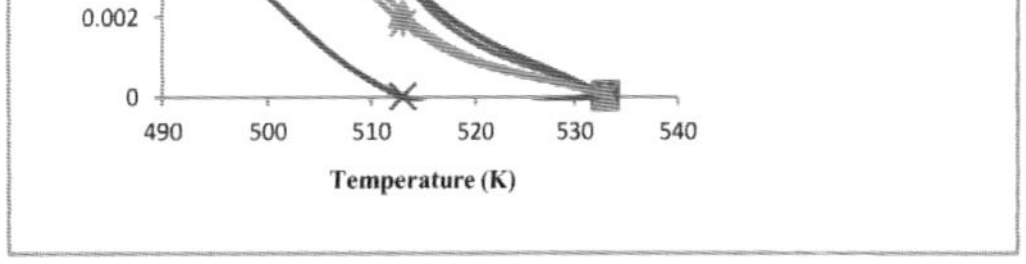

Figure 23. Liquid-to-vapor ratio for T = 493 K, 513 K and 533 K and P = 25 bar.

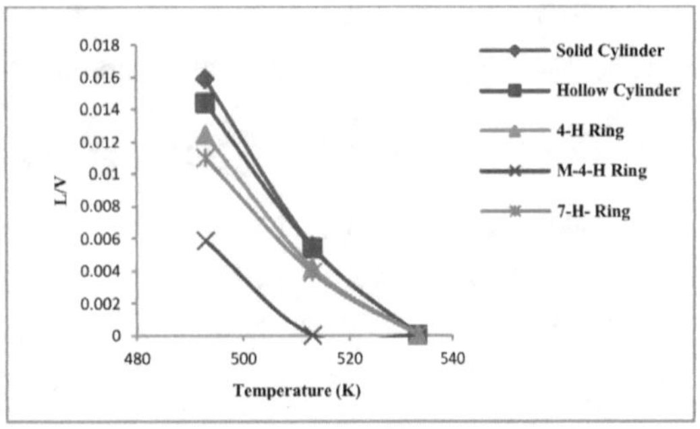

Figure 24. Liquid-to-vapor ratio for T = 493 K, 513 K and 533 K and P = 30 bar.

5. Summary and Conclusions

The transport-kinetic interactions for FTS in a 2-D catalyst pellet model using a detailed micro-kinetic model were successfully analyzed for the first time using COMSOL Multiphysics™. This study is an important extension to the previous work by Wang *et al.* (2008) who used the same micro-kinetic model to study the F-T reaction-diffusion mechanism in a spherical catalyst, but did not study the effect of catalyst shape nor report the inter-particle average diesel concentration. Also, the numerical technique used by Wang *et al.* (2008) was based upon a custom computer code that implemented the orthogonal collocation method to solve the two-point boundary value system of ODEs. The results in the current work will provide the basis for modeling the catalyst bed using the packed bed reactor model in COMSOL Multiphysics 5.0 as well other extensions that account for additional particle-scale transport effects.

6. Nomenclature

C_i	Concentration of species 'i' (mol/m^3)
$C_{i,bulk}$	Concentration of species 'i' in bulk phase (mol/m^3)
C_i^{avg}	$\frac{\int C_i dv}{V_p}$ (mol/m^3)
D_{ei}	Effective diffusivity of species *i* (m^2/s)
K_i	Equilibrium ratio
L	Number of moles in the liquid phase (moles)
L_p	Length of catalyst pellet (mm)
P	Pressure (bar)
P_{ic}	Critical pressure of component *i* (bar)
r	Radial coordinate (mm)
R_p	Radius of catalyst pellet (mm)
R_i	Internal radius of hollow cylinder (mm)
R_o	External radius of hollow cylinder (mm)
R_{ij}	Rate of component *i* in jth reaction (mol kg^{-1} s^{-1})
T	Temperature (K)
T_{ic}	Critical temperature of component *i* (K)
V	Number of moles in vapor phase
z_i	Number of moles of component i
Greek Letters	
α_g	Intra-particle vapor fraction
α_{ij}	Stoichiometric coefficient of component *i* in reaction *j*
η	Overall effectiveness factor $(\frac{\int \sum_{i,j} \alpha_{ij} R_{ij} dv}{(\sum_{i,j} \alpha_{ij} R_{ij})_{max} V_p})$
ρ_p	Density of catalyst pellet (kg/m^3)
ε	Catalyst pellet porosity
τ	Catalyst tortuosity

7. References

1. D. A. Wood, C. Nwaoha, and B. F. Towler, Gas-To-Liquids (GTL): A Review of an Industry Offering Several Routes for Monetizing Natural Gas, *Journal of Natural Gas Science and Engineering*, **9**, 196-208 (2012)

2. N. J. Mariani, C. Mocciaro, O. M. Martinez, and G. F. Barreto, Estimation of Effectiveness Factor for Arbitrary Particle Shape and Non-Linear Kinetics, *Industrial & Engineering Chemistry Research*, **48**, 1172-1177 (2009)

3. Yi-Ning Wang, Yuan-Yuan Xu, Hong-Wei Xiang, Yong-Wang Li, and Bi-Jiang Zhang, Modeling of Catalyst Pellets for Fischer-Tropsch Synthesis, *Industrial & Engineering Chemistry Research*, **40**, 4324-4335 (2001)

4. Y. N. Wang, Y. W. Li, L. Bai, Y. L. Zhao, and B. J. Zhang, Correlation for Gas-Liquid Equilibrium Prediction in Fischer-Tropsch Synthesis, *Fuel*, **78**, 911-917 (1999)

5. Yi-Ning Wang *et al.*, Kinetic Modelling of Fischer-Tropsch Synthesis Over an Industrial Fe-Cu-K Catalyst, *Fuel*, **82**, 195-213 (2003)

375527 Applications of Computational Chemistry and Molecular Simulation for Product and Process Development

Abstract:

Applications of computational chemistry and molecular simulation for product and process development at Air Products and Chemicals, Inc., a manufacturer of industrial gases and chemicals, will be discussed. The company serves customers in industrial, energy, technology, personal care, and electronics markets worldwide with a unique portfolio of atmospheric gases, process and specialty gases, performance materials, and equipment and services. Example applications of computational chemistry and molecular simulation will be selected from business areas including industrial-scale gas production, polymers, and electronics. Classical and quantum chemical methodologies, as well as microkinetic modeling and structure-activity correlation techniques, have been employed and will be discussed.

Force Field Assessment for Predicting Vapor-Liquid Equilibrium of Water in Ionic Liquids

Eliseo Marin-Rimoldi and Edward Maginn
Department of Chemical and Biomolecular Engineering
University of Notre Dame
Notre Dame, IN 46545 USA

Ionic Liquids (ILs) have emerged as potential substitutes for common organic solvents for industrial applications. ILs present many interesting thermodynamic and transport properties that suggest their utilization in carbon capture, catalysis or energy storage technologies. These properties can be tuned by means of modifying the constituent ions or by varying the concentration of the impurities present in the system. Many experimental and computational studies have pointed to the fact that ILs are hygroscopic in nature and that the presence of water can have a dramatic effect on IL properties such as density, viscosity or surface tension. In this work, Gibbs ensemble Monte Carlo simulations were used to calculate absorption isotherms of water in three imidazolium-based Ionic Liquids (ILs): 1-n-butyl-3-methylimidazolium hexafluorophosphate ([C_4MIM][PF_6]), 1-n-butyl-3-methylimidazolium bis(trifluoromethylsulfonyl) imide ([C_4MIM][TF2N]) and 1-n-butyl-3-methylimidazolium chloride ([C_4MIM][Cl]). The hydrophobicity of the selected ILs depends on the anion and decreases as [Tf2N] > [PF_6] > [Cl]. An evaluation of several modified water models and various IL force fields was conducted. The selected water models are the TIP5P, TIP4P2005 and a collection of three site models based on the SPC geometry whose charges are scaled to mimic different electrostatic environments. In the case of the IL force fields, partial charges were also scaled by either 1.0 (full charge) or 0.8 (scaled charge). Scaling partial charges is a known method to account for polarization and charge transfer. It is found that most water models tested in the full charge [C_4mim][PF_6] overestimate the solubility. The best agreement is found when the three site model whose charges were derived from water immersed in a low dielectric continuum is used. Moreover, the pure phase equilibria behavior of these modified water model is presented, showing a significant deviation from experimental results. When the scaled charge [C_4mim][PF_6] force field is used, SPC presents an underestimation of the solubility, suggesting that an agreement can be obtained using a scaling between 1.0 and 0.8. In the case of [C_4mim][Tf2N], a quantitative agreement is found when a full charge IL model is used along with the regular SPC and an underestimation is observed using a scaled charge force field. Finally, a good agreement is found for [C_4mim][Cl] if scaled charges are used along with the SPC model. The main conclusion of the study is that, to consistently model water absorption, force fields that include effects such as polarization are likely necessary.

390607 Accurate Calculations of Partition Coefficients (Log POW and Log PMW) with Atomistic Simulation Methods

Abstract:

Octanol/water and membrane/water partition coefficients of a substance measure its solubility in octanol/membrane compared to water. They can be seen as simple models for its solubility in biological membranes, which ultimately is a rough estimate for toxicity. If a substance is hardly soluble in octanol, it is practically impossible for it to enter (human) cells, and therefore is less likely to be toxic. On the other hand, for novel drugs it might be important to penetrate the cell through the membrane, or even integrate into it. While for simple substances the log POW is concentration-independent, this is not true for ionic liquids. Being able to determine log POW values a priori from molecular dynamics simulations is an important step in virtual material design. To do so, we firstly present our methods to semi-automatically derive robust and reliable force field parameters for soft matter components. Secondly, we present a method to accurately calculate log POW values from atomistic computer simulations, even for substances with concentration-dependent log POW . Our method is based on the calculation of solvation free energies in water and octanol using thermodynamic integration. By way of using a simple association-dissociation model for the concentration dependence of the chemical potentials, we were able to reproduce concentration-dependent log POW values for a commonly used ionic liquids like [CnMIM][NTf2], [N1888][NTf2], and [P666(16)][NTf2]. Additionally we present results for simulated Log PMW values of these ILs generated with the umbrella sampling technique.

References:

[1] M. Hülsmann, T. Köddermann, J. Vrabec, and D. Reith: “GROW: A Gradient-based optimisation workflow for the automated development of molecular models”, Computer Physics Communications 181, 499–513 (2010). DOI:10.1016/j.cpc.2009.10.024

[2] T. Köddermann, D. Reith, and R. Ludwig, "Force Field comparison on various model approaches – how to design the best model for the ionic liquid family [CnMIM][NTf2]," Chem. Phys. Chem. 14, 3368-3374 (2013). DOI: 10.1002/cphc.201300486

[3] T. Köddermann, D. Reith, and A. Arnold, "Calculating concentration-dependent Log Pow values from Atomistic Computer Simulations," J. Chem. Phys. B ,117, 10711-10718 (2013). DOI: 10.1021/jp405383f

We have built a specialized united atom model for Tween® 80 (Polyoxyethylene (20) sorbitan monooleate), based on the $GROMOS53A6_{oxy+D}$ force field, and applied it to several different Tween® 80 components, including two stereoisomers, three constitutional isomers, and three structures with 1, 2, and 3 tails, to represent components in the Tween® 80 commercial mixture. The structure of a preassembled micelle containing 60 molecules is determined, including the distribution of Tween® tail and ethylene oxide head groups and is found to be insensitive to stereo-isomerization, but sensitive to changes in relative lengths of the four polyoxyethylene head groups. The rate at which a Tween® 80 spherical micelle engulfs a thin oil layer is studied, including four stages through which equilibrium is reached. At the air-water and oil-water interfaces, the interfacial tension is significantly lower for the constitutional isomer with a shorter W head group, which attaches the tail to the sorbitan ring, and for Tween® 80 isomers with more than one tail group, than for the canonical Tween structure with a single tail and a W head group containing five oxyethylene monomers. The results indicate possible scope for improvement in the design of polyoxyethylene sorbitan oleates with improved surface tension reduction or better spreading at the oil-water interface.

372457 Computational Screening of Cellulosic Excipients for Drug Delivery

Abstract:

Solid dispersions of cellulosic polymers such as hydroxypropyl methylcellulose (HPMC) and hydroxypropyl methylcellulose acetate succinate (HPMCAS) are widely used in the oral delivery of poorly soluble active pharmaceutical ingredients (APIs). The presence of three substitution positions on each monomeric unit of cellulosic polymers facilitates the chemical design of excipients with varying degrees of hydrophobicity. However, the large design space of cellulosic polymers also poses a technological challenge of designing polymers with optimal properties for a given API. We use atomistic molecular dynamics simulations to study excipient-API interactions and compare the efficiency of different excipients for representative APIs. Simulations reveal formation of excipient-API complexes for some of the excipients and drugs, which results in a reduction of drug aggregation and crystallization. Beyond a certain excipient weight percent, the excipient forms a "gel"-like phase spanning the simulation box with API molecules encapsulated inside, which considerably slows API diffusion. Substantial differences are observed with variations in methyl, hydroxypropyl, acetate, and succinate substitution levels in the excipients. Further, the protonation state of succinate groups in HPMCAS substantially influence the complexation behavior, signaling possible excipient response to pH changes in the stomach and GI tract. Simulations are also performed with seed drug crystals to understand excipient effect on crystal nucleation and growth. Quantitative measures developed in this study are expected to prove useful as computational screening tools in excipient design.

APPLICATION OF MOLECULAR SIMULATIONS: MOLECULAR UNDERSTANDING OF ELECTROSPINNING

I. Nezbeda, F. Moucka, J. Jirsak
Faculty of Science
J. E. Purkinje University
Usti nad Labem, Czech Republic
ivonez@icpf.cas.cz

Nanospider technology is a method to produce nanofibers by electrospinning from the free surface of polymer solutions. Due to the lack of understanding of the process at the molecular level, the actual setup of the process (strength of the electric field, composition of the solution, temperature, etc.) has been based so far on the trial-and-error.

In the presentation the process will be briefly described, the main problems identified, and the role of molecular simulations (i) to understand what is going on in the system during the spinning and (ii) to reveal relations between various molecular and/or thermodynamic properties affecting the process detailed.

Regions accessible to simulations are identified along with potential methodologies.
We report first simulation results focusing on the interplay between the ions, electric field and water molecules, and on the onset of the surface waves, the origin of the Taylor cones, and stability of the jet. Furthermore, of a particular interest is the orientational arrangement of the water molecules in the interphase layer because this arrangement seems to be responsible for all the observed phenomena. All these results provide the necessary input information into semiempirical and theoretical macroscopic methods leading to improved efficiency of the production.

Vapor-Liquid Coexistence Data for p-cymene using Equation of State Methods and Monte Carlo Simulations

Madakashira Harini[†], Jhumpa Adhikari[†] and K. Yamuna Rani[*]

[†]*Department of Chemical Engineering, Indian Institute of Technology Bombay, Mumbai-400076*
**Chemical Engineering Division, Indian Institute of Chemical Technology, Hyderabad-500007*

P-cymene, a naturally occurring organic compound, is produced by distilling turpentine oil and is also obtained as a byproduct in the manufacture of sulphite paper pulp. It finds wide applications in the manufacture of fragrances, fine chemicals and pharmaceuticals either as reaction intermediate or solvent. In addition, it is used as solvent in recycling polystyrene, where p-cymene, itself is recycled after use by distillation. The experimental vapor liquid equilibria (VLE) data of p-cymene at high temperatures (above 523.15 K) is not reported in the literature.[1] As performing experiments at high temperatures is difficult and expensive, in this study we report the estimation of thermodynamic properties using theoretical methods such as equation of state (EoS) methods[2] combined with group contribution based parameters and Monte Carlo (MC) simulations.[3] Three EoS methods, Soave-Redlich-Kwong (SRK), Peng-Robinson (PR) and volume translated Peng-Robinson (VTPR); and two molecular simulation techniques, Gibbs ensemble Monte Carlo (GEMC) and grand canonical - transition matrix Monte Carlo (GC-TMMC) have been employed for prediction of the VLE. The estimates of properties, including vapor pressures, heats of vaporization (ΔH_{vap}), coexistence densities and critical properties from both the approaches have been compared with available experimental data.

The VLE curves have been determined by EoS employing the critical property data and acentric factor value as predicted by the Marrero-Gani based group contribution models.[4] The GEMC-NVT and GC-TMMC simulations have been performed using MCCCS *towhee* package[5]. The interactions in the p-Cymene system have been modelled utilizing the transferable potentials for phase equilibria -united atom (TraPPE-UA) force field; where an unknown torsion has been determined from the density functional theory analysis at the B3PW91/6-31++G(d,p) level of theory using *Gaussian03* package. The vapor pressures as predicted from PR and VTPR EoS, and GC-TMMC simulations are in good agreement with the experimental data (limited to temperatures up to 523.15 K). The GEMC-NVT predicted vapour pressures have been found to be in agreement with VTPR and PR EoS results with increase in temperature. The VTPR EoS predicted ΔH_{vap} data is in good agreement with the experimental data in literature; while the GEMC-NVT predicted ΔH_{vap} data lie between the predictions of VTPR and SRK EoS. The estimated coexistence liquid densities from GEMC-NVT and GC-TMMC coincide and lie between the predictions of PR and VTPR EoS at temperatures below 475 K, and are in better agreement with PR EoS predictions as temperature increases. However, the predicted coexistence vapor densities from GEMC-NVT and GC-TMMC are not in agreement. The

GC-TMMC predicted vapour densities match well with the predictions of SRK EoS. The GEMC-NVT results underestimate the vapour densities when compared to GC-TMMC, VTPR, PR and SRK EoS at high temperatures and the vapour density values lie close to the PR EoS predictions at temperatures below 475 K. It is to be noted that the experimental data on coexistence densities is not available for comparison. The estimated critical properties from GC-TMMC and GEMC-NVT simulations complement with experimental data with the exception of critical volume where only GC-TMMC prediction is in agreement.

References

1. Tsonopoulos, C.; Ambrose, D. *J. Chem. Eng. Data* ***40***, 547–558 (1995); Riddick, J.A., Bunger, W.B., Sakano, T.K.Wiley Interscience, New York 1986; Timmermans, J. Elsevier, New York, 1965.
2. Schmid, B. & Gmehling, J. *The J. Supercrit. Fluids* **55**, 438–447 (2010).
3. Frenkel, D.; Smit, B. Academic Press: San Diego, 1996.
4. Marrero, J. & Gani, R. *Fluid Phase Equilib.* **184**, 183–208 (2001).
5. MCCCS Towhee. Available at: http://towhee.sourceforge.net.

378053 Towards Transparent Aluminum

Abstract:

From the Stone Age, to the Bronze Age, to the Iron Age and beyond, it is often said that civilizations are defined by their materials. Today we are on the verge of a new Age, where simulation-based engineering and science will drive the syntheses of new types of materials that are designed - not simply selected - for specific applications. In the future, materials will be made on the spot, when desired, to precise specifications; that is, on demand. As simulators, we control the fate of this new Age of Materials on Demand because it is our tools that will make real-time material design and optimization possible. We discuss design strategies for materials on demand, and introduce new statistical mechanical simulation approaches based on computational alchemy.

376396 Structure-Conductance Relationships in Atomic-Scale Junctions: Insights from Reactive Molecular Simulation

Abstract:

Molecular electronics, i.e., self-assembled molecular-based electronic systems, have been proposed as a possible alternative to the traditional Si-based components used in the nano electronics industry. While several promising examples have been created, much is still not well understood in these systems, often as a result of the disconnect between experiment and computation. Experimentally, the mechanically controllable break-junction technique has served as a platform for studying the electron transport properties of low-dimensional metallic structures and individual molecules. Computational studies of these junctions have provided a wealth of atomic-level insight, enabling the identification of important structural mechanisms that influence the transport properties. However, the utility of these computational studies relies not only on the ability to accurately model the atomic interactions, but also on the ability to appropriately capture the non-ideality present in the experiments. Much of this non-ideality arises from thermal motion and the mechanical deformation process by which the junctions are formed, neither of which are typically considered in first-principles-based calculations of the molecular junctions.

In order to model the formation of junctions under conditions more representative of experiment, we use reactive molecular simulations, parameterized from first principles calculations, to model the spontaneous formation of single-molecule junctions of benzene-1,4-dithiolate (BDT) bridged between Au tips[1-3]. These simulations are used as input to high-fidelity first-principles conductance calculations to establish structure-conductance relationships of the more realistic molecular junctions[2,3]. This combined approach allows for significantly reduced computational cost, as compared to first principles calculations alone, enabling the accurate study of large system sizes, long time-scales, and many independent simulations, needed to fully understand the behavior of these junctions. Results detailing the role that structure, thermal motion, and stochasticity play on the conductance of break-junctions is presented, with an emphasis on the structural origin of "anomalous" conductance increases seen in experiment for Au-BDT-Au junctions[4].

[1] French WR, Iacovella CR, Cummings PT, ACS Nano, 6, pp 2779–2789, 2012
[2] French WR, Iacovella CR, Rungger I, Souza A, Sanvito S, Cummings PT, Nanoscale, 5 pp 3654-3659, 2013
[3] French WR, Iacovella CR, Rungger I, Souza A, Sanvito S, Cummings PT, J. Phys. Chem. Lett., 4 pp 887–891, 2013
[4] Bruot C, Hihath J, Tao, NJ, Nature Nano. 7, 35-40, 2012

376417 Number Crunching or Solving Problems? Peter's Impact on My Research Philosophy

Abstract:

Peter Cummings has made tremendous impact in the chemical engineering community because he has shown that massive computer simulations, when performed accurately and carefully, can explain experiemntal observations, innovate stagnant research fields, and even open up new research directions. In other words, he contributed enormously in making the simulation community relevant to our society. In this contribution I will discuss some of the research directions that originated when I was a post-doc in Peter's laboratory. These range from water desalination to nanoparticle design to lubrication. The goal is to elucidate Peter's impact on the research philosophy implemented in my group.

391908 Ionic Conductivities of Ionic Liquids and Solutions Containing Ionic Liquids

Abstract:

Ionic Conductivities of Ionic Liquids and Solutions Containing Ionic Liquids

Chaojun Shi, Liyuan Sun, Samuel Seo and Joan F. Brennecke*

Department of Chemical and Biomolecular Engineering

University of Notre Dame

Notre Dame, IN 46556

jfb@nd.edu

Ionic liquids have been considered for numerous electrochemical applications, ranging from electrolytes in lithium ion batteries and supercapacitors to media for electroplating. Their extremely low volatility presents the possibility of safer devices and lower environmental impact than aqueous acid and organic-based electrolyte solutions. Unfortunately, most ionic liquids have ionic conductivities that are too low for practical applications. Using a series of imidazolium and phosphonium ionic liquids with aprotic heterocyclic anions, we explore the key features that determine ionic conductivity. Conductivity, viscosity, density and ion diffusivity (from pulse gradient spin echo NMR spectroscopy) measurements indicate that both bulk viscosity and the association between the anion and the cation (as determined from Walden plots) determine the conductivity. We show how the cation-anion association that influences conductivity also influences the availability of the anion for reaction (with CO_2).

362160 Simulations of Thermodynamic and Transport Properties for Aqueous Mixtures

Abstract:

The development of accurate force fields and simulation methods for prediction of thermodynamic and transport properties of water and aqueous solutions are important topics to which Prof. Peter Cummings has made major contributions. This talk will focus on recent efforts in the speaker's group to model properties of water-electrolyte-CO_2 systems over a broad range of temperatures and pressures. The properties of interest include activity and osmotic coefficients of electrolyte solutions, the phase behavior at elevated pressures, and the diffusivity and viscosity of CO_2-rich and H_2O-rich phases at geological storage reservoir conditions. Deficiencies of common fixed-point-charge models for the prediction of mixture properties are identified and improved models are proposed.

Effects of Crowders, Inhibitors and Interfaces on A-beta Peptide Aggregation Using Coarse-Grained Simulations

In Alzheimer's Disease (AD), the amyloid β (Aβ) peptide aggregates in the human brain and forms oligomer and fibril structures, causing a decline in thought processes and memory, ultimately resulting in dementia and death. The intracellular and extracellular environments of the brain contain a variety of macromolecules and structures that can occupy anywhere from 7% to 40% of the total volume of their respective mediums with cellular cytoplasm typically falling within the 20% to 30% range. To examine the effect of crowding on protein aggregation, discontinuous molecular dynamics (DMD) simulations combined with an intermediate resolution protein model developed in the Hall group, PRIME20, were applied to a peptide/crowder system. The systems contained 192 Aβ(16-22) peptides and crowders of diameters 5Å, 20Å, and 40Å, represented here by simple hard spheres, at volume fractions of φ_c=0.00, 0.10, and 0.20. The effects of varying the crowder volume fraction and crowder diameter were examined. Results show that aggregation is promoted when the crowder volume fraction is increased because of an increase in effective concentration. Additionally, aggregation is promoted when the crowder diameter decreases at a constant volume fraction. Increasing crowder volume fraction and decreasing crowded diameter leads to more rapid oligomer formation and fibril seeding, ultimately eliminating the lag phase of aggregation and altering the mechanism of aggregation. The aggregation mechanism shifts from the reaction-limited dock-lock mechanism in dilute solution to a mechanism more characteristic of hydrophobic collapse followed by subsequent rearrangement into fibrils in a crowded environment. Special attention was paid to the amount of dimers, trimers,tetramers, pentamers, and hexamers formed during the simulations because of the recent focus on small oligomers as toxic agents in AD. We show that 40Å crowders increase the amount of dimers present in our simulations and 5Å crowders promote the formation of larger oligomers, in this case hexamers. Our analysis shows that the size of the crowder relative to the peptide dictates what type of oligomer is preferentially formed. We also show preliminary results for the effects of adding aggregation inhibiting molecules to the simulation or imposing an oil-water interface in the simulation cell.

376325 Molecular Modeling Insights into the Mechanisms for Viscosity Trends in Ionic Liquids

Abstract:

Ionic liquids, defined as salts that in their pure state melting at a temperature below 100 ^{o}C, have come under intense study over the past decade due to their unique properties such as negligible vapor pressure, ionic conductivity and excellent solvation properties. One of the major drawbacks of ionic liquids, however, is that they tend to be more viscous than conventional organic solvents. In this work, we report the results of molecular dynamics simulations in which the viscosity of pure ionic liquids and those mixed with low viscosity diluents is computed. We show that the characteristic time associated with local ion exchange correlates well with the viscosity of the liquid. Using this concept, we propose several structural modifications that can be made which should lower the viscosity of ionic liquids.

On the nature of the active sites for HCOOH decomposition on supported Au catalysts

Sha Li, Suyash Singh, Ronald Carrasquillo-Flores, Ana C. Alba-Rubio, James A. Dumesic, Manos Mavrikakis*

Department of Chemical & Biological Engineering

University of Wisconsin-Madison

Formic acid (HCOOH) is a major byproduct in biomass processing[1] and a promising liquid hydrogen carrier that can be used for *in situ* hydrogen production. Well-dispersed Au nanoparticles are found to be highly reactive for HCOOH decomposition with 100% selectivity toward CO_2 and H_2.[2-3] In this work, an integrated approach that combines state-of-the-art density functional theory (DFT) with reactivity experiments and microkinetic modeling is presented to unravel the detailed reaction mechanism and the nature of active site on Au catalysts. DFT calculations were performed on Au(111), Au(100) and Au(211) model surfaces to obtain initial guesses for the surface reaction energetics needed for the microkinetic model. Then by adjusting the DFT derived binding energies and activation energies in the microkinetic model, good agreement is achieved between the model predicted reaction rates and reaction orders with the experimental results collected on Au/SiC catalysts. The reaction is predicted to proceed entirely through the formate (HCOO) mediated route which is comprised of three elementary steps: HCOOH*→HCOO**+H*, HCOO**→CO_2*+H*, H*+H*→H_2. The large adjustments made to the DFT derived parameters in order to obtain good fits with the experimental data suggest that none of Au(111), Au(100) and Au(211) provide an accurate representation of the active site for this reaction. In fact, the active site might be an even more under-coordinated Au site with coordination number ≤ 6. Then, *ab initio* molecular dynamics simulations were carried out to determine the structures of small gold clusters which generally contain a high concentration of under-coordinated sites. Theoretical investigations of HCOOH decomposition on several Au cluster models further confirm our conclusion regarding the importance of highly under-coordinated Au sites in this chemistry.

References:

(1) Bond JQ, Alonso DM, Wang D, West RM, Dumesic JA. *Science* **2010,** *327*, 1110.

(2) Ojeda M, Iglesia E. *Angew. Chem. Int. Ed.* **2009,** *48*, 4800.

(3) Yi N, Saltsburg H, Flytzani-Stephanopoulos M. *Chemsuschem* **2013**, 6, 816.

First-Principles Based Kinetic Monte Carlo Simulation of Catalytic Systems over Metal Particles

Eric Dybeck and Matthew Neurock
University of Virginia
AICHE Annual Meeting November 17, 2014

1

Background

Macro Scale:

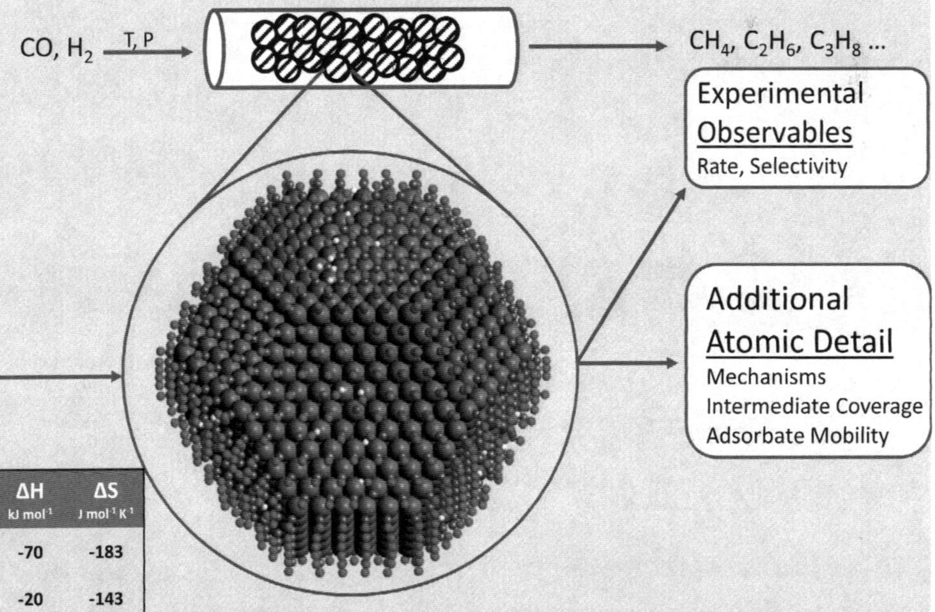

Micro Scale:

Event	ΔH kJ mol^{-1}	ΔS J mol^{-1} K^{-1}
CO -> CO*	-70	-183
H2 -> 2H*	-20	-143
CO*+H* -> HCO*	60	0
HCO*+H*->CH*+OH*	9	0

2

2D-Surface KMC 3D-Particle KMC

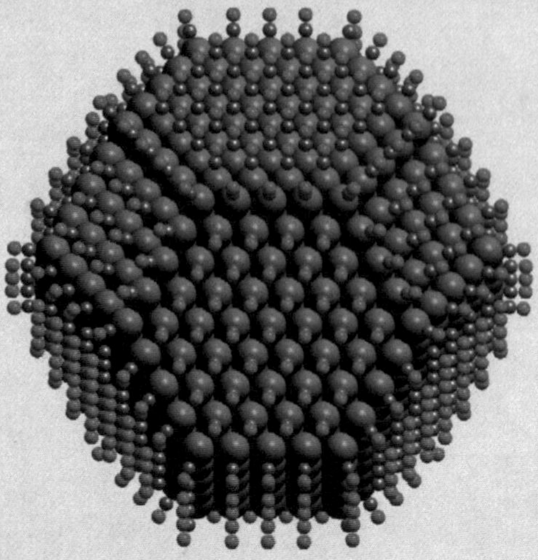

3

2D-Surface KMC 3D-Particle KMC

Adlayers on surfaces are constricted to perpendicular orientation

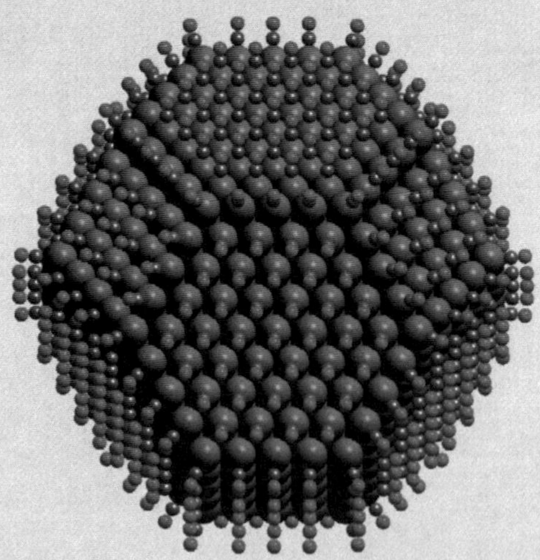

Adlayers on particles can realign to reduce repulsion

4

2D-Surface KMC 3D-Particle KMC

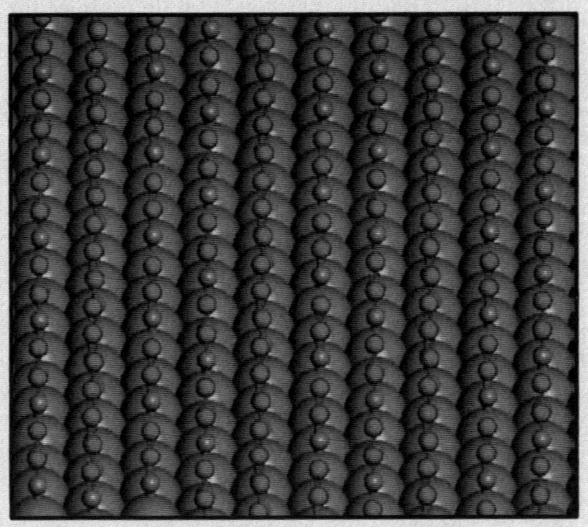

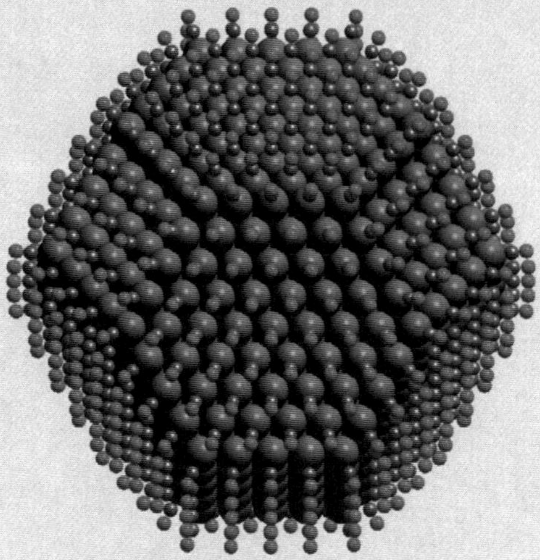

Adlayers on surfaces are constricted to perpendicular orientation

$\theta_{CO} < 0.75$ ML

Adlayers on particles can realign to reduce repulsion

$\theta_{CO} > 1.0$ ML

5

Coordinatively Unsaturated Atoms

<u>**Example Reactions**</u>
Fischer-Tropsch Synthesis
Ammonia Synthesis
Methane Reformation
CO Oxidation

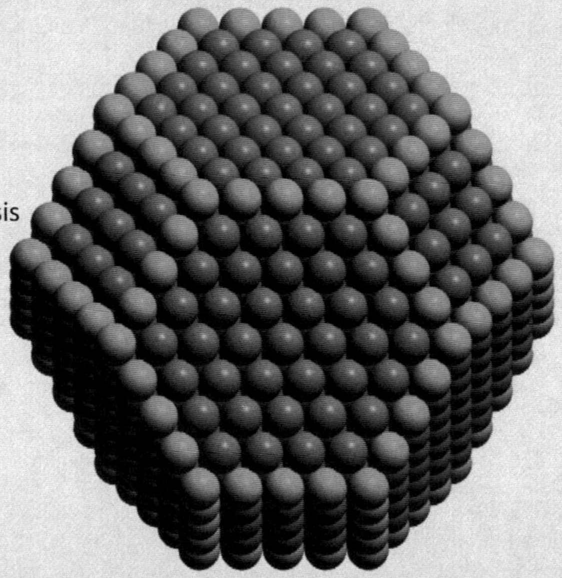

Particles contain coordinately unsaturated sites which exhibit markedly different activity and binding properties.

6

Coordinatively Unsaturated Atoms

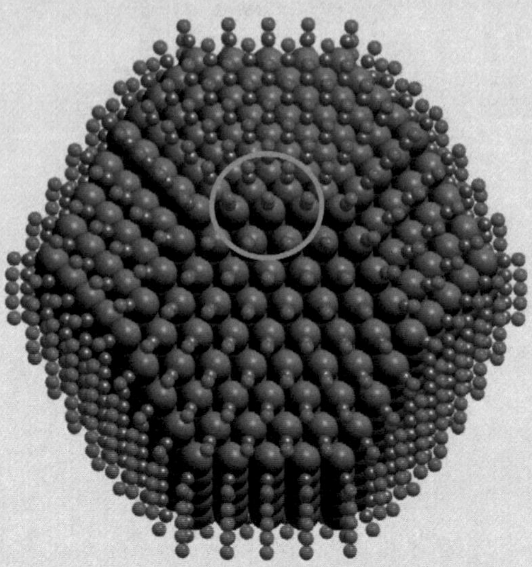

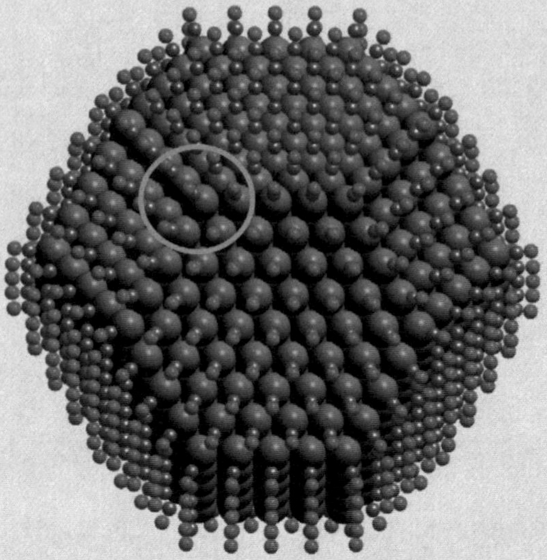

Unsaturated sites on corners and edges can bind to more than one adsorbate

7

Coordinatively Unsaturated Atoms

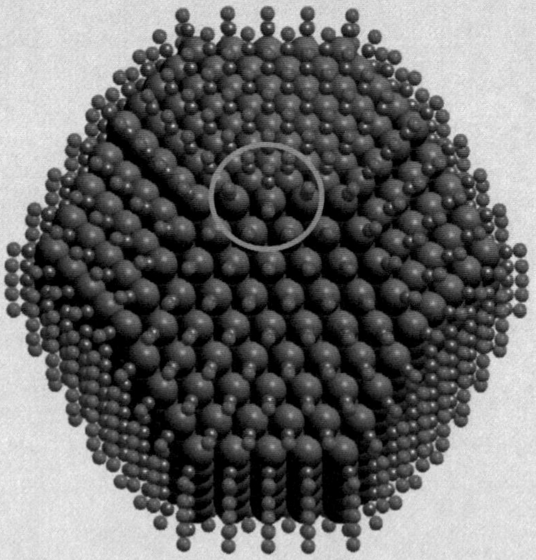

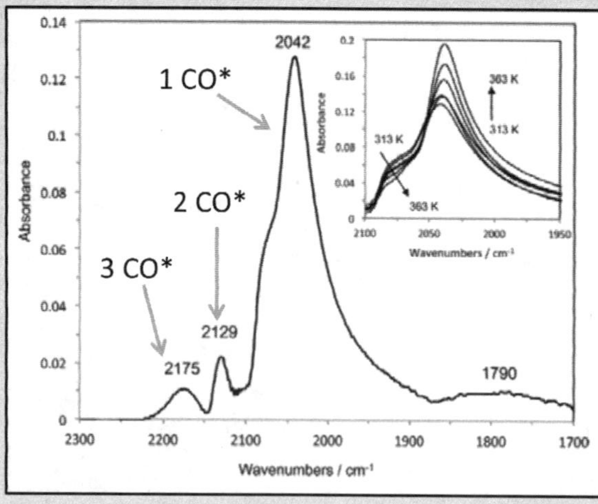

Unsaturated sites on corners and edges can bind to more than one adsorbate

Loveless, B, et al *Journal of the American Chemical Society* (2013) **135** (16) 6107-6121

8

Building the Particle

9

Bond Order Conservation (Surface)

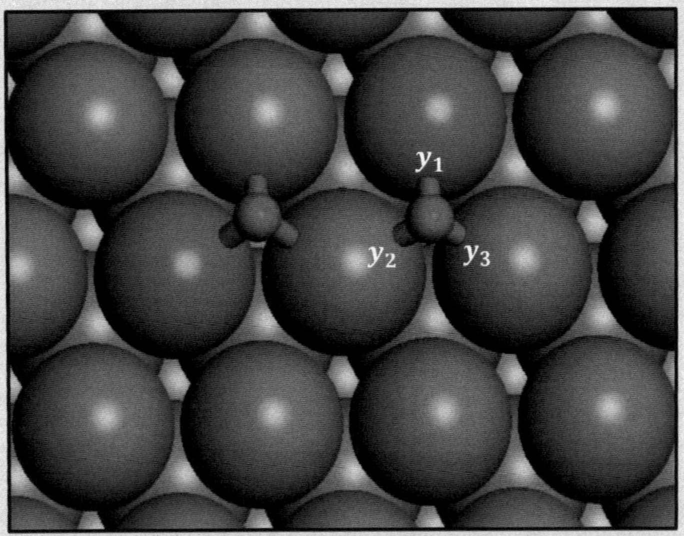

Adsorbate bonds are conserved to unity

$$Q_{A,M} = Q_o(A,M)(2y - y^2)$$

$$1 = \sum y_i$$

Shustorovich E., Sellers H., *Surface Science Reports* (1998) **31** (1) 0167-5729

10

Bond Order Conservation (Particle)

Functionality extended to coordinatively unsaturated atoms.

$$y_i = 1 - \frac{\prod_{j=1}^{n} Q_o\left(A_j, M_j, C_j\right)^{1-\delta_{ij}}}{\sum_{k=1}^{n}\prod_{p=1}^{n} Q_o\left(A_p, M_p, C_p\right)^{1-\delta_{kp}}}$$

Shustorovich E., Sellers H., *Surface Science Reports* (1998) **31** (1) 0167-5729

11

Intermolecular Interactions

Merck Molecular Force Field 94

$$E_{int\,ij} = \varepsilon_{ij}\underbrace{\left(\frac{1.07R_{IJ}^{*}}{R_{ij}+0.07R_{ij}^{*}}\right)^{7}\left(\frac{1.12R_{IJ}^{*\,7}}{R_{ij}^{\;7}+0.12R_{IJ}^{*\,7}}-2\right)}_{\text{Van der Waals}}+\underbrace{\frac{332.716\mathrm{q_i q_j}}{\mathrm{D}\left(\mathrm{R_{ij}}+\delta\right)^{n}}}_{\text{Electrostatic}}$$

Halgren, T. A. *Journal of Computational Chemistry* (1996), **17**, 490-519

12

Kinetic Monte Carlo Algorithm

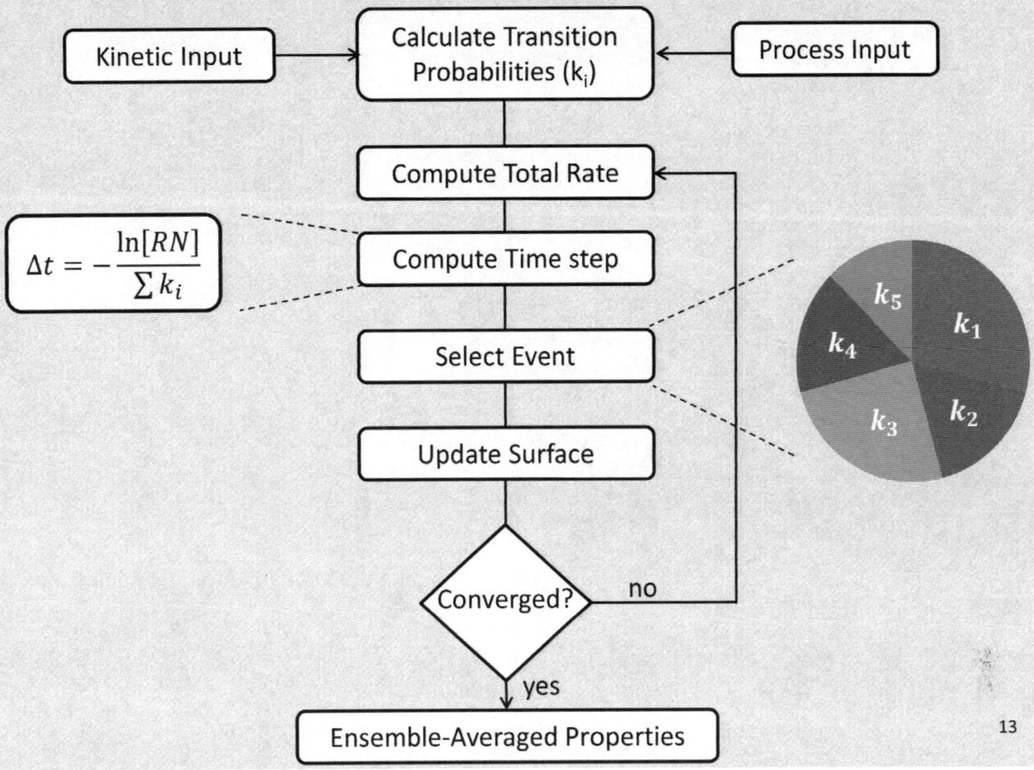

13

Kinetic Monte Carlo Simulation

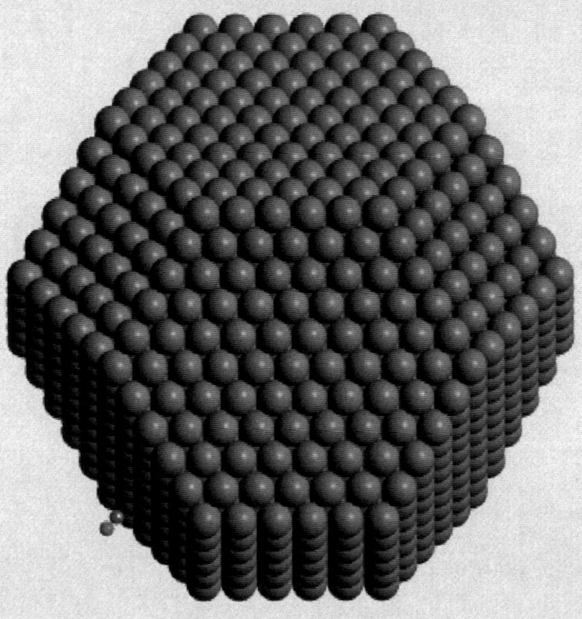

14

CO Oxidation on Platinum Nanoclusters

$$CO + * \leftrightarrows CO^*$$

$$O_2 + CO^* + * \leftrightarrows CO_2 + O^*$$

$$O^* + CO^* \leftrightarrows CO_2 + 2*$$

Allian, A., et al. *Journal of the American Chemical Society* (2011), **133** (12) 4498-4517

15

Model Binding Affinity Fitted to DFT

CO*

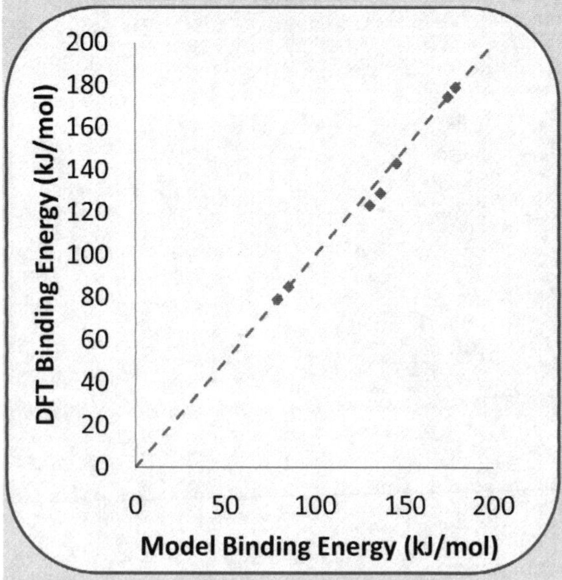

O*

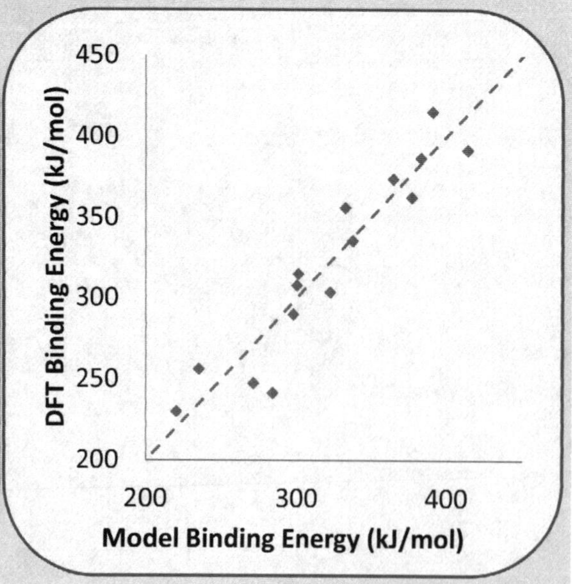

16

Simulated Pressure Effects

$$r_{CO2} = k_{eff}\frac{P_{O2}}{P_{CO}}$$

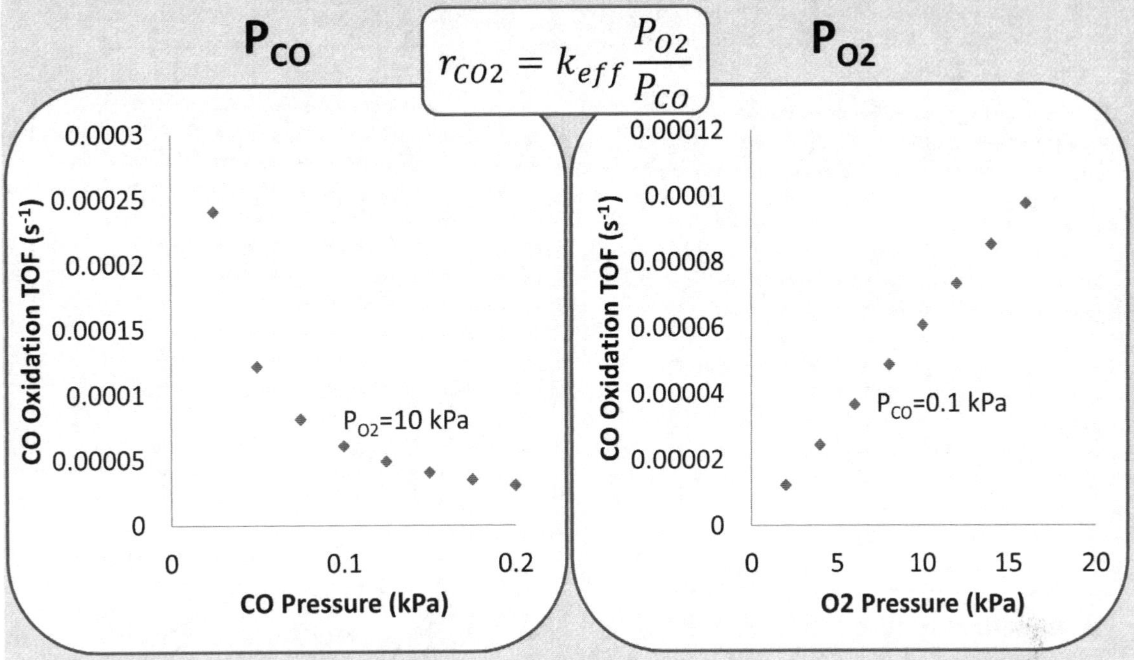

Reaction decreases with CO pressure and increases with O_2 pressure due to high coverage

17

CO Relaxation Effect (1.8 nm)

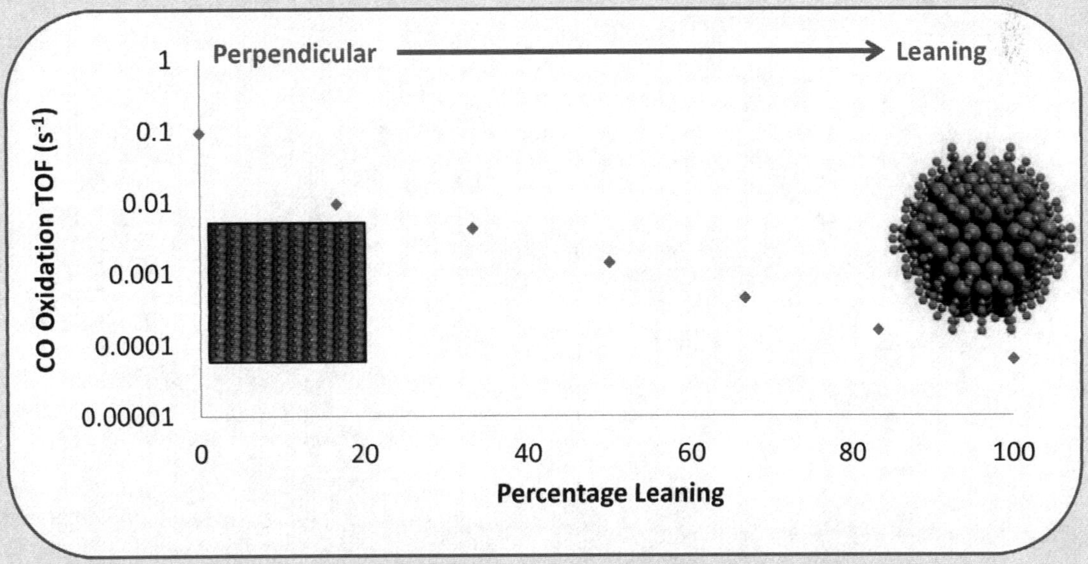

The effect of the CO adlayer relaxing on the particle alters the reaction rate by as much as 3 orders of magnitude on small particles.

18

Particle Size Effects (Simulation)

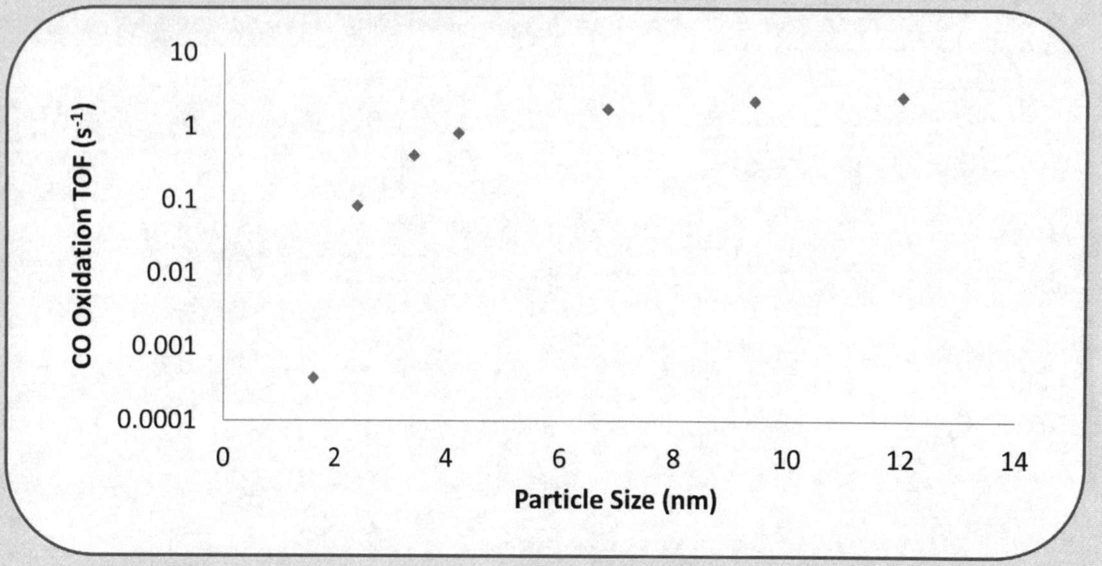

Without di-carbonyl sites, the reaction rate increases sharply with particle size.

19

Poisoning Leads to Structure Sensitivity

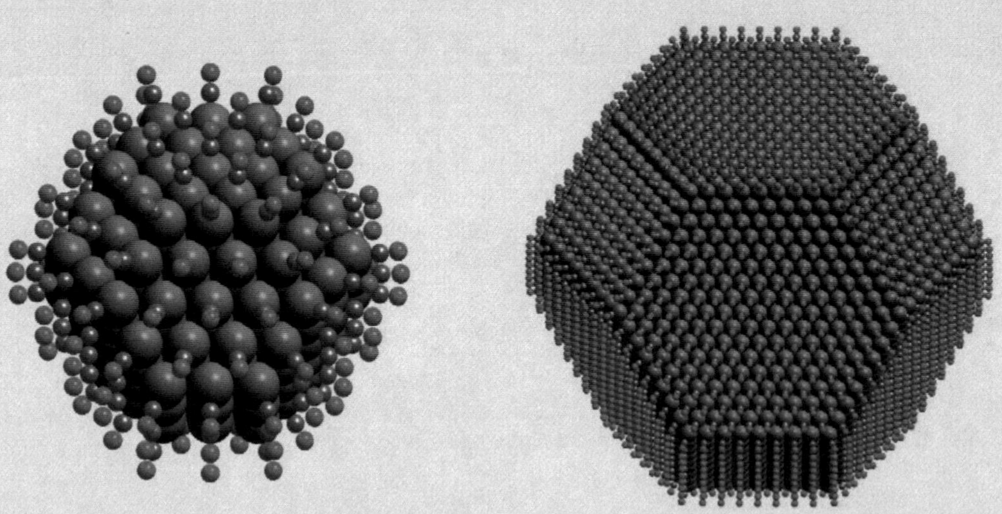

This structure sensitivity arises because the corners and edges are poisoned during the reaction, and because the CO adlayer has more room to fan out on smaller particles.

20

Incorporating Di-carbonyl Groups

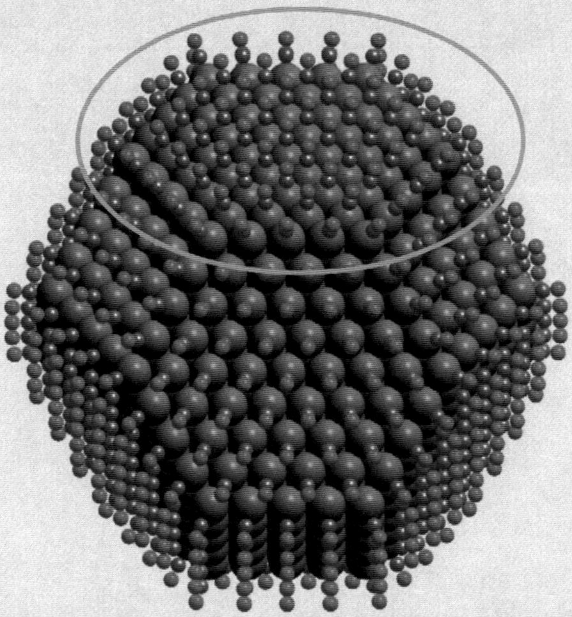

The formation of di-carbonyl groups destabilizes the CO* on the terrace

21

Incorporating Di-carbonyl Groups

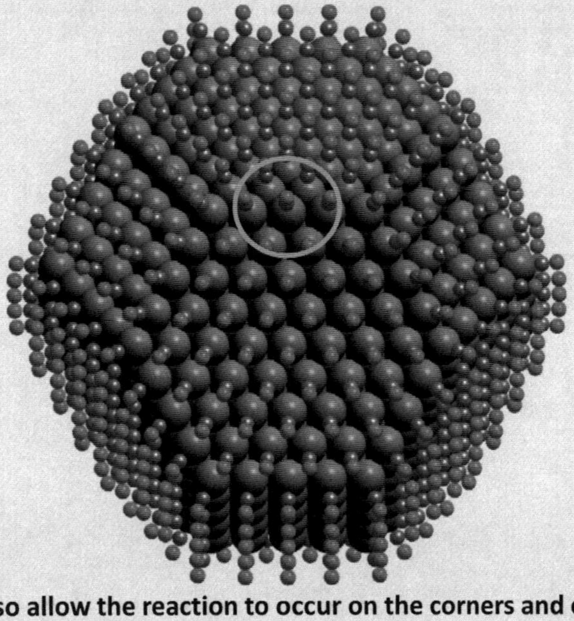

These extra sites also allow the reaction to occur on the corners and edges

22

Particle Size Effects (Simulation)

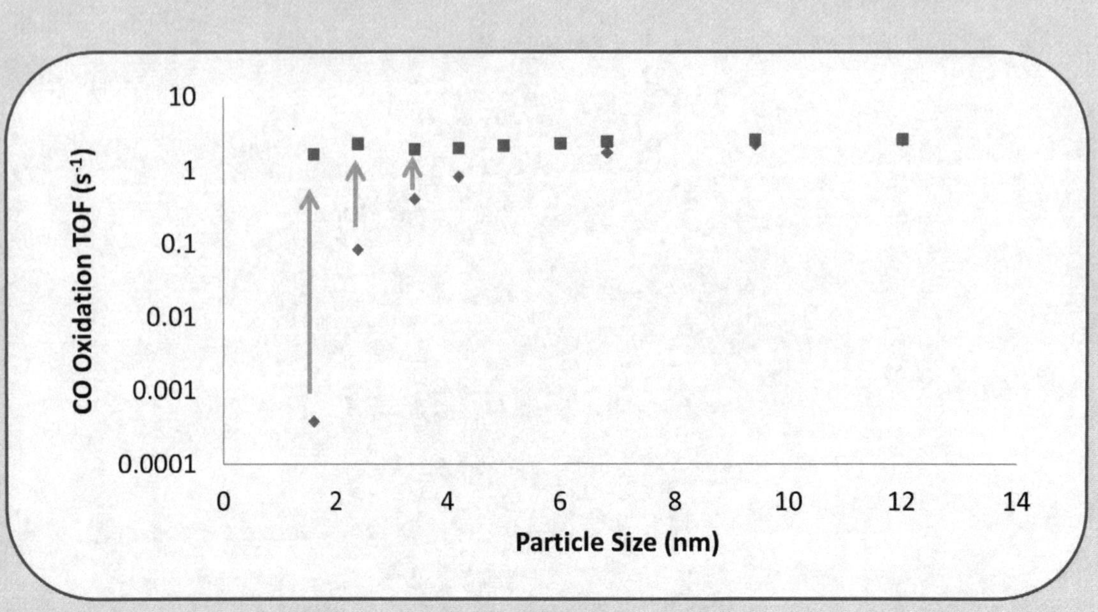

Without di-carbonyl sites, the reaction rate increases sharply with particle size.

However, this structure sensitivity largely disappears when dicarbonyl sites are included

23

Particle Size Effects (Experiments)

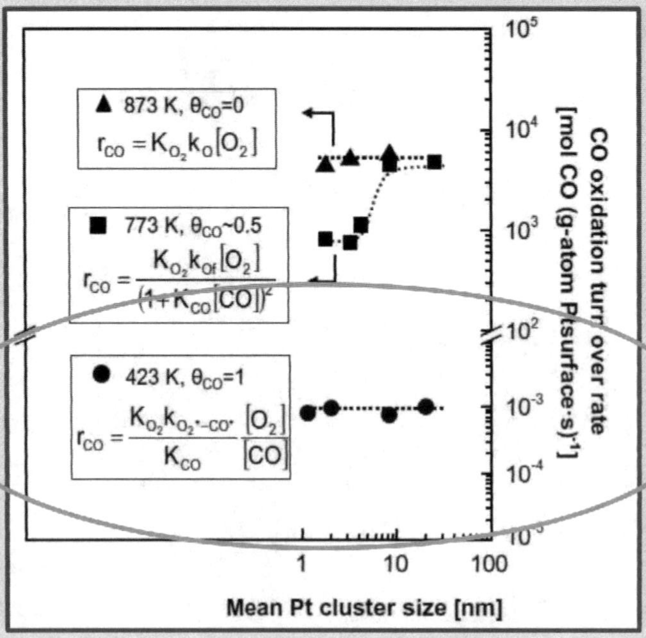

Garcia-Dieguez, M., Iglesia, E. *Journal of Catalysis* (2013) **301** 198-209

24

Summary

Particle Curvature Effects

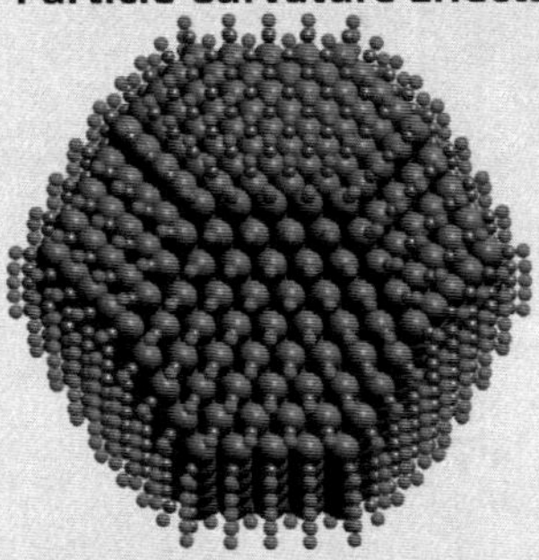

Coordinatively Unsaturated Sites

Generic Catalyst Surface

Extended Bond Order Conservation

25

Acknowledgements

Neurock Research Group

bp

BP XC[2] Program

26

379663 Ab Initio Kinetic Monte Carlo Simulations of H2 Spillover Control on a Bimetallic Catalyst

Abstract:

The heterogeneous catalysis of hydrogenation reactions plays a significant role within numerous areas such as the petrochemical, pharmaceutical and food industries. Precious metals including Pt, Pd and Ru are often employed as hydrogenation catalysts owing to their high reactivity towards hydrogen. Using such metals on a commercial scale can prove costly and so cheaper alternatives are being sought after.

Experiment has shown that doping a less reactive, cheaper host metal such as Cu, with a low concentration of a more reactive metal, such as Pd, increases the surface reactivity towards hydrogen with respect to the pure host metal [i]. The dopant metal provides a low energy pathway for hydrogen dissociation and facilitates spillover onto the host metal. Furthermore, control over the spillover, and hence the surface coverage, may be established through the selective binding of a spectator CO molecule at the minority metal site. The low energy pathway is selectively blocked, thereby inhibiting the dissociation or re-association of hydrogen, trapping the latter on the surface. This phenomenon has been termed the 'molecular cork effect' and may have implications in mediating the kinetics of hydrogen uptake in bimetallic catalysis. Though observed in experiment, the underlying cause of the molecular cork effect is still under investigation [ii].

First-principles kinetic Monte Carlo (KMC) was employed to elucidate the origin and underlying kinetics of the molecular cork effect. This multiscale theoretical approach entails a comprehensive molecular-level analysis at the nanoscale before extending to the catalyst level at the microscale. Density functional theory (DFT) has been used in order to calculate the ground state energies of a variety of hydrogen and CO binding configurations on the (111) surface of Pd, Cu and a Pd/Cu alloy. The binding energies computed support the observation that hydrogen binding at the minority Pd site within the Pd/Cu alloy is more stable than binding to pure Cu. Furthermore the affinity of CO for Pd during 'corking' is also investigated, with CO binding more strongly to Pd than Cu within the alloy. Additionally, Nudged Elastic Band (NEB) calculations were performed to compute the transition state energy for hydrogen dissociation and diffusion across the Pd/Cu alloy surface. The barriers calculated corroborate the experimental findings; diffusion barriers were found to be low, thereby promoting facile spillover. Moreover, the dissociation barriers specifically at the Pd minority site were found to be lower than on pure Cu or elsewhere on the alloy surface.

The binding and lateral interaction energies along with NEB barrier heights and gaseous formation energies are taken from our DFT calculations to parameterise KMC simulations. Subsequently, first-principles KMC was able to reproduce the observations seen in experiment, with the resulting data extracted from the simulations providing valuable insight into an unknown mechanism and also aiding in the design of improved hydrogenation catalysts.

---

[i] Tierney, H. L., Baber, A. E., Kitchin, J. R., Sykes, E. C. H., *Phys. Rev. Lett.*, 2009, **103**, 246102.

[ii] Marcinkowski, M. D., Jewell, A. D., Stamatakis, M., Boucher, M. B., Lewis, E. A., Murphy, C. J., Kyriakou, G.., Sykes, E. C. H., *Nat. Mater.*, 2013, **12**, 523.

Density Functional Theory (DFT) simulations have been used to successfully describe heterogeneously catalyzed reaction mechanisms in the gas phase, but describing liquid phase heterogeneously catalyzed reactions is still hindered due to a lack of understanding of the relationship between the solvent and the surface-bound species. This relationship has at least three components: 1) liquid-phase molecules interact with the surface species through dispersion interactions and/or chemical binding, 2) liquid molecules can adsorb to the catalyst, altering its chemical composition, and 3) the liquid environment can alter the adsorbate surface interaction[1,2,3,4]. These effects can also influence the reactant surface coverage on the catalyst by stabilizing adsorbates through hydrogen bonding or dispersion forces[5,6], thus increasing adsorbate chemical potentials and coverages. Both solvent molecules and coadsorbed molecules can influence catalytic activity through interactions with the reactants, which can alter binding energies and activation energies[7]. Therefore, solvation and surface coverage are intimately linked, and this link must be defined in order to understand catalysis in the liquid phase and, more importantly, to effectively design catalysts for liquid-phase reactions. Given that many catalysts for liquid-phase reactions are carried out over expensive transition metal catalysts, catalyst design focused on incorporation of cheaper catalyst materials would enable more widespread utilization of systems such as fuel cells.

In this work, we study the link between solvation and surface coverage in terms of Pt-catalyzed CO oxidation with OH, which is thought to be the rate-limiting step at direct methanol fuel cell (DMFC) anodes[8], and which is also important in the aqueous phase reforming of glycerol[9,10]. Simulations are conducted using the Vienna Ab-Initio Simulation Package[11] (VASP) with Projector-Augmented Wave[12] (PAW) pseudopotentials and the Perdue, Burke, and Erzenhof[13] (PBE) exchange/correlation functional. The DFT-D2[14] method is used to compute dispersion forces. Kinetic calculations are performed using the climbing image-nudged elastic band (ci-NEB) method[15]. We compare interaction energies between several explicit solvation models at various adsorbate coverages of CO and OH to determine the optimal solvation model. We consider a 2-D ice-like monolayer, a 3-D Ih ice lattice, a liquid-like 3-D water environment formed by "melting" the ice lattice at various temperatures using molecular dynamics (MD), and a 3-D water environment formed by randomly placing water molecules in the simulation box to form a vapor-like model. We also consider adsorbate coverages up to 1/3rd of a monolayer (ML).

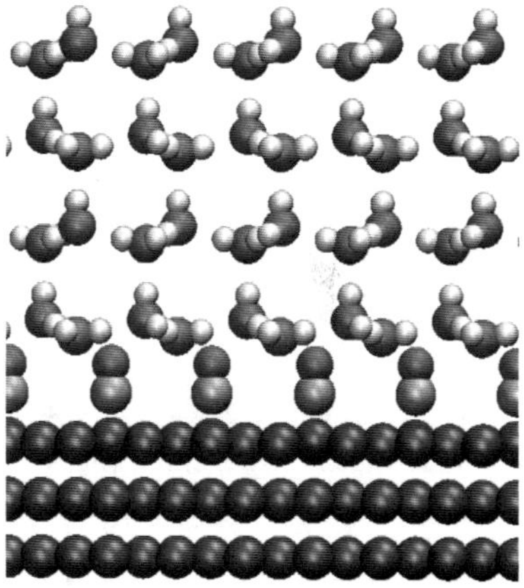
Figure 1: CO adsorbed on the Pt(111) surface under the 3D Ih ice lattice at a coverage of 1/9th ML.
Color code: Pt=grey, C=teal, O=red, and H=white.

Results show that adsorption site, solvation model, water configuration, adsorbate coverage, and dispersion forces all heavily influence the water-adsorbate interaction energies, which we calculate using

$$\text{Pt (aq)} + \text{Pt-CO (vacuum)} \rightarrow \text{Pt (vacuum)} + \text{Pt-CO (aq)}$$

Other important findings are that

1. Interaction energies cannot be correlated solely to the adsorbate dipole. For example, the interaction energies for FCC- and HCP-bound CO are 2.0 eV stronger than for atop-bound CO, despite that the dipole for atop-bound CO is more than three times larger (7.7 Debye versus 2.2 Debye).
2. Coverage has a remarkable effect on the interaction energies. For example, the interaction energies for CO adsorbates decrease by up to 1.0 eV as the coverage increases from 1/9th to 1/3rd ML.

3. Including dispersion corrections has a significant effect on the calculated interaction energies. For example, the interaction energy of atop CO at 1/9th ML increases by 0.8 eV when the D2 method is turned on.

We summarize our findings by providing suggestions about how to explicitly include solvation in computational catalysis at moderate computational expense.

References

1. Zope, B. N.; Hibbitts, D. D.; Neurock, M.; Davis, R. J.; *Science*, 2010, **330**:74-78.
2. Wang, H. F.; Liu, Z. P.; *J. Phys. Chem. C*, 2009, **113**:17502-17508.
3. Okamoto, Y.; Sugino, O.; Mochizuki, Y.; Ikeshoji, T.; Morikawa, Y.; *Chem. Phys. Lett.*, 2003, **377**:236-242.
4. Behtash, S.; Lu, J.; Faheem, M.; Heyden, A.; *Green Chemistry*, 2014, **16**:605–616.
5. Nørskov, J.; Rossmeisl, J.; *J. Phys. Chem. B*, 2004, 17886–17892.
6. Taylor, C. D.; Neurock, M.; *Current Opinion in Solid State and Materials Science*, 2005, **9(1-2)**, 49–65.
7. Getman, R. B.; Schneider, W. F.; Smeltz, A. D.; Delgass, W. N.; Ribeiro, F. H. *Phys. Rev. Lett.*, 2009.
8. H.A. Gasteiger, N. Markovic, J.P.N. Ross, E.J. Cairns, *J. Phys. Chem.*,1994, **98**:617.
9. Huber, *G. W.;* Shabaker, J. W.; Dumesic, J. A.; *Science* 2003, **300**, 2078.
10. Cortright, R. D.; Davda, R. R.; Dumesic, *J. A.; Nature* 2002, **418**, 964.
11. Kresse, G.; Furthmüller, J.; *Comput. Mat. Sci.*, 1996, **6**:15.
12. Blochl, P. E.; *Phys. Rev. B*, 1994, **50**:17953.
13. Perdew, J. P.; Burke K.; Ernzerhof, M.; *Phys. Rev. Lett.*, 1996, **77**:3865.
14. Grimme, S.; *J. Comp. Chem.* 2006, **27**:1787.
15. Henkleman, G.; Uberuaga, B. P.; Jonsson, H.; *J. Chem. Phys.*, 2000, **113**:9901.

376366 Ethylene Glycol Reforming over Pt (111): Modeling of C-C Versus C-O Bond Cleavage Selectivity in Vapor and Aqueous Environments

Abstract:

Catalytic conversion of biomass-derived oxygenates to fuels and value-added chemicals is a promising strategy in the search for renewable and sustainable energy sources. Decomposition chemistry on a metal catalyst is a complex network of multiple series and parallel pathways leading to formation of hydrogen, alkanes, and lighter oxygenates. The final product distribution ultimately depends on the sequence and competition of C-C, C-O, C-H, and O-H bonds scissions. Ethylene glycol (EG) is the simplest such molecule to contain all of C-C, C-O, C-H, and O-H bonds and a C/O stoichiometry of 1:1, and has received significant attention as a model molecule of larger biomass-derived polyols. While the reaction mechanism of EG reforming is to some degree understood at the metal-gas interface, lack of a well-established methodology for describing the influence of a complex liquid phase on a reaction across a solid-liquid interface has hindered similar theoretical studies in an aqueous environment.

In this talk, we present an application of our recently developed iSMS[1] methodology to study the mechanism of EG reforming over Pt (111) in water. Reaction free-energies and free-energy barriers for all elementary reactions in vapor phase are computed within the framework of periodic planewave DFT calculations. The effect of water is then included through a combination of Gaussian-type orbital calculations and COSMO-RS calculations as required by iSMS. Finally, detailed microkinetic models are developed for both vapor and aqueous phases to provide insights into the effect of an aqueous environment on the activity and selectivity of the metal catalyst.

1. Faheem, M.; Suthirakun, S.; Heyden, A. *J. Phys. Chem. C* **2012,** *116*, 22458-22462.

Semi-empirical thermochemical property estimation for biomass derivatives on transition metal surfaces

Density functional theory (DFT) has become an essential tool in estimating thermochemistry and activation barriers in heterogeneous catalysis. For small molecules, we have been able to elucidate reaction mechanisms and parameterize kinetic models that reproduce experimental data with high accuracy[1]. A shift towards modeling larger biomass derivatives interacting with metal surfaces introduces new levels of complexity, including longer computational time, multiple adsorption conformations, and a greater number of intermediates and elementary steps due to multiple pathways encountered in this chemistry. For example, furfural, an important biomass derivative, can undergo hydrogenation, hydrogenolysis, decarbonylation, and even ring opening depending on the metal catalyst[2], making it impractical to compute all possible intermediates, conformations, and reaction barriers with DFT, even on a single metal surface.

In this work, we present a DFT-parameterized semi-empirical scheme capable of predicting thermochemical properties of furan-related intermediates on metal surfaces. This model extends prior group additivity for open-chain molucles[3] to account for furan ring corrections, such as ring deformation. The properties include the heat of formation, entropy, heat capacity, and the dispersion contribution to the heat of adsorption, estimated using dispersion-corrected DFT (D3) and statistical mechanics formulae. For the proposed model, the vibrational and rotational contributions of gas-phase-like groups are estimated using G4-level calculations. We find that the group additivity scheme developed for open-chain molecules is inaccurate for furanics. We report 17 new groups, involving heteroatom-nearest neighbor interactions and 4 new corrections that account for furan ring deformation and the level of ring hydrogenation. Furthermore, we extend the thermochemistry scheme so the thermochemistry estimation can be applied to multiple metal catalysts as a generalization of the linear scaling relations[4] introduced for small molecules. We demonstrate how to construct energy diagrams for various catalysts by combining these methods with linear free energy relations. Comparison of the pathways of furfural chemistry on a select number of metals will be shown.

References

1. Salciccioli *et al.*, *Chemical Engineering Science* **2011,** *66* (19), 4319-4355.
2. Sitthisa, Resasco, *Catalysis Letters* **2011,** *141* (6), 784-791.
3. Salciccioli *et al.*, *The Journal of Physical Chemistry C* **2012,** *116* (2), 1873-1886.
4. Abild-Pedersen *et al.*, *Physical Review Letters* **2007,** *99* (1), 016105.

372257 CO Oxidation on O-Rich Pd(111): How Does Adlayer Structure Affect Reactivity?

Abstract:

The CO oxidation reaction is crucial in the function of catalytic converters found in personal vehicles. Despite its simplicity, several aspects of this reaction are still under investigation. Thus, a recent experimental study by Nakai *et al.* [J. Chem. Phys. 124, 224712 (2006)] proposed that CO oxidation on O-precovered Pd(111) exhibits markedly different rates at different temperatures as a result of different phases in the O overlayer exhibiting disparate reactivities. Thus, the *p*(2x2) phase appears as inert, whereas the (√3x√3)R30° and *p*(2x1) are reactive and exhibit distinct apparent activation energies and reaction orders with respect to O coverage. It still unclear however, whether there is a causal relation between the changes in the overlayer structure and the different reactivity, and if this is the case what is the mechanism giving rise to this effect.

To shed light into this question we perform first principles-based kinetic Monte Carlo (KMC) simulations using *Zacros*, our software package implementing the Graph-Theoretical KMC approach. These simulations incorporate a detailed cluster expansion Hamiltonian and Brønsted-Evans-Polanyi relations, in order to explicitly account for lateral interactions and their impact on the activation energies of elementary processes. Our simulations reproduce quantitatively the main features of the experimental measurements, and in particular the temperature dependence of CO adsorption and surface reaction rates. The ordering of the adsorbate layer is found to strongly depend on the strength of the lateral interactions but does not seem to have a significant role on the catalytic properties of the system. By analysing the appearance and realisation statistics of the elementary CO oxidation processes, we show that only a small subpopulation of events with rather low activation energies contribute to the observed rate, in agreement with previous observations on a different system, namely the NO oxidation reaction on Pt(111) [Wu et al. J. Catal. 286, 88 (2012)].

385736 Computational Screening of Metal-Organic Frameworks for Hexane Isomer Separation

Abstract:

Hexane isomers differ greatly in their octane number, and better separation of these isomers could lead to better processes for the production of high-quality gasoline. Current adsorbents can differentiate linear alkanes from mono- and di-branched alkanes via molecular sieving. However, the desired operation is to efficiently separate the di-branched isomers, which have the highest octane number, from the linear and mono-branched alkanes. Recently, a highly stable metal-organic framework (MOF) with triangular pores was synthesized by Long and co-workers, and it was demonstrated that the separation of hexane isomers could be done efficiently based on their shape[1].

It has been hypothesized that the exceptional separation capability of the MOF with triangular pores for hexane isomers arises from the shape and size of the pore, which can differentiate between different isomers. In order to test the generality of this hypothesis and to investigate the molecular-level details, we have created and computationally screened a database of thousands of existing metal-organic frameworks. The database contains a structurally diverse set of MOFs with various pore sizes, pore shapes, and other structural properties. The screening was carried out using a hierarchical strategy. First, thousands of MOFs were characterized by a descriptor-based classification of their pore shapes, and the Henry's constants were calculated for 5 different hexane isomers in all MOF structures. Next, grand canonical Monte Carlo simulations were performed to calculate the full adsorption isotherms for a subset of structures, and ideal adsorbed solution theory was used to predict the mixture behavior. Finally, full mixture simulations were performed for the most promising structures. The results were used to find correlations among structural properties, Henry's constants, and the selectivity and capacity in mixtures of hexane isomers.

References

1. Herm, Z. R.; Wiers, B. M.; Mason, J. A.; van Baten, J. M.; Hudson, M. R.; Zajdel, P.; Brown, C. M.; Masciocchi, N.; Krishna, R.; Long, J. R., Separation of Hexane Isomers in a Metal-Organic Framework with Triangular Channels. *Science* **2013**, 340 (6135), 960-964

376181 Applications of DFT Derived Force Fields for Olefin/ Paraffin Separation in MOFs

Abstract:

Grand Canonical Monte Carlo (GCMC) simulations have been widely used to predict adsorption uptakes in nanoporous structures, but the accuracy of these simulations is highly dependent on the quality of the force field used for modeling the intermolecular interactions. Significant advances have been made in recent years in the development of force fields for nanoporous materials from first principles calculations[1]. In this work, we will describe a DFT-based force field development approach for modeling olefin and paraffin interactions in MOFs. The results show that our DFT consistent force fields successfully describe adsorption of linear alkanes and alkenes in MIL-47 (V) and HKUST-1 MOFs. We will discuss the transferability of our HKUST-1 force field to other Cu open metal site MOFs and discuss the potential for using calculations based on these force fields to screen MOFs for olefin/ paraffin separations.

[1] H. Fang, H. Demir, P. Kamakoti, and D. S. Sholl, J. Mater. Chem. A 2 (2014) 274

Computation Ready and DFT Optimized MOF Crystallographic Structure Data

Dalar Nazarian and David S. Sholl*

School of Chemical and Biomolecular Engineering, Georgia Institute of Technology, 311 Ferst Drive NW, Atlanta, Georgia 30332-0100

*Corresponding author: david.sholl@chbe.gatech.edu

Metal Organic Frameworks (MOFs) have demonstrated highly selective adsorption properties and ultra-high surface areas. Thousands of already synthesized MOF structures are available in the Cambridge Structure Database (CSD). High-throughput computational screening of MOFs is a powerful tool for identifying top MOF candidates for target adsorption applications. Such screening requires a database of computation ready crystallographic structure data of MOFs. This work will describe a systematic procedure for assessing the quality of the experimentally refined MOF structure data available in the CSD. We will discuss DFT functional performance for predicting MOF ground state structure and compare the DFT-based ground state structures of a large set of MOFs to experimental structure data. We will also demonstrate the assignment of point charges to atoms comprising MOFs to make the structural data suitable for computations involving polar molecules while accurately representing the electrostatic potential in the pores of the materials as determined from quantum chemistry calculations.

375857 Grand Canonical Monte Carlo (GCMC) Simulations on the Effect of Pore Size on Electrosorption Capacitance during Capacitive Deionization (CDI)

Abstract:

Securing clean and potable water for future generations presents the technological challenge of developing energy efficient and cost effective desalination processes. Ninety percent of the water in the planet is either brackish or sea water.

A promising water desalination technology trying to take the leap from bench top to commercial application is capacitive deionization (CDI). During CDI, a low electrical potential (± 1.5 V maximum) is applied to an array of highly porous electrodes to enable the electrosorption of oppositely charged ions within the electrical double layer (EDL) formed at the surface of the electrodes. The separation process is reversible, and part of the electrical energy invested in separating the ions can be recovered during the discharge cycle.

Grand Canonical Monte Carlo (GCMC) simulations were used in this work to explore the effect of pore size on the equilibrium electrosorption capacity of sodium chloride ions during CDI at conditions of operation for desalination of sea water and brackish water. The scientific question driving this study was: does the structure of the electrode, basically pore size, contribute to the low separation efficiencies observed in CDI? Reports on experimental work state that only between fifteen and fifty percent of the BET available surface area of the electrodes is used during CDI.

It was found that lower uptake of chloride ions within the EDL occurs at slit-type pores with widths equal or larger than 12 nm, at all applied potentials explored. This observation implies that larger electrode pores do not contribute to the desalination process in the same extent as smaller pores, therefore negatively affecting the performance of CDI processes.

Molecular Modelling and Simulation of Wetting on Solid Surfaces

Stefan Becker, Martin Horsch and Hans Hasse

Laboratory of Engineering Thermodynamics (LTD), University of Kaiserslautern, Germany

E-mail: martin.horsch@mv.uni-kl.de

Wetting of solids is important for many processes. Molecular simulations are useful for investigating wetting and yield insight with atomistic resolution. The results solely depend on the choice of the force fields that are used for describing the intermolecular interactions. The simulations yield a consistent description of wetting and adsorption. They enable systematic studies of the influence of the molecular interactions and the surface structural properties on wetting.

In the present study, wetting phenomena are studied by molecular dynamics (MD) simulations with the massively parallel simulation code *ls1 mardyn* on the basis of classical force fields. Droplets resting on a solid substrate are simulated in a model system interacting via the truncated and shifted Lennard-Jones potential. The contact angles are obtained by evaluation of the density profile. The temperature is varied over a wide range of the vapour-liquid coexistence of the fluid. The dispersive fluid-solid potential strength is varied and the entire range between complete wetting ($\theta = 0°$) and dewetting ($\theta = 180°$) is observed. A correlation is given that reproduces the results quantitatively: The contact angle cosine varies linearly with the reduced fluid-solid potential strength and a power law term accounts for the temperature dependence. The study is extended for different models of the wall and the expected strong impact of the solid density on the contact angle is found and characterized. A variation of the droplet size shows a decreasing solid liquid surface tension for very small droplets which is due to the perturbation of the liquid in vicinity to the wall.

The pinning of the three phase contact line is examined by MD simulations of droplets of different radii resting on pillars of different shapes. The transition from pinning to depinning is found to be continuous on the molecular scale. Atomistically, the three phase contact is not a line but has a width of a few nanometers. The precursor layer has an important influence on the line pinning. It brings about the full development of the solid-gas interfacial tension.

388317 Atomistic Simulations of Structure and Mechanical Behavior of Primed and Unprimed Epoxy-Alumina Interfaces

Abstract:

The nature of the adhesive bonding between crosslinked thermosetting resins and inorganic or metallic substrates plays a critical role in a variety of materials applications including fiber or particle reinforced composites and bonding of metal parts in the aerospace and automotive industries. In the simpler cases, bonding between the organic and inorganic or metal components occurs via van der Waals dispersive and electrostatic interactions. However, frequently, additional primer compounds are introduced to form chemical bonds at the interface between the major system components.

In recent work, we have performed a series of investigations of the properties of crosslinked thermosets, including studies of the effect of resin architecture on small strain elastic constants and gelation studies in polyester and epoxy systems (1), which show good quantitative agreement with available experimental data. Accordingly we now perform extensions of these calculations to include interfaces with a solid substrate. Specifically, the present work focuses on model adhesive type systems in which crosslinked epoxy interacts with an alumina substrate similar to that produced by anodization pretreatment of surfaces prior to assembly using adhesives. Two types of model systems are considered, with the first containing only crosslinked epoxy and alumina, and the second containing in addition the commonly-used primer gamma-aminopropyltriethoxysilane in which the ethoxysilane moieties have been pre reacted with the alumina substrate prior to performing the amine-based crosslinking with the epoxy resin and curing agent.

Following the model building, the epoxy-alumina interface systems are equilibrated using the MedeA®-LAMMPS simulation environment (2, 3) before being to subjected to structural analysis to elucidate the nature of packing at the interface, accompanied by mechanical property calculation to compare and contrast the behavior of the systems when subjected to small and large deformations. Finally, we report on possible interface failure mechanisms by analyzing the model behavior at large deformations.

1. D. Rigby, C.M. Freeman, P.W. Saxe and B. Leblanc, *Computational Prediction of Mechanical Properties of Glassy Polymer Blends and Thermosets*, Proceedings of the 143rd Annual TMS Meeting, San Diego, CA, Feb 16-20, 2014.
2. Materials Design, Inc.; http://www.materialsdesign.com/medea
3. S. Plimpton, *Fast Parallel Algorithms for Short-Range Molecular Dynamics*, J Comp Phys, **117**, 1-19 (1995); http://lammps.sandia.gov

376871 Miming Acid Oil Component Interactions with Carbonate Reservoirs: A First Principles View on Low Salinity Recovery Mechanisms

Abstract:

Enhanced oil recovery by low salinity water (LSW) injection in carbonate reservoirs constitutes one of the main interests for petroleum industry. Lately, a molecular mechanism involving acid oil components and the replacement of surface Ca atom by Mg atom has been proposed. In this work, we studied the thermodynamic feasibility of this mechanism by the calculation of the propionic acid adsorption upon calcite (10-14) and Mg-calcite(10-14) surfaces, in the last one a Mg atom replace a Ca surface atom. We employed first-principles calculations based on Density Functional Theory, in vacuum and in solvent. For the last case we used a continuum solvent model implemented in the Car Parinello approach. In vacuum, we observed that the propionic acid adsorption energy upon both surfaces is negative, not in favor of the proposed mechanism. As a first approach to describe the reactivity in solvent, we include a water monolayer (ML) and we observe a decrease on the adsorption energy due to a change on the surface oxygen atom involved in the acid-surface hydrogen bond, but the acid remains stable upon both surfaces. Additionally, we explore the effect of adding an aqueous media on the adsorption energy. We employ the continuum solvent model and we obtain substantial changes on the energy adsorption due to the increasing of the acid-surface hydrogen bond. In addition, and in order to account for explicit interactions, we add an explicit water ML to this scheme finding a minor energy variation. In both scenarios the acid becomes unstable upon Mg while remains stable upon Ca, in favor of the proposed mechanism. Furthermore, the effect of salinity, temperature and pressure are explored in an effective way through the variation of the solvent dielectric constant (50-90). It can be inferred from our results that if Ca-Mg replacement occurs, this mechanism is thermodynamically feasible for the whole dielectric constant range studied.

376168 Glycoside Hydrolase Processivity from a Thermodynamic and Molecular Perspective

Abstract:

Cell wall polysaccharides represent a vast source of organic carbon on Earth. To access these carbohydrate polymers as food sources, micro-organisms have evolved cocktails of glycoside hydrolase enzymes to depolymerize individual chains from insoluble substrates employing a host of different mechanisms to locate and hydrolyze glycosidic bonds. These enzymes are key components of many commercial biomass conversion processes and are thus primary targets for protein engineering. My group uses molecular simulation to understand the fundamental mechanisms driving deconstruction of recalcitrant polysaccharides by carbohydrate-active enzymes.

Here, I will focus on two distinct vignettes related to how cellulases degrade cellulose. First, we will focus on our recent discovery of additional linker domain functionality. Many cellulases are multi-modular consisting of a carbohydrate binding module (CBM) connected to a catalytic domain by a flexible, glycosylated linker. These linkers have long been thought to simply serve as a tether between structured domains or to act in an inchworm-like fashion during catalytic action. With molecular dynamics simulations of *Trichoderma reesei* Family 6 and Family 7 cellulases bound to cellulose, we predicted that glycosylated linkers directly bind to cellulose. Our prediction was experimentally examined by measuring binding affinity of the isolated CBM and the glycosylated CBM-linker construct; an order of magnitude enhancement was observed as a result of the linker. Together, these results suggest glycosylated linkers in carbohydrate-active enzymes, which are intrinsically disordered proteins in solution, aid in dynamic binding during enzymatic deconstruction of plant cell walls. Secondly, I will focus on the processive mechanism of cellulases whereby an individual carbohydrate polymer chain is decrystallized and hydrolyzed along the chain without substrate dissociation. We developed a theoretical relationship to describe glycoside hydrolase processivity in terms of experimentally measureable parameters and apply enhanced sampling free energy calculations to examine our model in Family 7 cellulases. Overall, our approaches represent significant steps toward the development of structure-activity relationships in these industrially and environmentally important enzymes.

376154 Computational Cluster-Based Methods for Molecular Thermodynamics

Abstract:

Cluster-based methods in statistical mechanics express fluid properties in terms of power series in density or other state variables. Series coefficients can be determined rigorously from molecular models, and consequently these approaches can produce thermodynamic models that exactly describe the behavior that corresponds to the molecular model, over some range of conditions. This capability provides a unique route for connecting molecular-based treatments, including first-principles methods, to experiment, and in doing so it can guide the formulation of molecular models, methods, and theories; alternatively, this capability provides a modeling tool that can serve a wide range of applications in science and engineering, including diverse fields such as nanotechnology, combustion engineering, environmental science, separations, and more.

We provide an overview of work we and others have performed in recent years to make computational cluster-based methods into a viable tool for molecular thermodynamics. We consider methods for computing cluster integrals, treatment of complications such as molecular flexibility, nuclear quantum effects, and multibody interactions, and methods to extend the range of application of the general approach. We also show ways that these methods can be used in conjunction with experiment and standard molecular simulations to improve interpretation of results obtained there.

Open-system balance equations about "black-box" units, with flows of heat, work, and material streams possessing mass, energy, and entropy, yield thermodynamically rigorous connections among the inputs and outputs of proposed or existing processes. Quantitative process descriptions are reliable when accurate property models provide values for conceptual properties, such as enthalpies and entropies, in terms of measurable properties, such as temperature, pressure, and composition. Consistency can be checked by the sum of surroundings energy flows and entropy productions of individual sections reproducing results from separate analysis of the complete system.

This methodology can be used to determine what, if any, process configuration(s) would operate with minimum energy effects by imposing reversibility – no entropy generation - for the specified constraints. The same relations and models can also characterize the impacts of irreversibilities, either overall or locally, by including various (positive) nonzero values of entropy generation.

Results obtained so far have revealed process results for potential reversible processes, found entropy generation for known processes or some constructed with process simulation software, shown inconsistencies in property models for complex chemical systems, yielded comparisons of energy effects using different property models, and demonstrated process sensitivities to entropy generation values.

The talk will summarize the fundamental equations, illustrate applications to systems of contemporary interest, and suggest extensions.

Imperial College London

MOLECULAR SYSTEMS ENGINEERING

Computing the pressure and interfacial tension of inhomogenenous nanoscopic systems including drops and confined anisotropic phases: The Devil is in the Virial.

AIChE Annual Meeting
Session in Honour of Peter T. Cummings
Atlanta, Georgia, USA, 17 November 2014

George Jackson
Department of Chemical Engineering
Molecular Systems Engineering/Centre for Process Systems Engineering

Peter T Cummings

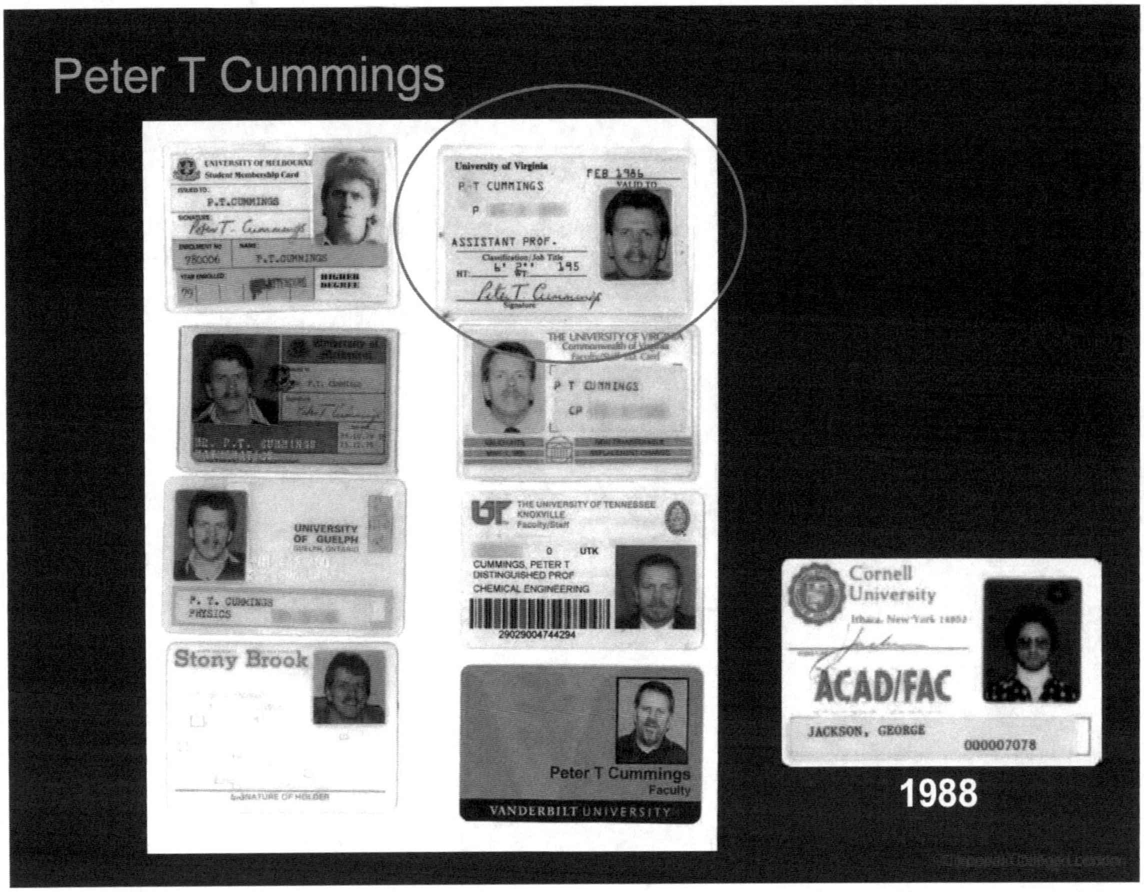
Peter T Cummings
UNIVERSITY OF MELBOURNE
Student Membership Card
P.T.CUMMINGS
HIGHER DEGREE
University of Virginia
FEB 1986
P T CUMMINGS
ASSISTANT PROF.
THE UNIVERSITY OF VIRGINIA
P T CUMMINGS
UNIVERSITY OF GUELPH
P. T. CUMMINGS
PHYSICS
THE UNIVERSITY OF TENNESSEE
KNOXVILLE
Faculty/Staff
UTK
CUMMINGS, PETER T
DISTINGUISHED PROF
CHEMICAL ENGINEERING
29029004744294
Stony Brook
Peter T Cummings
Faculty
VANDERBILT UNIVERSITY
Cornell University
ACAD/FAC
JACKSON, GEORGE
000007078
1988

Peter T Cummings
John S. Rowlinson

Peter T Cummings

Acknowledgements

- **Collaborators**:

  José G. Sampayo (Imperial)
  Enrique de Miguel (Huelva)
  Alexandr Malijevský (Prague)
  Erich A. Müller (Imperial)

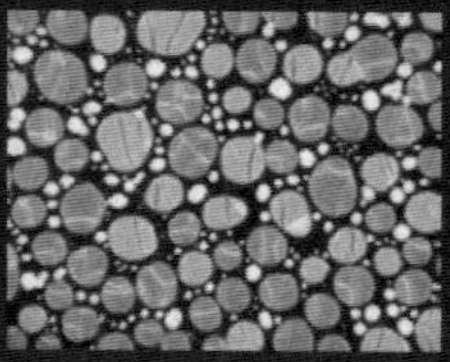

  Gabriel V. Lau (Imperial)
  Ian J. Ford (UCL)
  Patricia A. Hunt (Imperial)
  Erich A. Müller (Imperial)

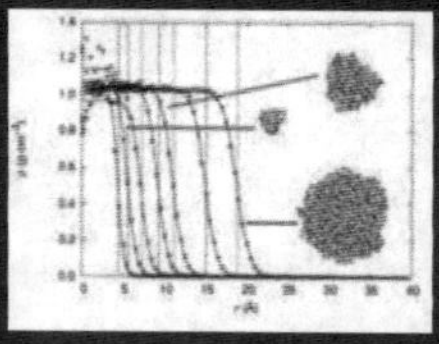

- **Funding**:
  EPSRC, EU, Royal Society, Shell

Outline

- Brief review of thermodynamics of interfaces
- Molecular simulation of the interfacial tension
  - ✓ Mechanical approach - Pressure tensor (Kirkwood & Buff, 1949)
  - ✓ Thermodynamic approach - Free energy difference (Bennett, 1976)
  - ✓ Finite-size scaling approaches - Landau free energy (Binder, 1982)
  - ✓ Test area (TA) method - Free energy perturbation (Gloor et al., 2005)
- Issues with the computation of the interfacial tension of liquid drops (curved interfaces)
- Applications to drops of LJ fluid and TIP4P/2005 water
- Conclusions

Mechanical Definition of Pressure

Clausius (1822-1888)

XVI. *On a Mechanical Theorem applicable to Heat.*
By R. Clausius*.

In a treatise which appeared in 1862, on the mechanical theory of heat†, I advanced a theorem which, in its simplest form, may be thus expressed:—*The effective force of heat is*

* Translated from a separate impression communicated by the Author, having been read before the Niederrheinischen Gesellschaft für Natur- und Heilkunde, on June 13, 1870.
† Phil. Mag. S. 4. vol. xxiv. pp. 81, 201; The Mechanical Theory of Heat, p. 215.

- Pressure Virial - Phil. Mag. **40**, 122-127 (1870)

$$\Sigma \frac{m}{2}\overline{v^2} = -\tfrac{1}{2}\Sigma\overline{(Xx+Yy+Zz)}.$$

$$P = \langle \rho kT \rangle + \left\langle \frac{1}{3V}\sum_i \sum_{j>i} \vec{r}_{ij}\cdot\vec{f}_{ij} \right\rangle$$

$$= \rho kT - \frac{2\pi}{3}\rho^2 \int r \frac{\partial u(r)}{\partial r} g(r) r^2 dr$$

Planar fluid interfaces

- Mechanical (pressure tensor) route to the tension

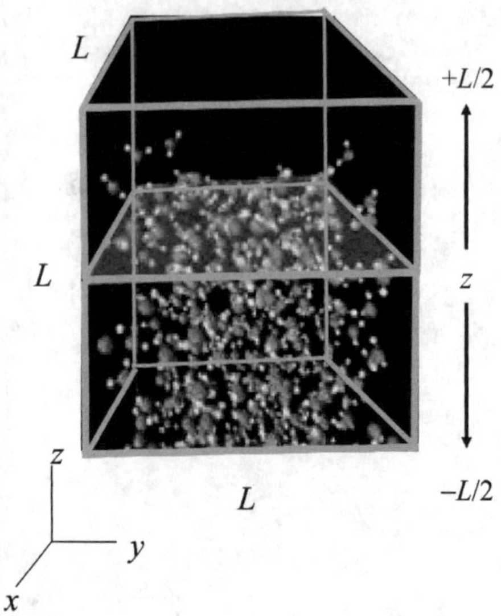

Surface area

$$A = L^2$$

Pressure tensor

$$\mathbf{P}(\mathbf{r}) = \begin{pmatrix} P_{xx}(\mathbf{r}) & 0 & 0 \\ 0 & P_{yy}(\mathbf{r}) & 0 \\ 0 & 0 & P_{zz}(\mathbf{r}) \end{pmatrix}$$

Tangential & normal components

$$\nabla \cdot \mathbf{P}(\mathbf{r}) = 0$$

$$P_{xx}(\mathbf{r}) = P_{yy}(\mathbf{r}) = P_T(z)$$

$$P_{zz}(\mathbf{r}) = P_N(z) = P$$

Planar fluid interfaces

- Work of deforming the interface

Infinitesimal deformation of area

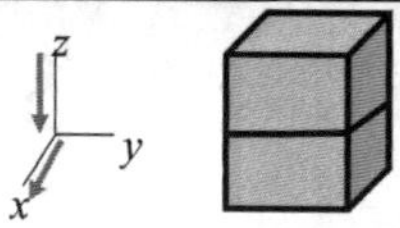

$$A_1 = A_0 + \delta A \qquad L_{x,1} = L + \frac{\delta A}{L} \qquad L_{z,1} = L - \frac{\delta A}{L} \qquad V_1 = V_0 = L^3$$

work done due to deformation in x direction

$$\delta W_T(z) = -F_{xx}(z)\delta L_x = -\left(P_{xx}(z)\,\mathrm{d}z\,L\right)\times\left(\frac{\delta A}{L}\right) = -\delta A\,P_T(z)\,\mathrm{d}z$$

$$\delta W_T = -\delta A \int_{-L/2}^{+L/2} P_T(z)\,\mathrm{d}z$$

tangential contribution

work done due to deformation in z direction

$$\delta W_N(z) = -F_{zz}(z)\delta L_z = -\left(P_{zz}(z)\,L^2\right)\times\left(-\frac{\delta A}{L}\right) = \delta A\,P_{zz}(z)\,L$$

$$\delta W_N = \delta A\,P_N(z)\,L = \delta A\,P\,L$$

normal contribution

Planar fluid interfaces

- Mechanical definition of interfacial tension

Total work done

$$\delta W = \delta W_T + \delta W_N = -\delta A \int_{-L/2}^{+L/2} P_T(z)\,\mathrm{d}z + \delta A \int_{-L/2}^{+L/2} P_N(z)\,\mathrm{d}z$$

$$= \delta A \int_{-\infty}^{\infty} [P_N(z) - P_T(z)]\mathrm{d}z = \delta A \int_{-\infty}^{\infty} [P - P_T(z)]\mathrm{d}z$$

Interfacial tension – mechanical expression

$$\gamma = \left(\frac{\delta W}{\delta A}\right)_V = \int_{-\infty}^{\infty} [P - P_T(z)]\mathrm{d}z$$

$\gamma \sim 75\,\mathrm{mN\,m^{-1}}$ $\quad w \sim 1\,\mathrm{nm}$ $\quad \langle P - P_T(z)\rangle_w \sim 7.5\times10^7\,\mathrm{N\,m^{-2}} = 750\,\mathrm{bar}$

$P \sim 0.03\,\mathrm{bar}$ $\quad \therefore \quad \langle P_T(z)\rangle_w \sim -750\,\mathrm{bar}$ **water 25°C**

Molecular Simulation of Interfacial Tension

- Mechanical route of Kirkwood & Buff (1949)

Interfacial tension

$$\gamma = \int_{-\infty}^{\infty} [P_N(z) - P_T(z)]\mathrm{d}z = \frac{1}{A}\left\langle \sum_i \sum_{j>i} \left(\frac{x_{ij}^2 + y_{ij}^2}{2r_{ij}} - \frac{z_{ij}^2}{r_{ij}} \right) \frac{du(r_{ij})}{dr_{ij}} \right\rangle$$

$$= \frac{1}{A}\left\langle \sum_i \sum_{j>i} \left(1 - \frac{3z_{ij}^2}{r_{ij}^2} \right) r_{ij} \frac{du(r_{ij})}{dr_{ij}} \right\rangle$$

Lee & Barker (1974), Lui (1974)
Chapela, Saville & Rowlinson (1975)
Rao & Levesque (1976)

- Kirwood-Buff relation is natural choice in MD simulation.
- Accuracy can be poor (large values of pressure tensor).
- Not straight forward for discontinous potentials.

Drops of liquid

- Mechanical expression

Interfacial tension in terms of pressure virials

$$\nabla \cdot \mathbf{P}(\mathbf{r}) = 0$$

$$P_T(r) = P_N(r) + \frac{r}{2}\frac{\partial P_N(r)}{\partial r}$$

$$\gamma^3 = -\frac{1}{8}\left(P_l - P_g\right)\int_0^\infty \frac{\partial P_N(r)}{\partial r} r^3 \mathrm{d}r$$

Buff (1955)
Thomson *et al.* (1984)

Schofield and Henderson (1982)
Non-unique nature of pressure tensor, tension and surface of tension for drops

Fluid Interfaces

- Thermodynamic definition of interfacial tension

Change in Helmholtz free energy

$$dF = SdT - PdV + \sum_i \mu_i dN_i + \gamma d\mathrm{A}$$

$$\gamma = \left(\frac{dF}{dA}\right)_{N_i VT}$$

Total Helmholtz free energy

$$F = -PV + \sum_i \mu_i N_i + \gamma A$$

Surface free energy

$$F^s = F - F^\alpha - F^\beta = \sum_i \mu_i N_i^s + \gamma A$$

$$\gamma = \frac{F^s}{A} \quad \text{when} \quad N_i^s = 0 \quad \forall_i$$

Molecular Simulation of Interfacial Tension

- Thermodynamic route - free energy perturbation

Free energy difference of two systems 0 and 1

$$\Delta F = F_1 - F_0 = -kT\ln\frac{Q_1}{Q_0} = -kT\ln\frac{Z_1}{Z_0}$$

$$\frac{Z_1}{Z_0} = \frac{\int d\mathbf{r}^N \exp(-U_1/kT)}{\int d\mathbf{r}^N \exp(-U_0/kT)} = \frac{\int d\mathbf{r}^N \exp(-U_0/kT)\exp(-\Delta U/kT)}{\int d\mathbf{r}^N \exp(-U_0/kT)}$$

$$\frac{Z_1}{Z_0} = \langle \exp(-\Delta U/kT)\rangle_0 \qquad \Delta U = U_1 - U_0$$

System 1 is a perturbation of 0

Molecular Simulation of Interfacial Tension

- Thermodynamic route - free energy perturbation

Free energy difference (Zwanzig, 1954)

$$\Delta F = -kT\ln\frac{Z_1}{Z_0} = -kT\ln\langle \exp(-\Delta U/kT)\rangle_0$$

Longuet-Higgins (1951) – mixtures of conformal fluids
Barker (1951), Pople(1952) – polar fluids

Molecular simulation of free energy derivatives

- Other free energy perturbation methods

Chemical potential – Test particle method

Widom (1963)

$$\mu = \left(\frac{\partial F}{\partial N}\right)_{VT} = -kT \lim_{N\to\infty} \ln \frac{Q_{N+1}}{Q_N} \approx -kT \ln\left[(N+1)\Lambda^3 / V\right] - kT \ln\langle \exp(-\Delta U / kT)\rangle_N$$

Pressure – Test volume method (isotropic)

Eppenga & Frenkel (1984)

Harismiadis *et al.* (1996); Vörtler & Smith (2000)

$$P = -\left(\frac{\partial F}{\partial V}\right)_{NT} = kT \lim_{\Delta V\to 0} \ln \frac{Q_{V_1}}{Q_{V_0}} = kT \lim_{\Delta V\to 0} \ln\left\langle (1+\Delta V/V)^N \exp(-\Delta U/kT)\right\rangle_0$$

Pressure tensor – Test volume method (anisotropic)

de Miguel & Jackson (2006)

Brumby et al. (2010)

Free energy perturbations - Kofke & Cummings (1997)

Molecular Simulation of Interfacial Tension

- Thermodynamic route - free energy perturbation

Test Area (TA) perturbation approach (Gloor et al., 2005)

Area of system 0

A_0

$L_{x,0} = L_{y,0}$

$L_{z,0}$

Area of system 1

$A_1 = A_0(1+\Delta A)$

$L_{x,1} = L_{y,1} = L_{x,0}\sqrt{1+\Delta A}$

$L_{z,1} = \dfrac{L_{z,0}}{1+\Delta A}$

$V_0 = V_1$

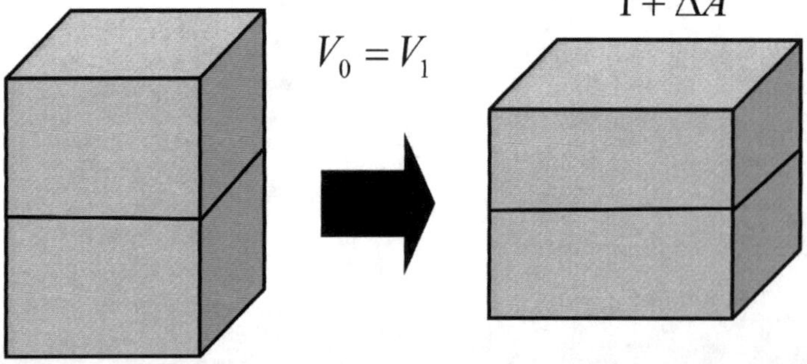

Molecular Simulation of Interfacial Tension

- Thermodynamic route - free energy perturbation

Test area (TA) free energy perturbation

$$\Delta F = -kT \ln \langle \exp(-\Delta U / kT) \rangle_0$$

Interfacial tension

$$\gamma = \left(\frac{\partial F}{\partial A} \right)_{NVT} = \lim_{\Delta A \to 0} \frac{\Delta F}{\Delta A} = \lim_{\Delta A \to 0} \left\{ \frac{-kT \ln \langle \exp(-\Delta U / kT) \rangle_0}{\Delta A} \right\}$$

forward or central difference

Gloor, Jackson, Blas, de Miguel, JCP, 123, 134703 (2005)

- as accurate than mechanical route.
- only requires simulation of a single system.
- useful for complex atomistic potentials.
- easy to extend to other types of interface/mixtures.
- can be used for discontinous potentials (with care).

Molecular Simulation of Planar Fluid Interfaces

- Thermodynamic route - TA method

Planar vapour-liquid interface of Lennard-Jones fluid

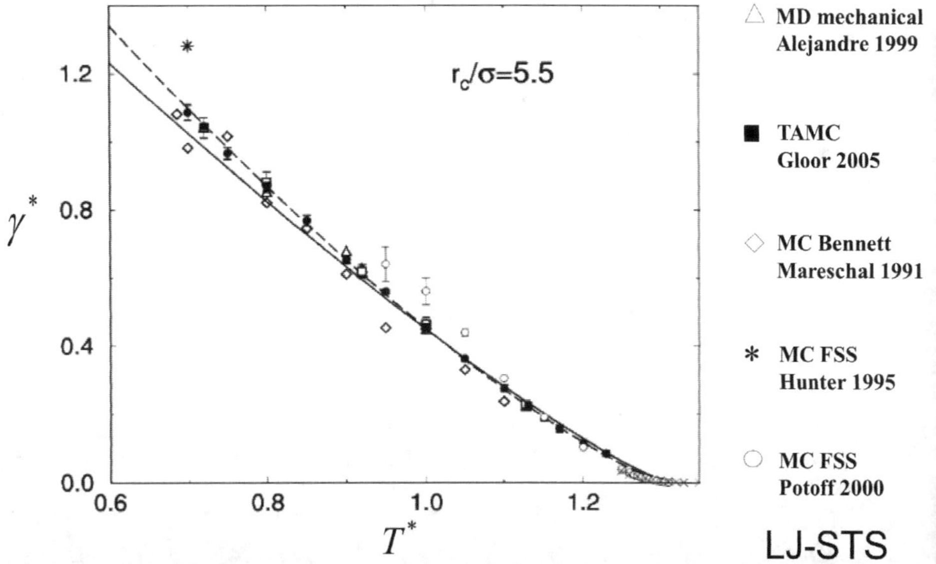

LJ-STS

A drop of liquid

- Rowlinson, *J. Phys.: Condens. Matter*, **6,** A1 (1994):

  "What could apparently be simpler than a drop of liquid? … [the] system throws up a set of mechanical, thermodynamic and statistical mechanical problems that are still matters of acute controversy".

Onda et al., *Langmuir*, **12**, 2125 (1996)

What do we know about liquid drops?

- **Classic Text**:
  J. S. Rowlinson and B. Widom
  Molecular Theory of Capillarity
  (Clarendon Press, Oxford, 1982)

- **Reviews**:
  J. R. Henderson
  "Statistical mechanics of spherical interfaces"
  in *Fluid Interfacial Phenomena*
  edited by C. A. Croxton (Wiley, New York, 1986)

  A. Malijevsky and G. Jackson
  A perspective on the interfacial properties of nanoscopic liquid drops
  J. Phys.: Condens. Matter, 24, 464121 (2012)

Molecular Simulation of Drops

- Thermodynamic route - free energy perturbation

Reference system 0

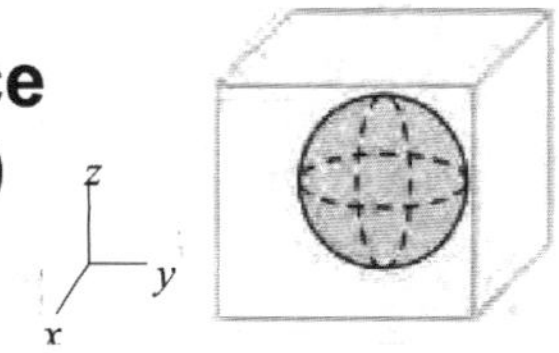

$A_0 = 4\pi R_e^2$

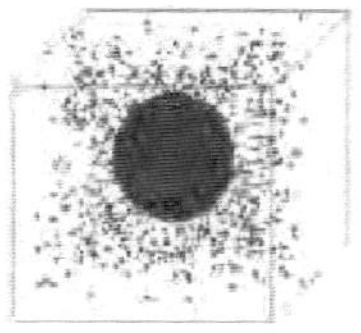

J. G. Sampayo et al., *J. Chem. Phys.*, 132 141101 (2010)

Molecular Simulation of Drops

- Thermodynamic route - free energy perturbation

Test Area (TA) perturbation

Perturbed system 1

Reference system 0

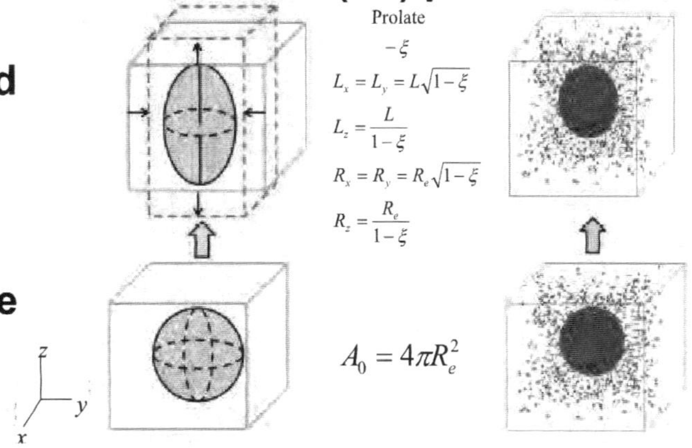

$\Delta A > 0$

J. G. Sampayo et al., *J. Chem. Phys.*, 132 141101 (2010)

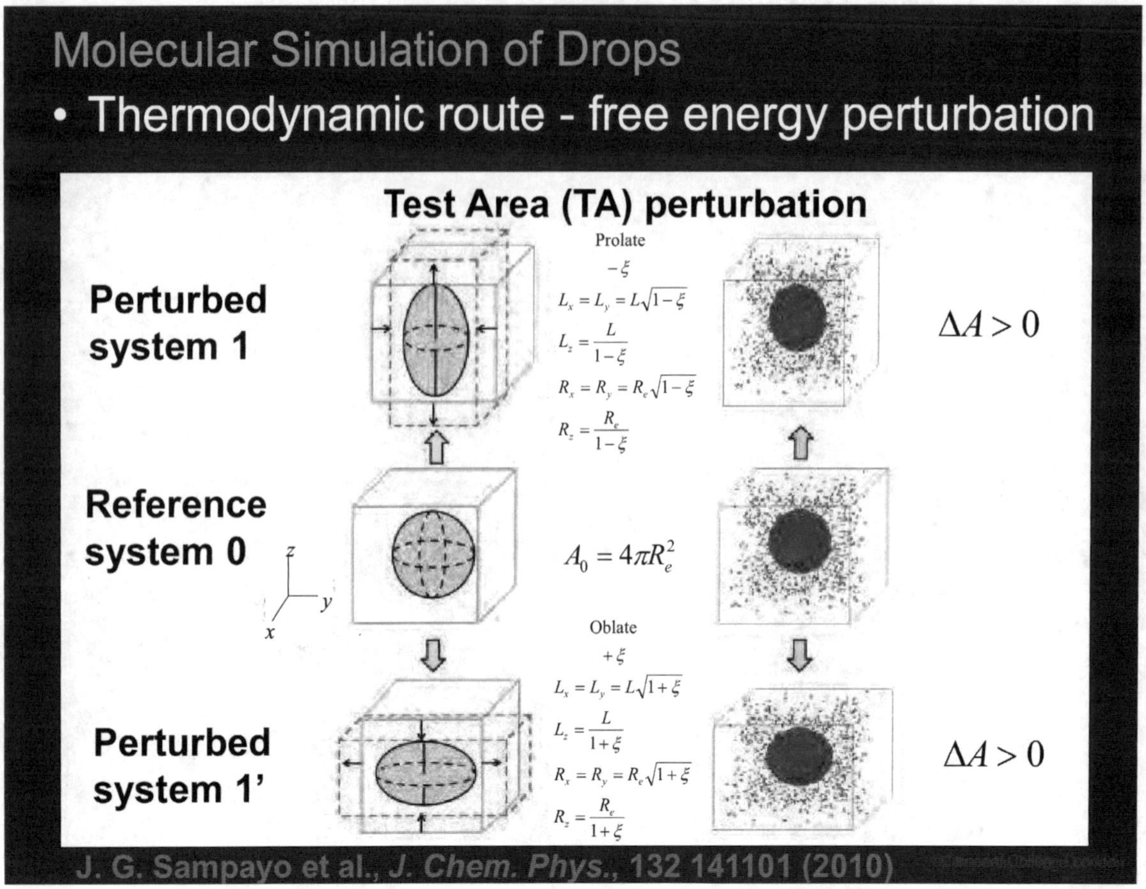
Molecular Simulation of Drops
• Thermodynamic route - free energy perturbation
Test Area (TA) perturbation
Perturbed system 1
Prolate
$-\xi$
$L_x = L_y = L\sqrt{1-\xi}$
$L_z = \frac{L}{1-\xi}$
$R_x = R_y = R_e\sqrt{1-\xi}$
$R_z = \frac{R_e}{1-\xi}$
$\Delta A > 0$
Reference system 0
$A_0 = 4\pi R_e^2$
Oblate
$+\xi$
$L_x = L_y = L\sqrt{1+\xi}$
$L_z = \frac{L}{1+\xi}$
$R_x = R_y = R_e\sqrt{1+\xi}$
$R_z = \frac{R_e}{1+\xi}$
Perturbed system 1'
$\Delta A > 0$
J. G. Sampayo et al., J. Chem. Phys., 132 141101 (2010)

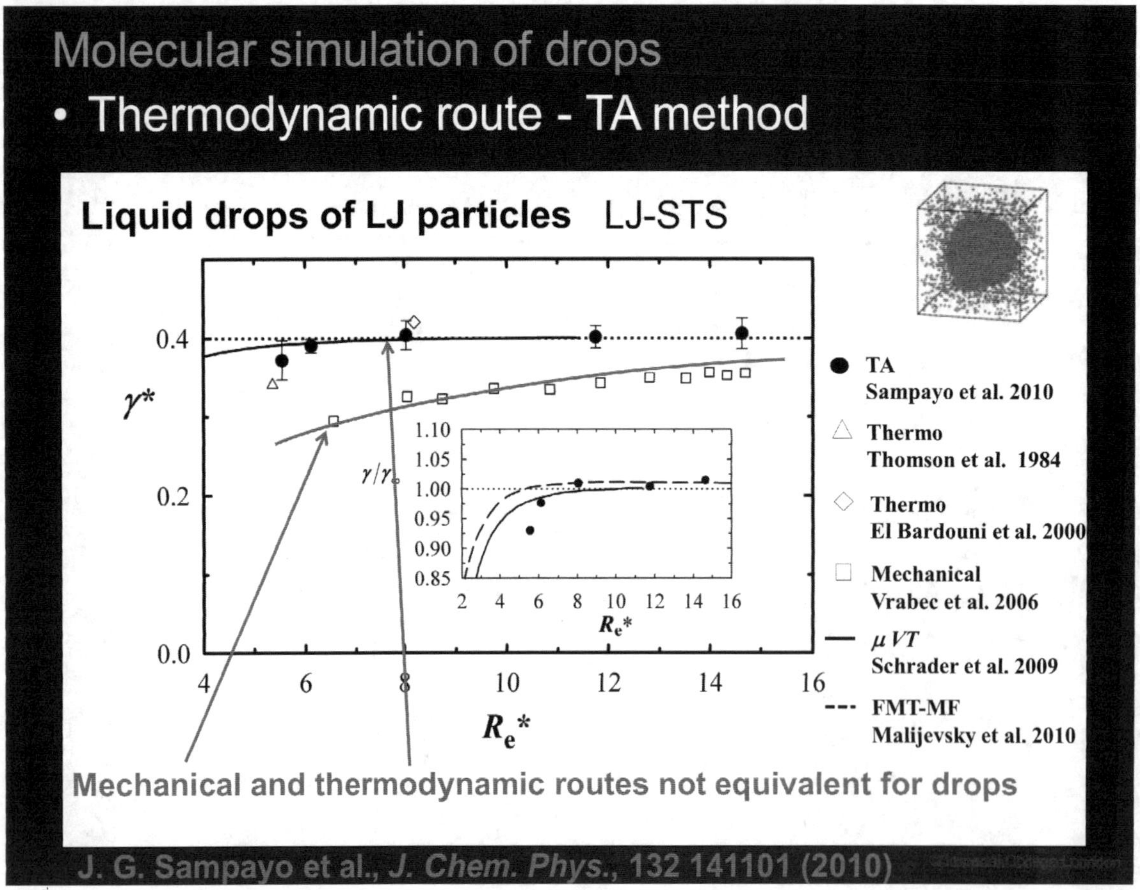
Molecular simulation of drops
• Thermodynamic route - TA method
Liquid drops of LJ particles LJ-STS
γ^*
R_e^*
γ/γ_0
TA Sampayo et al. 2010
Thermo Thomson et al. 1984
Thermo El Bardouni et al. 2000
Mechanical Vrabec et al. 2006
μVT Schrader et al. 2009
FMT-MF Malijevsky et al. 2010
Mechanical and thermodynamic routes not equivalent for drops
J. G. Sampayo et al., J. Chem. Phys., 132 141101 (2010)

Molecular simulation of interfacial tension

- Thermodynamic route - TA method

Analyse different terms of free energy perturbation:

$$\Delta F = -\frac{1}{\beta}\ln\left\langle \exp(-\beta \Delta U)\right\rangle_0 \qquad \textbf{expand exponential} \quad e^x = \sum_{i=0}^{\infty}\frac{x^i}{i!}$$

$$= -\frac{1}{\beta}\ln\left\langle \left(1-\beta\,\Delta U + \frac{1}{2!}[\beta\,\Delta U]^2 - \frac{1}{3!}[\beta\,\Delta U]^3 + \ldots\right)\right\rangle_0$$

$$= -\frac{1}{\beta}\ln\left(1-\beta\langle \Delta U\rangle_0 + \frac{1}{2}\beta^2\langle \Delta U^2\rangle_0 - \frac{1}{6}\beta^3\langle \Delta U^3\rangle_0 + \ldots\right)$$

We now expand the logarithm $\ln(1+x) = x - \frac{1}{2}x^2 + \frac{1}{3}x^3 - \ldots$

Molecular simulation of interfacial tension

- Thermodynamic route - TA method

$$\Delta F = -\frac{1}{\beta}\left(-\beta\langle \Delta U\rangle_0 + \beta^2\langle \Delta U^2\rangle_0 - \frac{1}{6}\beta^3\langle \Delta U^3\rangle_0 + \ldots\right)$$

$$+\frac{1}{2\beta}\left(-\beta\langle \Delta U\rangle_0 + \beta^2\langle \Delta U^2\rangle_0 - \frac{1}{6}\beta^3\langle \Delta U^3\rangle_0 + \ldots\right)^2$$

$$-\frac{1}{3\beta}\left(-\beta\langle \Delta U\rangle_0 + \beta^2\langle \Delta U^2\rangle_0 - \frac{1}{6}\beta^3\langle \Delta U^3\rangle_0 + \ldots\right)^3 + \ldots$$

Collecting the terms we obtain the so-called high-temperature expansion:

$$\Delta F = \underbrace{\langle \Delta U\rangle_0}_{\Delta F_1} \underbrace{-\frac{1}{2}\beta\left(\langle \Delta U^2\rangle_0 - \langle \Delta U\rangle_0^2\right)}_{\Delta F_2} \underbrace{+\frac{1}{6}\beta^2\left(\langle \Delta U^3\rangle_0 - 3\langle \Delta U^2\rangle_0\langle \Delta U\rangle_0 + 2\langle \Delta U\rangle_0^3\right)}_{\Delta F_3} + \ldots$$

leading-order contribution

$$\gamma = \left(\frac{\partial F}{\partial A}\right)_{NVT} = \lim_{\Delta A\to 0}\frac{\Delta F}{\Delta A} = \lim_{\Delta A\to 0}\left\{\frac{\Delta F_1}{\Delta A} + \frac{\Delta F_2}{\Delta A} + \frac{\Delta F_3}{\Delta A}\right\}$$

© Imperial College London

Molecular simulation of interfacial tension

- Thermodynamic route - TA method

To leading order: $\Delta F = \Delta F_1 = \langle \Delta U \rangle_0$

$$x_1 = x_0\sqrt{1+\xi} = x_0\left(1+\tfrac{1}{2}\xi - \ldots\right) \quad y_1 = y_0\sqrt{1+\xi} = y_0\left(1+\tfrac{1}{2}\xi - \ldots\right)$$

$$z_1 = z_0/\left(1+\xi\right) = z_0\left(1-\xi+\ldots\right) \quad V_1 = V_0 \quad \Delta A = A\xi$$

For infinitesimal deformations: $\xi \to 0 \quad \Delta U \to dU$

$$dU = \sum_{i=1}^{N}\left(\frac{\partial U}{\partial x_i}dx_i + \frac{\partial U}{\partial y_i}dy_i + \frac{\partial U}{\partial z_i}dz_i\right)$$

$$= \sum_{i=1}^{N}\left(\frac{\partial U}{\partial x_i}\frac{x_i\xi}{2} + \frac{\partial U}{\partial y_i}\frac{y_i\xi}{2} - \frac{\partial U}{\partial z_i}z_i\xi\right)$$

$$\langle \Delta U \rangle_0 = \langle dU \rangle_0 = N\xi\left\langle \tfrac{1}{2}x\frac{\partial U}{\partial x} + \tfrac{1}{2}y\frac{\partial U}{\partial y} - z\frac{\partial U}{\partial z_i}\right\rangle_0$$

Molecular simulation of interfacial tension

- Thermodynamic route - TA method

The interfacial tension to leading order:

$$\gamma = \left(\frac{dF}{dA}\right)_{NVT} = \lim_{\Delta A\to 0}\frac{\Delta F}{\Delta A} = \lim_{\Delta A\to 0}\frac{\langle \Delta U \rangle_0}{\Delta A} = \frac{N}{A}\left\langle \tfrac{1}{2}x\frac{\partial U}{\partial x} + \tfrac{1}{2}y\frac{\partial U}{\partial y} - z\frac{\partial U}{\partial z_i}\right\rangle$$

For pair-wise additive potentials: $U = \sum_{i=1}^{N}\sum_{j>i} u_{ij}(r_{ij})$

$$\gamma = \frac{N(N-1)}{2A}\left\langle \left(\tfrac{1}{2}x_{12}^2 + \tfrac{1}{2}y_{12}^2 - z_{12}^2\right)\frac{1}{r_{12}}\frac{du_{12}(r_{12})}{dr_{12}}\right\rangle$$

$$= \frac{1}{A}\left\langle \sum_{i}\sum_{j>i}\left\{\tfrac{1}{2}(x_{ij}^2 + y_{ij}^2) - z_{ij}^2\right\}\frac{1}{r_{ij}}\frac{du(r_{ij})}{dr_{ij}}\right\rangle$$

Lekner and Henderson (1977)
At leading order the thermodynamic route is equivalent to the mechanical route of Kirkwood and Buff.

Molecular simulation of interfacial tension

- Thermodynamic route - TA method

Planar vapour-liquid interface of LJ particles LJ-STS

$$\Delta F_1 = \langle \Delta U \rangle_0 \qquad \Delta F_2 = -\frac{1}{2}\beta\left(\langle \Delta U^2 \rangle_0 - \langle \Delta U \rangle_0^2\right)$$

Molecular simulation of interfacial tension

- Thermodynamic route - TA method

<u>Planar</u> vapour-liquid interface of LJ particles LJ-STS

$$\Delta F_1 = \langle \Delta U \rangle_0 \qquad \Delta F_2 = -\frac{1}{2}\beta\left(\langle \Delta U^2 \rangle_0 - \langle \Delta U \rangle_0^2\right)$$

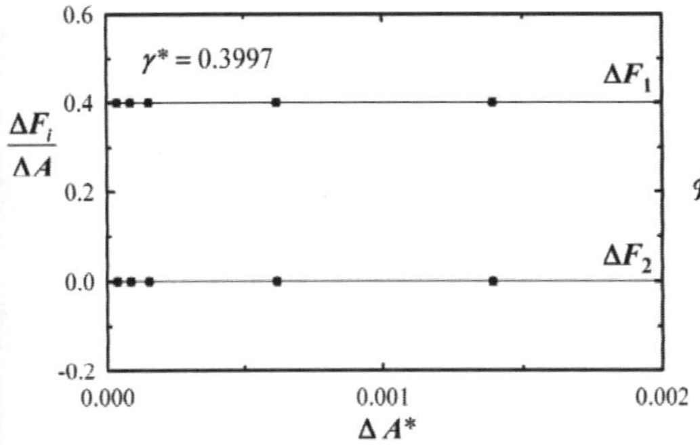

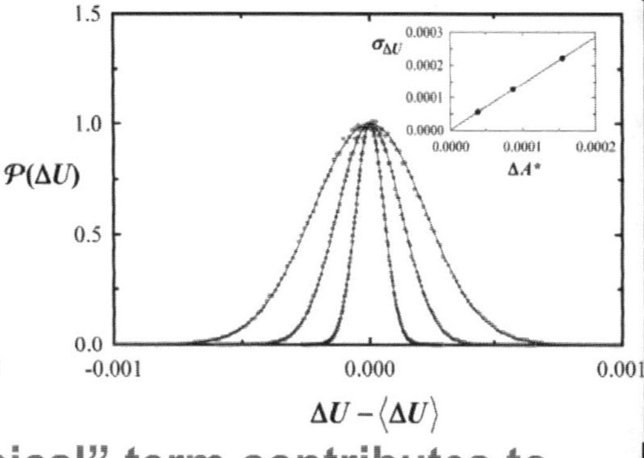

Only leading order "mechanical" term contributes to the interfacial tension. No appreciable fluctuations.

Thermodynamic and mechanical routes equivalent.

Molecular Simulation of Interfacial Tension

- Thermodynamic route - TA method

Liquid drops of LJ particles LJ-STS

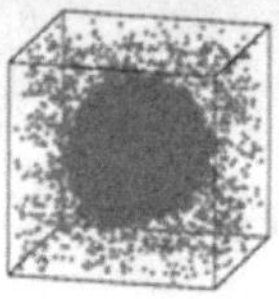

$$\Delta F_1 = \langle \Delta U \rangle_0 \qquad \Delta F_2 = -\frac{1}{2}\beta\left(\langle \Delta U^2 \rangle_0 - \langle \Delta U \rangle_0^2\right)$$

Molecular Simulation of Interfacial Tension

- Thermodynamic route - TA method

Liquid drops of LJ particles LJ-STS

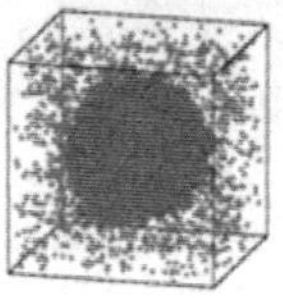

$$\Delta F_1 = \langle \Delta U \rangle_0 \qquad \Delta F_2 = -\frac{1}{2}\beta\left(\langle \Delta U^2 \rangle_0 - \langle \Delta U \rangle_0^2\right)$$

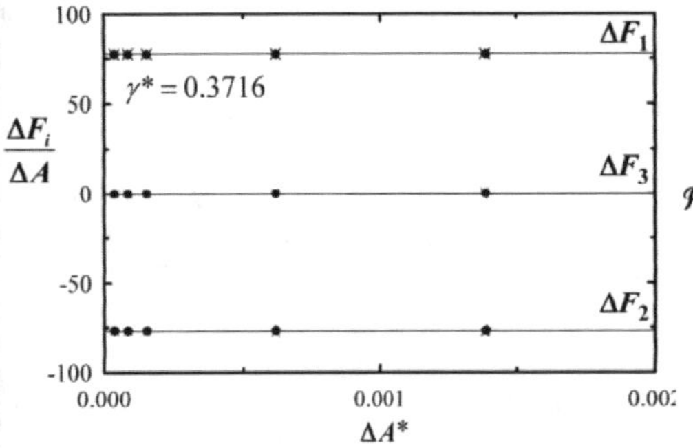

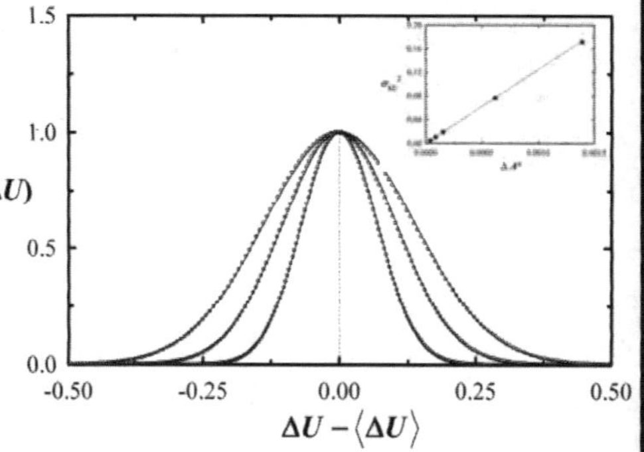

Second order "fluctuation" term now contributes markedly to the interfacial tension.

Thermodynamic and mechanical routes NOT equivalent!

J. G. Sampayo et al., *J. Chem. Phys.*, 132 141101 (2010)

Molecular Simulation of Interfacial Tension

- Thermodynamic route - TA method

Liquid drops of TIP4P/2005 Water

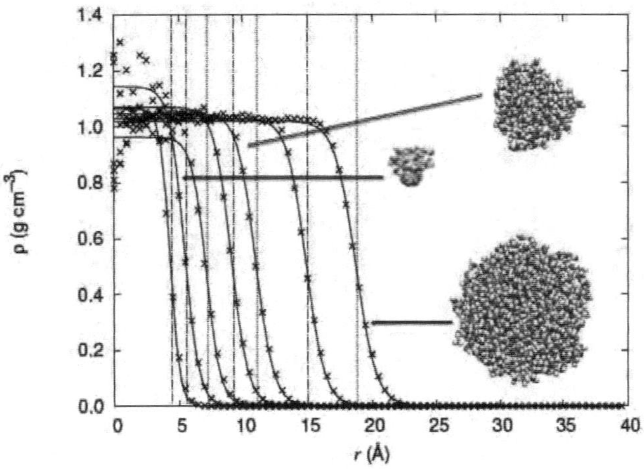

Small clusters down to *N* ~ 16 particles are mechanically stable

G. V. Lau et al., *J. Chem. Phys.*, submitted (2014)

Molecular Simulation of Interfacial Tension

- Thermodynamic route - TA method

Liquid drops of TIP4P/2005 Water

N	R_e/Å	γ/mNm^{-1}
16	4.40	0.1 ± 13.5
32	5.60	17.3 ± 9.8
64	7.25	39.0 ± 7.9
128	9.24	53.6 ± 7.6
216	11.11	55.8 ± 5.8
512	14.97	64.0 ± 9.8
1000	18.85	66.4 ± 9.7

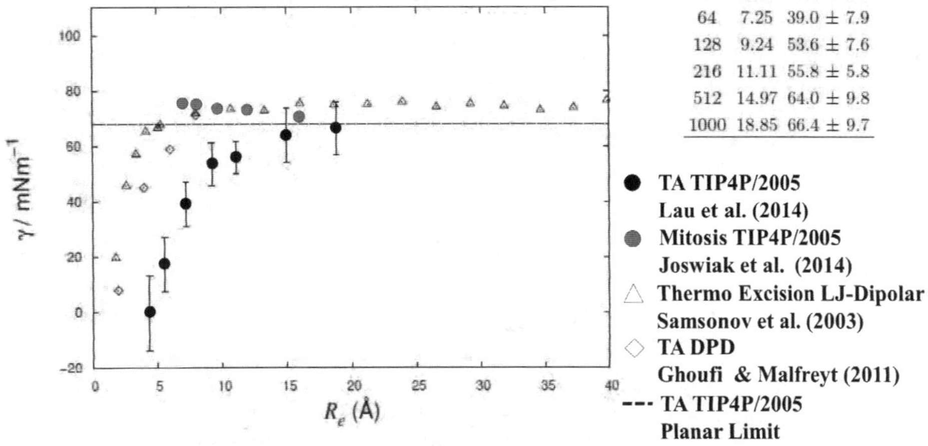

Inconsistent findings for tension of drops of water

G. V. Lau et al., *J. Chem. Phys.*, submitted (2014)

Conclusions

- First-order mechanical (virial) contribution characterises the planar interfacial tension.
- Large energetic fluctuations (second order) contribute to the interfacial tension of nanoscale drops in addition to mechanical contribution.
- Additional entropic contribution to the free energy of drops.
- The first-order mechanical (virial) route to the tension of drops is inappropriate.
- The correct “virial” expression must contain second-order terms of the type.

$$\left\langle x_\alpha \frac{dU}{dx_\alpha}\right\rangle\left\langle x_\beta \frac{dU}{dx_\beta}\right\rangle \quad \text{and} \quad \left\langle x_\alpha x_\beta \frac{dU}{dx_\alpha}\frac{dU}{dx_\beta}\right\rangle$$

- Implications for classical nucleation theory (CNT)

Thank you Peter!

Molecular Simulation of Interfacial Tension

- Thermodynamic route - free energy difference

Helmholtz free energy from partition function

$$F = -kT \ln Q \qquad Q = \frac{1}{N!\Lambda^{3N}} \int d\mathbf{r}^N \exp(-U/kT) = \frac{Z}{N!\Lambda^{3N}}$$

Free energy difference of two systems 0 and 1 (Bennett, 1976)

$$\Delta F = F_1 - F_0 = -kT \ln \frac{Q_1}{Q_0} = -kT \ln \frac{Z_1}{Z_0} = -kT \ln \frac{\langle W \exp(-U_1/kT) \rangle_0}{\langle W \exp(-U_1/kT) \rangle_1}$$

Interfacial tension

$$\gamma \approx \frac{F^s - F^\alpha - F^\beta}{A} = \frac{\Delta F}{A}$$

Miyazaki, Barker, Pound (1976)
Salomons & Mareschal(1991)
Moody & Attard (2004)

- Requires simulation of at least two systems.
- Care has to be taken with N_s contribution.

Molecular Simulation of Interfacial Tension

- Finite size scaling (Binder, 1982)

Grand partition function (Allen & Hansen 2002)

$$\Xi = \sum_{N=0}^{\infty} \frac{1}{N!\Lambda^{3N}} \exp(\mu N/kT) \int d\mathbf{r}^N \exp(-U/kT) = \sum_{N=0}^{\infty} \exp(\mu N/kT) Q(N)$$

Restricted partition function for systems with N^*

$$\Xi(N^*) = \sum_{N=0}^{\infty} \exp(\mu N/kT) Q(N) \delta(N - N^*) = \exp(\mu N^*/kT) Q(N^*)$$

Landau free energy (restricted grand potential)

$$\Omega(N^*) = -kT \ln \Xi(N^*) = -kT \ln \wp(N^*) + kT \ln \Xi \quad \text{where} \quad \wp(N^*) = \frac{\Xi(N^*)}{\Xi}$$

$$\Delta\Omega_L = \Omega(N^{\alpha\beta}) - \Omega(N^\alpha) = -kT \ln \frac{\wp_{\min}(N^{\alpha\beta})}{\wp_{\max}(N^\alpha)}$$

Potoff & Panagiotopoulos (2004)

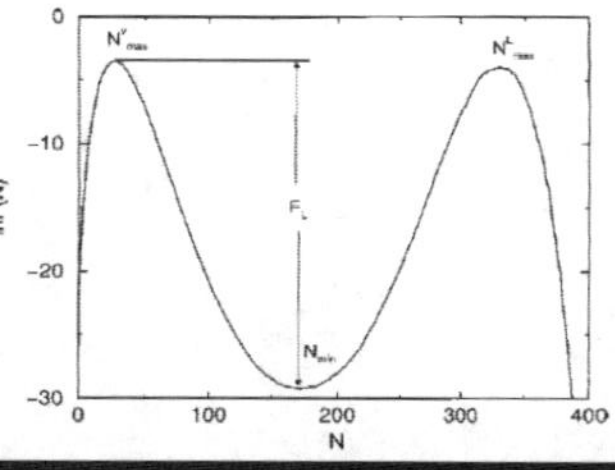

Molecular Simulation of Interfacial Tension

- Finite size scaling (Binder, 1982)

Finite size effects (Binder, 1986)

$$\lim_{L\to\infty}\frac{\wp_{\min}(N^{\alpha\beta})}{\wp_{\max}(N^{\alpha})}=\lim_{L\to\infty}\{\exp(-\Delta\Omega_L/kT)\}\propto L^x\exp(-\Omega^s_\infty/kT)$$

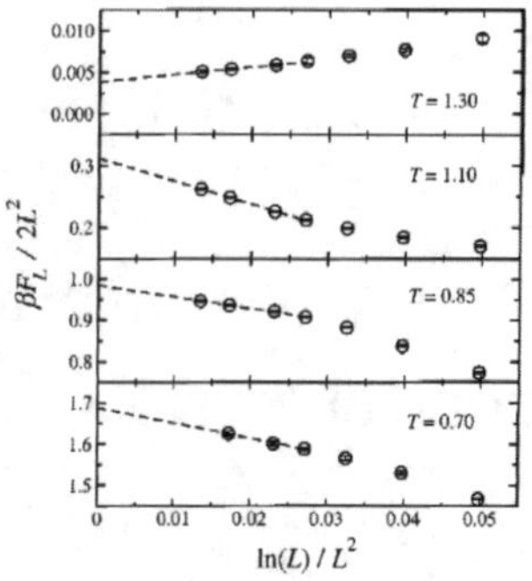

Interfacial tension

$$\gamma=\frac{\Omega^s_\infty}{A}\qquad \gamma_L=\frac{\Delta\Omega_L}{A}=\gamma-kTx\left(\frac{\ln L}{2L^2}\right)-\frac{kTC}{2L^2}$$

Hunter & Reinhardt (1995), Errington (2003)
Singh, Kofke, Errington (2003)

- Useful for critical region
- Progressively difficult at lower temperatures

What do we know about liquid drops?

Young (1805)

Laplace (1806)

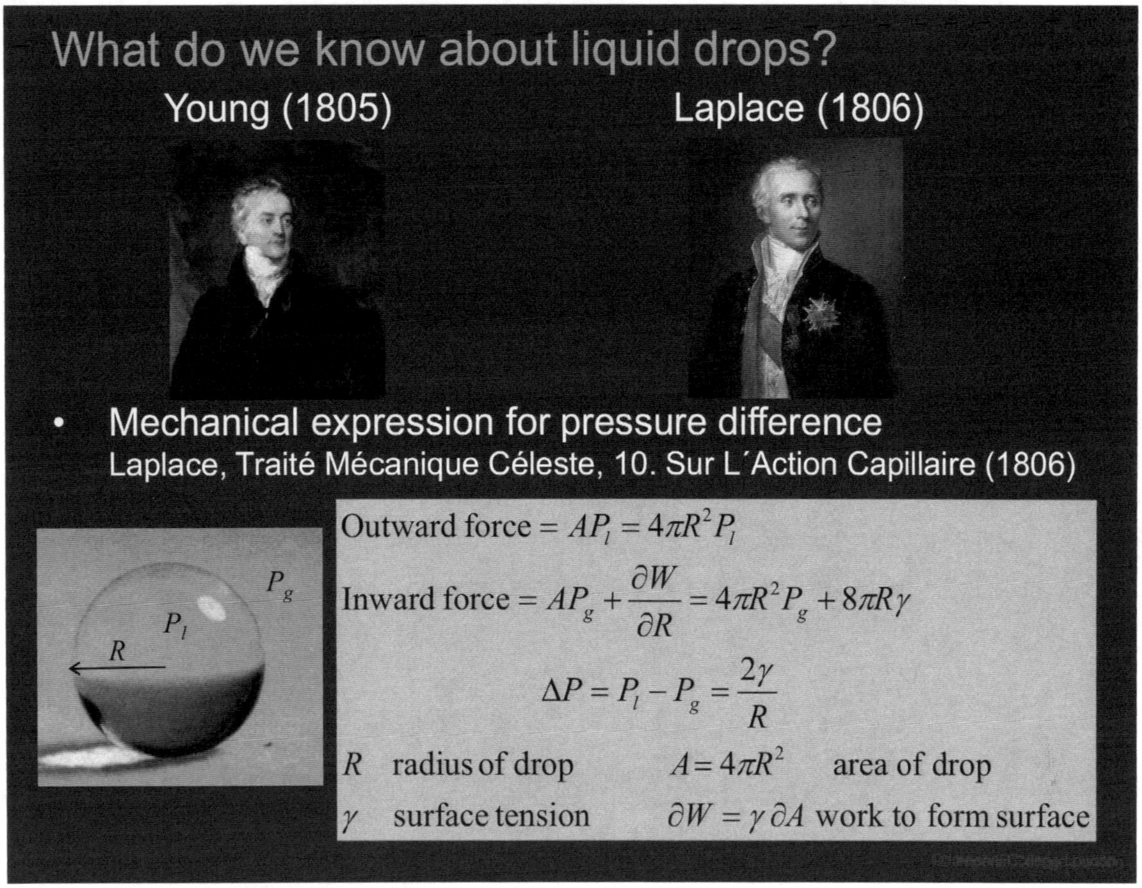
What do we know about liquid drops?
Young (1805)
Laplace (1806)
• Mechanical expression for pressure difference
Laplace, Traité Mécanique Céleste, 10. Sur L´Action Capillaire (1806)
P_g
P_l
R
Outward force $= AP_l = 4\pi R^2 P_l$
Inward force $= AP_g + \frac{\partial W}{\partial R} = 4\pi R^2 P_g + 8\pi R\gamma$
$\Delta P = P_l - P_g = \frac{2\gamma}{R}$
R radius of drop $A = 4\pi R^2$ area of drop
γ surface tension $\partial W = \gamma\,\partial A$ work to form surface

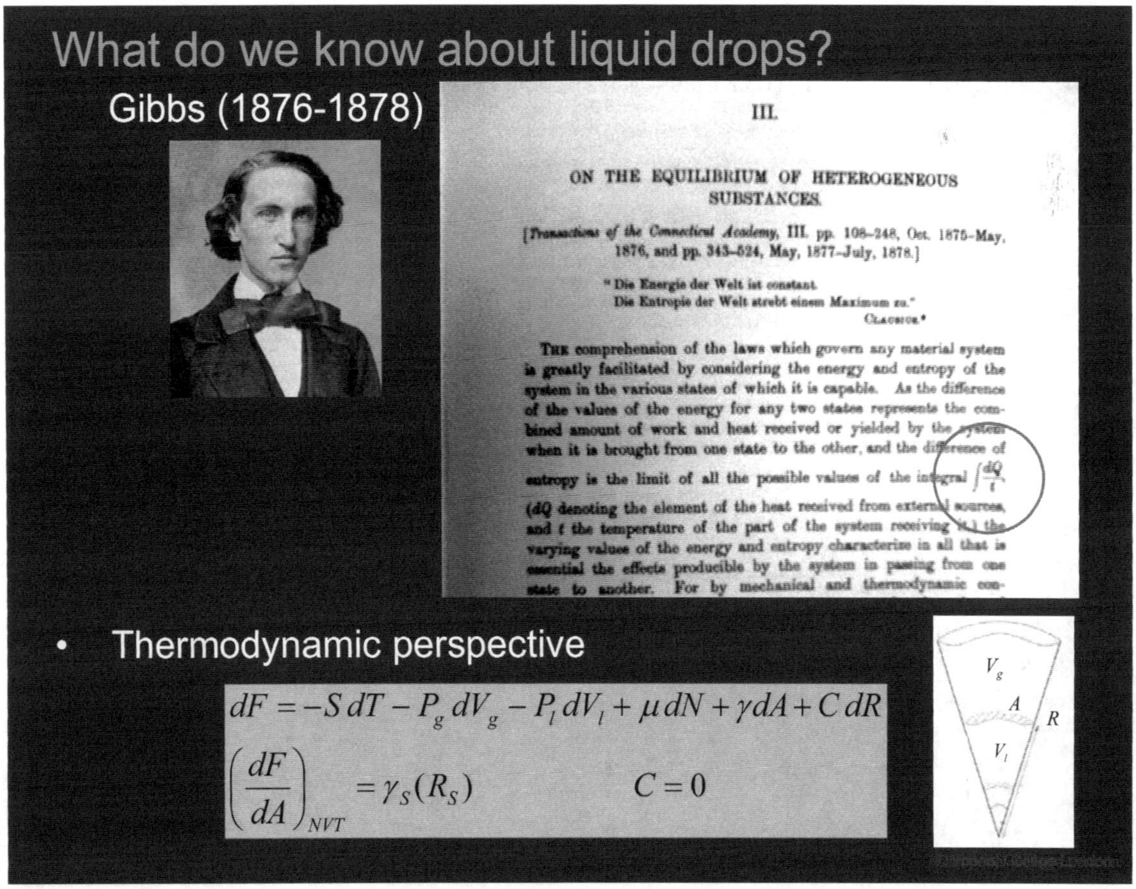
What do we know about liquid drops?
Gibbs (1876-1878)
III.
ON THE EQUILIBRIUM OF HETEROGENEOUS SUBSTANCES.
[Transactions of the Connecticut Academy, III. pp. 108–248, Oct. 1875–May, 1876, and pp. 343–524, May, 1877–July, 1878.]
"Die Energie der Welt ist constant.
Die Entropie der Welt strebt einem Maximum zu."
CLAUSIUS.*
THE comprehension of the laws which govern any material system is greatly facilitated by considering the energy and entropy of the system in the various states of which it is capable. As the difference of the values of the energy for any two states represents the combined amount of work and heat received or yielded by the system when it is brought from one state to the other, and the difference of entropy is the limit of all the possible values of the integral $\int \frac{dQ}{t}$, (dQ denoting the element of the heat received from external sources, and t the temperature of the part of the system receiving it,) the varying values of the energy and entropy characterize in all that is essential the effects producible by the system in passing from one state to another. For by mechanical and thermodynamic con-
• Thermodynamic perspective
$dF = -S\,dT - P_g\,dV_g - P_l\,dV_l + \mu\,dN + \gamma dA + C\,dR$
$\left(\frac{dF}{dA}\right)_{NVT} = \gamma_S(R_S) \qquad C = 0$
V_g
A
R
V_l

What do we know about liquid drops?

Tolman (ca. 1932 Caltech)

What do we know about liquid drops?

Tolman (1949)

THE JOURNAL OF CHEMICAL PHYSICS VOLUME 17, NUMBER 3 MARCH, 1949

The Effect of Droplet Size on Surface Tension

RICHARD C. TOLMAN§
California Institute of Technology, Pasadena, California

(Received September 23, 1948)

The effect of droplet size on surface tension is given theoretical consideration with the help of results of the Gibbs thermodynamic theory of capillarity and of previous results of the author as to the sign and magnitude of superficial densities. It is concluded that surface tension can be expected to decrease with decrease in droplet size over a wide range of circumstances. In addition, approximate figures are obtained for the rate at which such decreases may be expected. The decreases become significant for very small drops. The results are of interest in view of the important role of surface tension in determining the behavior of small droplets.

§ died September 5, 1948

What do we know about liquid drops?

Tolman (1949)

THE JOURNAL OF CHEMICAL PHYSICS VOLUME 17, NUMBER 3 MARCH, 1949

The Effect of Droplet Size on Surface Tension

RICHARD C. TOLMAN[§]
California Institute of Technology, Pasadena, California
(Received September 23, 1948)

The effect of droplet size on surface tension is given theoretical consideration with the help of results of the Gibbs thermodynamic theory of capillarity and of previous results of the author as to the sign and magnitude of superficial densities. It is concluded that surface tension can be expected to decrease with decrease in droplet size over a wide range of circumstances. In addition, approximate figures are obtained for the rate at which such decreases may be expected. The decreases become significant for very small drops. The results are of interest in view of the important role of surface tension in determining the behavior of small droplets.

§ died September 5, 1948

- Tolman curvature correction

$$\gamma(R) = \gamma_\infty \left(1 - 2\delta_\infty \frac{1}{R} + \ldots\right)$$

γ_∞ surface tension of planar interface $\qquad \delta_\infty$ Tolman length

$$\Delta P = P_l^* - P_g = \frac{2\gamma_S(R_S)}{R_S}$$

$\gamma_S(R_S)$ surface of tension

What do we know about liquid drops?

Post-Tolman developments

What do we know about liquid drops?

$$\gamma(R) = \gamma_\infty\left(1 - 2\delta_\infty \frac{1}{R} + \ldots\right)$$

Reliable estimates of the Tolman length

- Penetrable sphere model – Hemingway et al. (1981)

$$\delta_\infty = -\frac{\sigma}{2}$$

mean-field theory exactly solvable at zero temperature

- Ising model/curvature expansion – Wortis and Fisher (1984)

$$\delta_\infty = 0$$

symmetrical density profile

$$\delta_\infty = -0.02\sigma$$

mean-field theory appropriate near critical point

- Hard spheres against spherical wall/FMT – Bryk et al. (2003)

$$\delta_\infty = -\left(\frac{1-\eta}{2+\eta}\right)\frac{\sigma}{2}$$

always negative for physical values of packing fractions

Conflicting findings for LJ liquid drops

$$\gamma(R) = \gamma_\infty\left(1 - 2\delta_\infty \frac{1}{R} + \ldots\right)$$

Conflicting findings for LJ liquid drops

$$\gamma(R) = \gamma_\infty \left(1 - 2\delta_\infty \frac{1}{R} + \ldots\right)$$

- Square Gradient Theory (SGT)

Rayleigh (1892), van der Waals (1893), Cahn and Hilliard (1958)

$$F[\rho(\mathbf{r})] = \int d\mathbf{r} \left\{ f^{\text{homo.}}(\rho(\mathbf{r})) + \frac{C}{2}(\nabla \rho(\mathbf{r}))^2 \right\}$$

$$C = \frac{kT}{6} \int d\mathbf{s}\, s^2 c(s; \rho(\mathbf{r}))$$

Conflicting findings for LJ liquid drops

$$\gamma(R) = \gamma_\infty \left(1 - 2\delta_\infty \frac{1}{R} + \ldots\right)$$

- Square Gradient Theory (SGT)

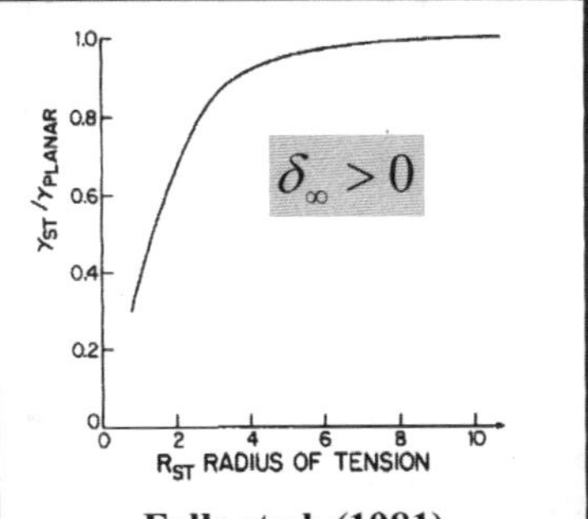

Falls et al. (1981)

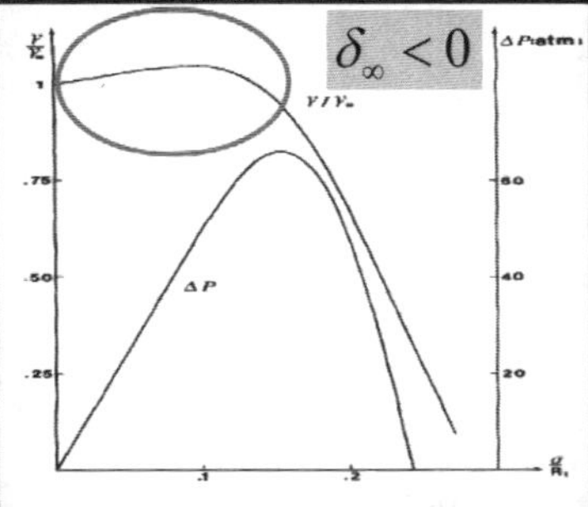

Guermeur et al. (1985)

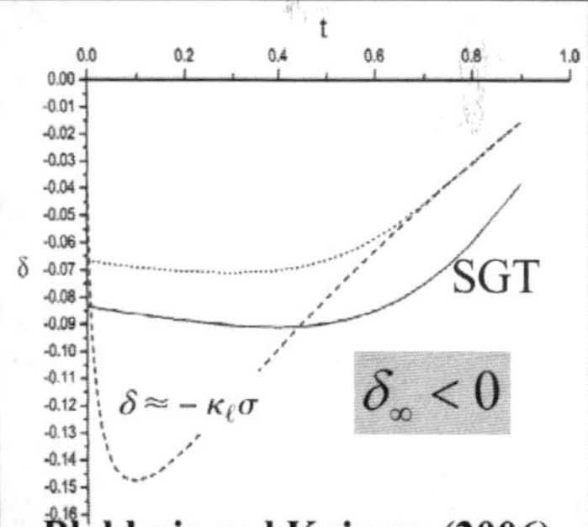

Blokhuis and Kuipers (2006)

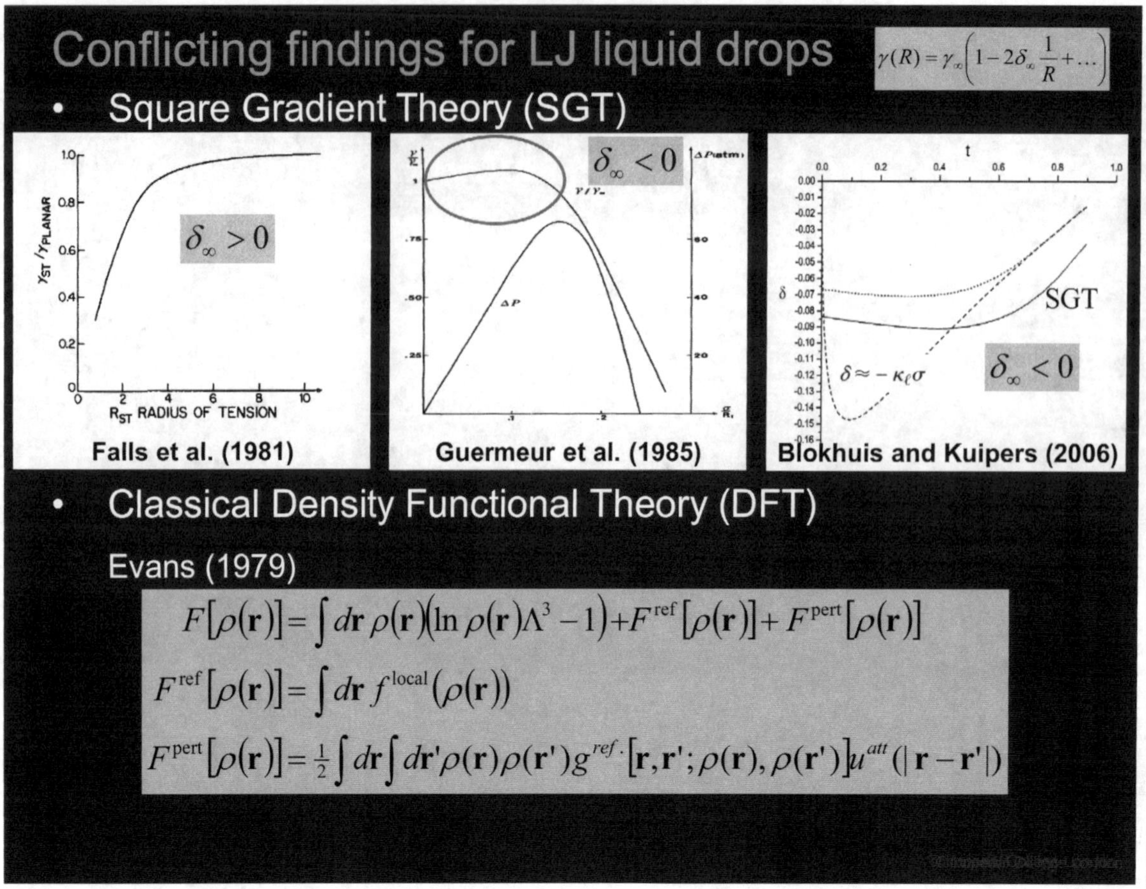
Conflicting findings for LJ liquid drops
$\gamma(R)=\gamma_\infty\left(1-2\delta_\infty\frac{1}{R}+\ldots\right)$
• Square Gradient Theory (SGT)
$\delta_\infty>0$
$\gamma_{ST}/\gamma_{PLANAR}$
R_{ST} RADIUS OF TENSION
Falls et al. (1981)
$\delta_\infty<0$
ΔP
Guermeur et al. (1985)
t
SGT
$\delta\approx-\kappa_\ell\sigma$
$\delta_\infty<0$
Blokhuis and Kuipers (2006)
• Classical Density Functional Theory (DFT)
Evans (1979)
$F[\rho(\mathbf{r})]=\int d\mathbf{r}\,\rho(\mathbf{r})(\ln\rho(\mathbf{r})\Lambda^3-1)+F^{\text{ref}}[\rho(\mathbf{r})]+F^{\text{pert}}[\rho(\mathbf{r})]$
$F^{\text{ref}}[\rho(\mathbf{r})]=\int d\mathbf{r}\,f^{\text{local}}(\rho(\mathbf{r}))$
$F^{\text{pert}}[\rho(\mathbf{r})]=\frac{1}{2}\int d\mathbf{r}\int d\mathbf{r}'\rho(\mathbf{r})\rho(\mathbf{r}')g^{ref.}[\mathbf{r},\mathbf{r}';\rho(\mathbf{r}),\rho(\mathbf{r}')]u^{att}(|\mathbf{r}-\mathbf{r}'|)$

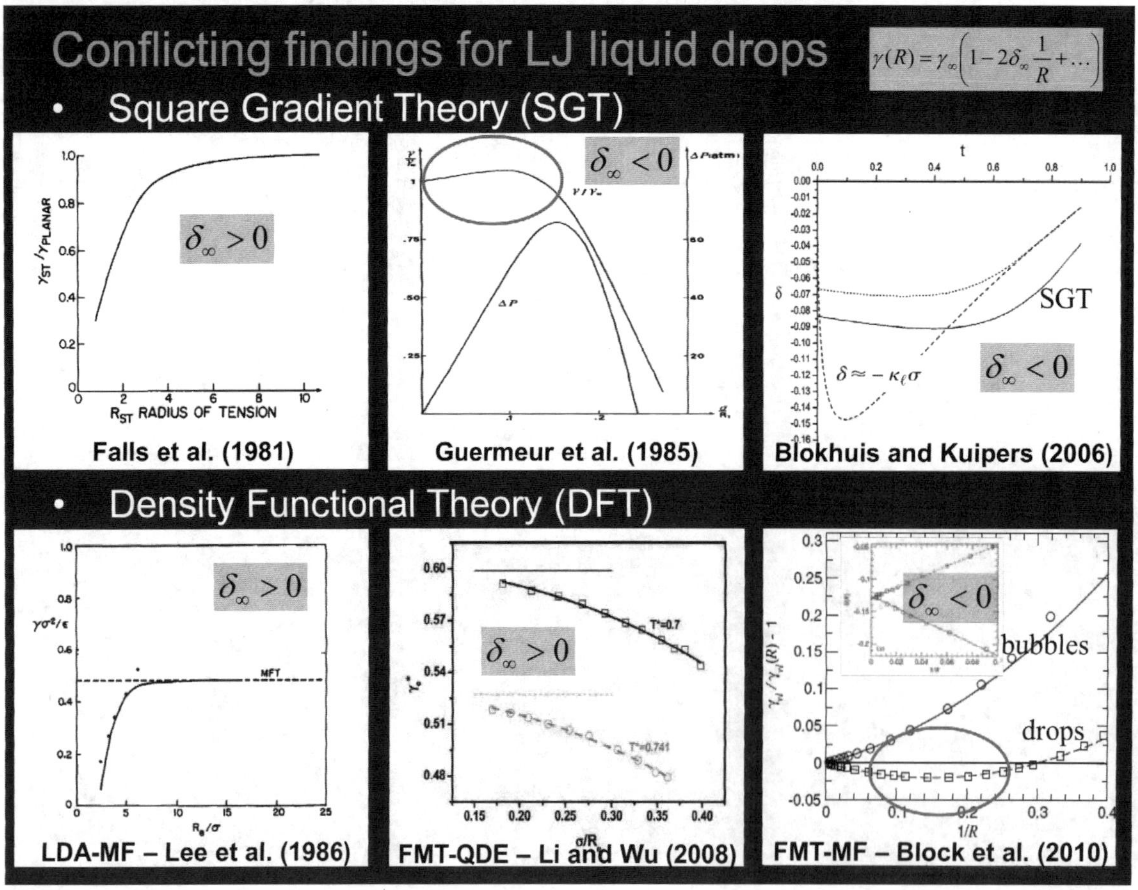
Conflicting findings for LJ liquid drops
$\gamma(R)=\gamma_\infty\left(1-2\delta_\infty\frac{1}{R}+\ldots\right)$
• Square Gradient Theory (SGT)
$\delta_\infty>0$
$\gamma_{ST}/\gamma_{PLANAR}$
R_{ST} RADIUS OF TENSION
Falls et al. (1981)
$\delta_\infty<0$
ΔP
Guermeur et al. (1985)
t
SGT
$\delta\approx-\kappa_\ell\sigma$
$\delta_\infty<0$
Blokhuis and Kuipers (2006)
• Density Functional Theory (DFT)
$\delta_\infty>0$
MFT
LDA-MF – Lee et al. (1986)
$\delta_\infty>0$
T*=0.7
T*=0.741
σ/R
FMT-QDE – Li and Wu (2008)
$\delta_\infty<0$
bubbles
drops
1/R
FMT-MF – Block et al. (2010)

Conflicting findings for LJ liquid drops

$$\gamma(R) = \gamma_{\infty}\left(1 - 2\delta_{\infty}\frac{1}{R} + \ldots\right)$$

- Molecular Simulation

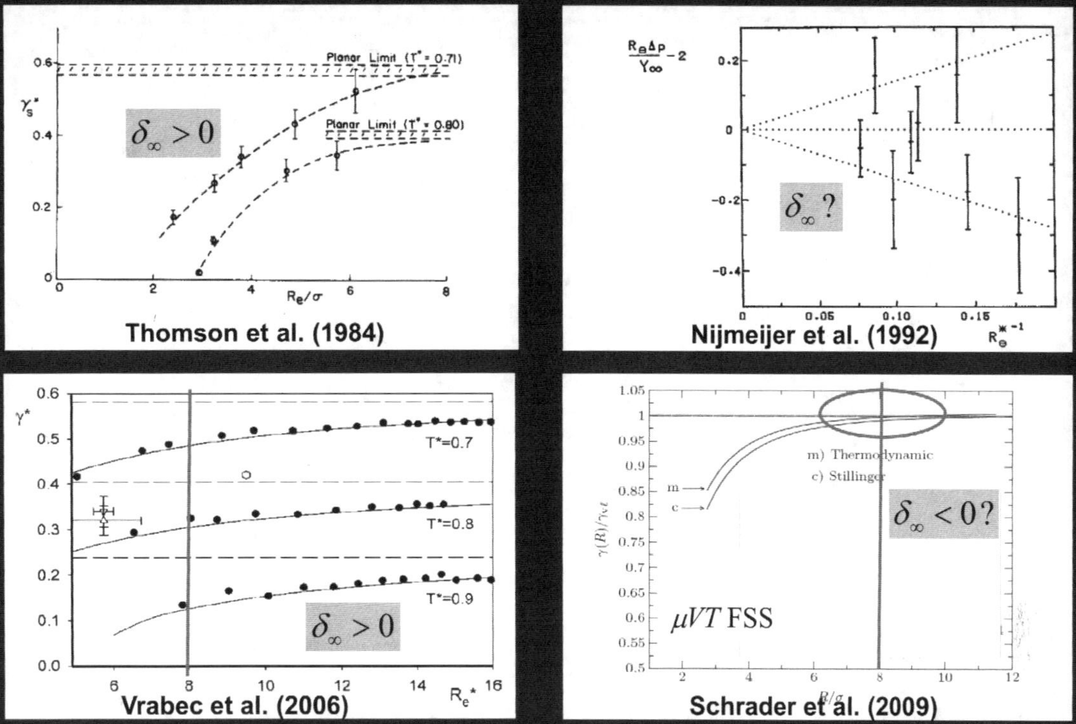

Simulation of confined non-spherical particles

Lekkerkerker and co-workers (1997, 2006)
Allen and co-workers (1999-2002)
Dijkstra, van Roij & Evans (2000)
Cleaver and co-workers (2001-2004)
Schoen and co-workers (2004-date)
Schilling and co-workers (2005-date)
Galindo and co-workers (2007)
Rull and co-workers (2007-date)

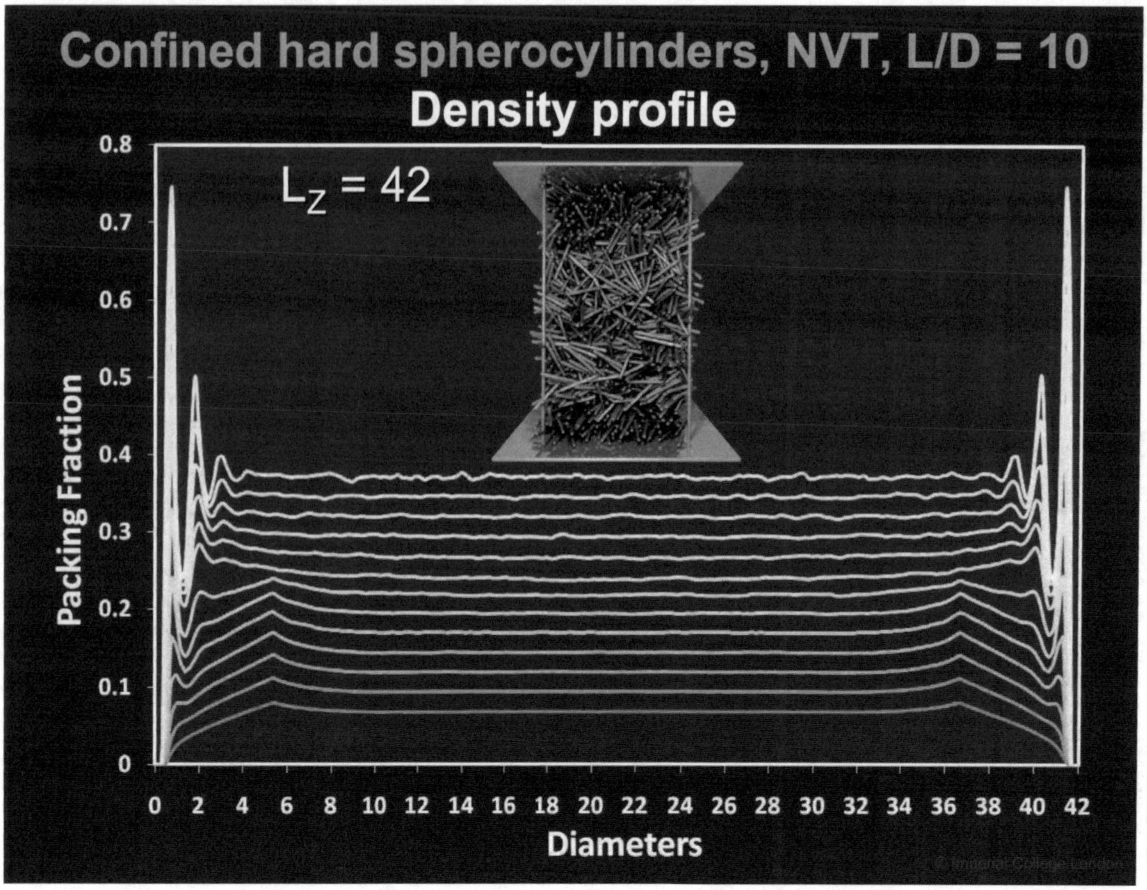
Confined hard spherocylinders, NVT, L/D = 10
Density profile
$L_Z = 42$
Packing Fraction
0.8
0.7
0.6
0.5
0.4
0.3
0.2
0.1
0
0 2 4 6 8 10 12 14 16 18 20 22 24 26 28 30 32 34 36 38 40 42
Diameters
© Imperial College London

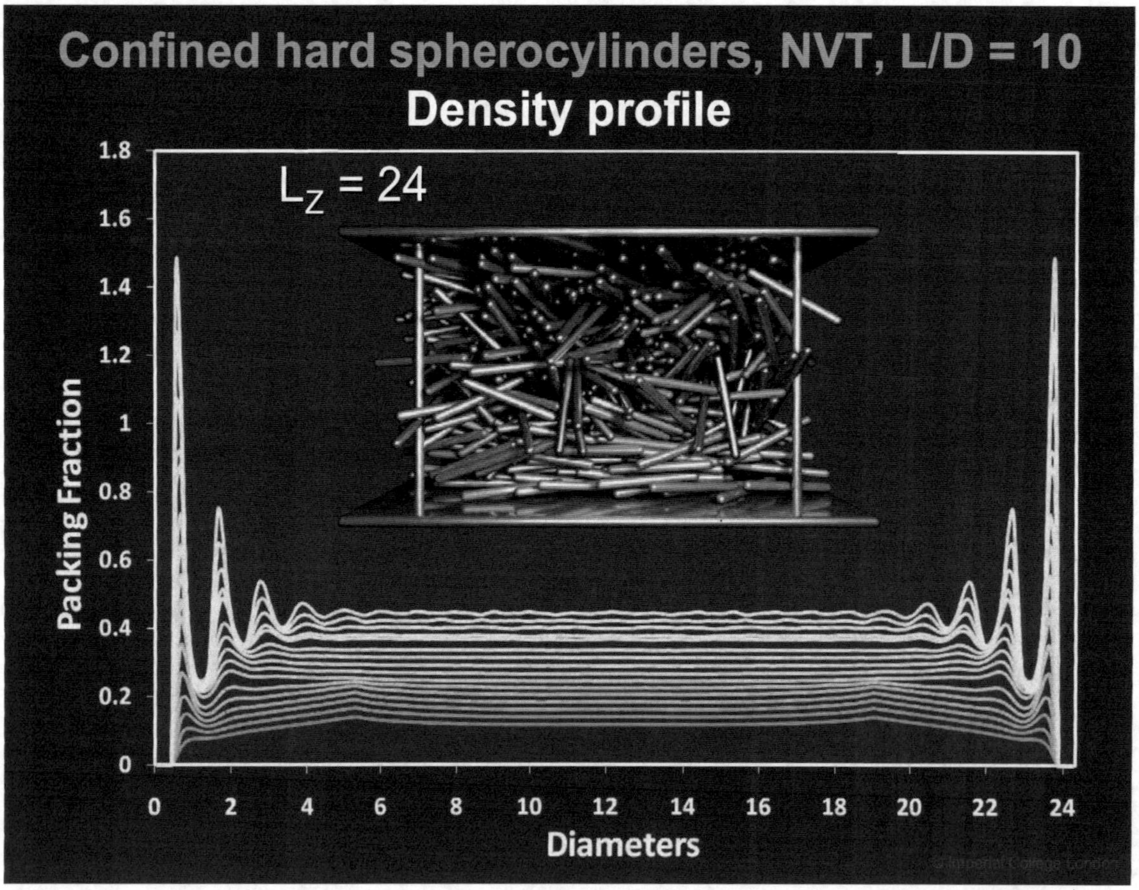
Confined hard spherocylinders, NVT, L/D = 10
Density profile
$L_Z = 24$
Packing Fraction
1.8
1.6
1.4
1.2
1
0.8
0.6
0.4
0.2
0
0 2 4 6 8 10 12 14 16 18 20 22 24
Diameters
© Imperial College London

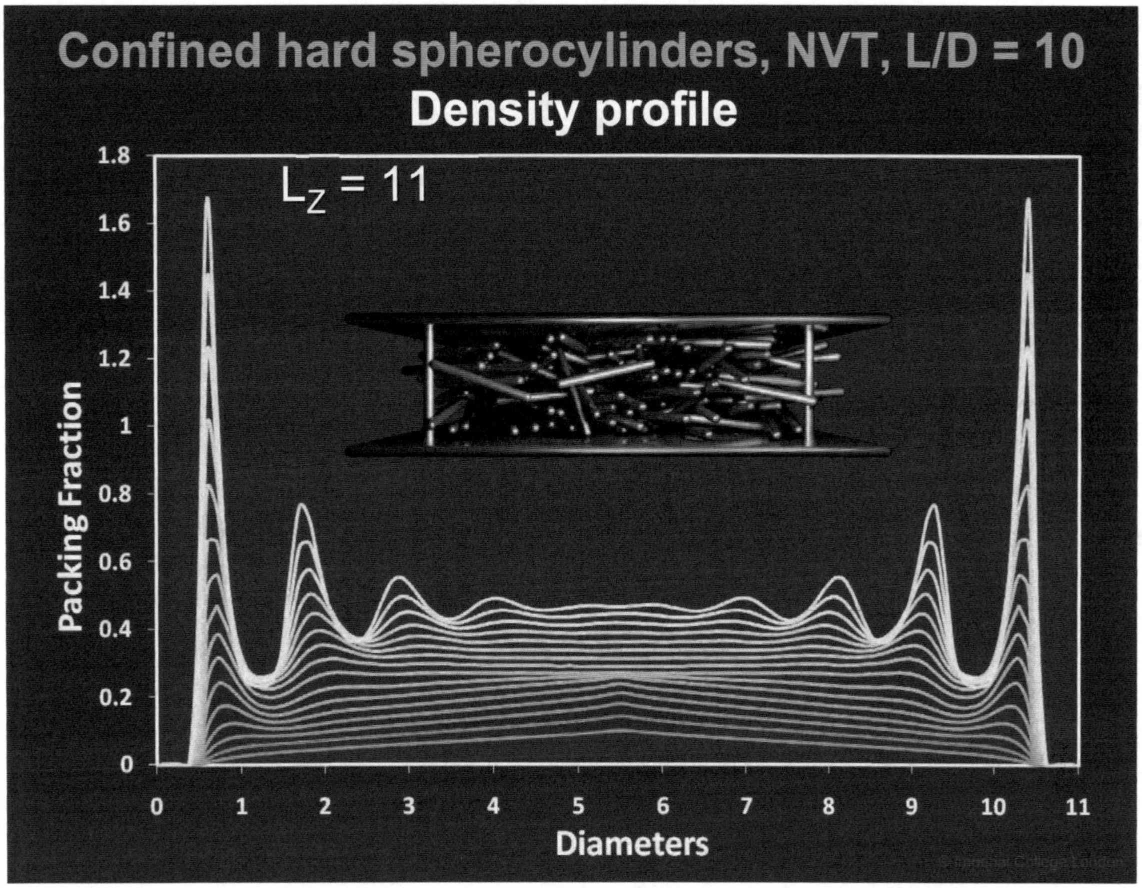
Confined hard spherocylinders, NVT, L/D = 10
Density profile
L_Z = 11
Packing Fraction
Diameters
0 1 2 3 4 5 6 7 8 9 10 11
0 0.2 0.4 0.6 0.8 1 1.2 1.4 1.6 1.8

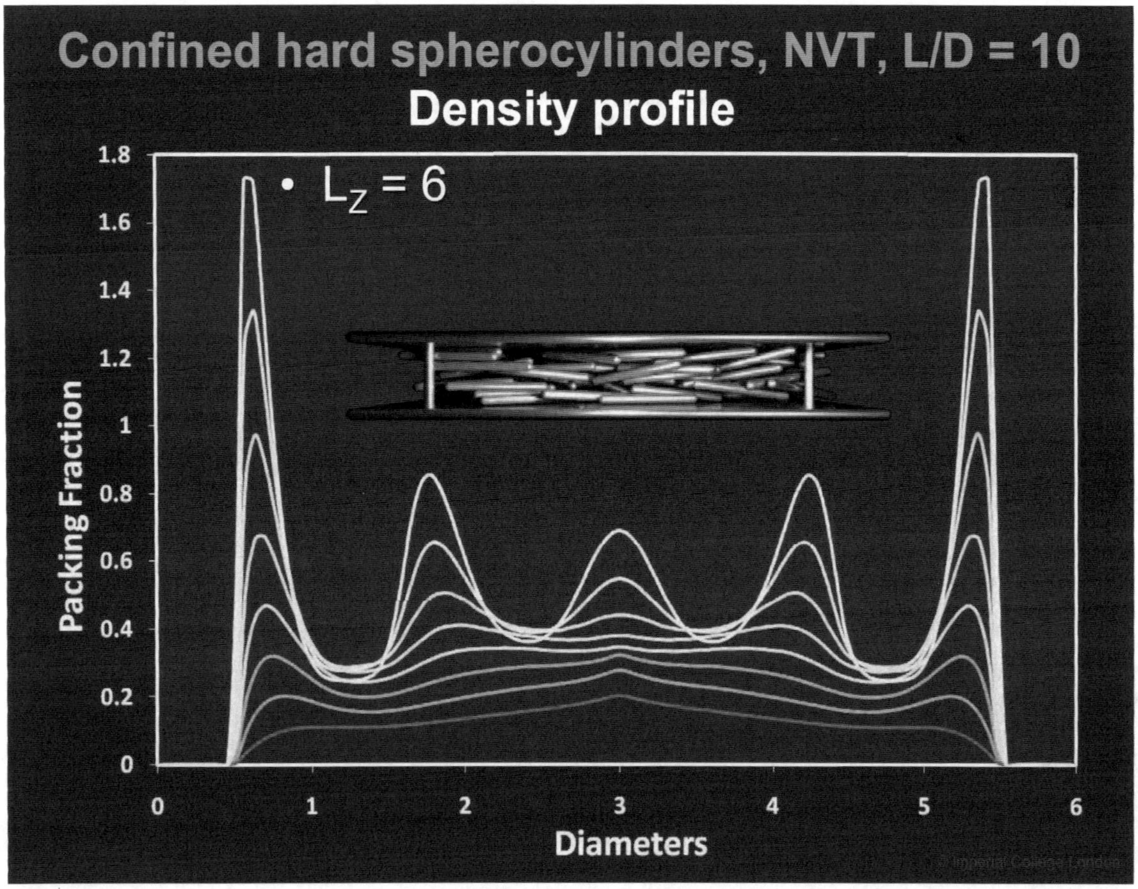
Confined hard spherocylinders, NVT, L/D = 10
Density profile
• L_Z = 6
Packing Fraction
Diameters
0 1 2 3 4 5 6
0 0.2 0.4 0.6 0.8 1 1.2 1.4 1.6 1.8

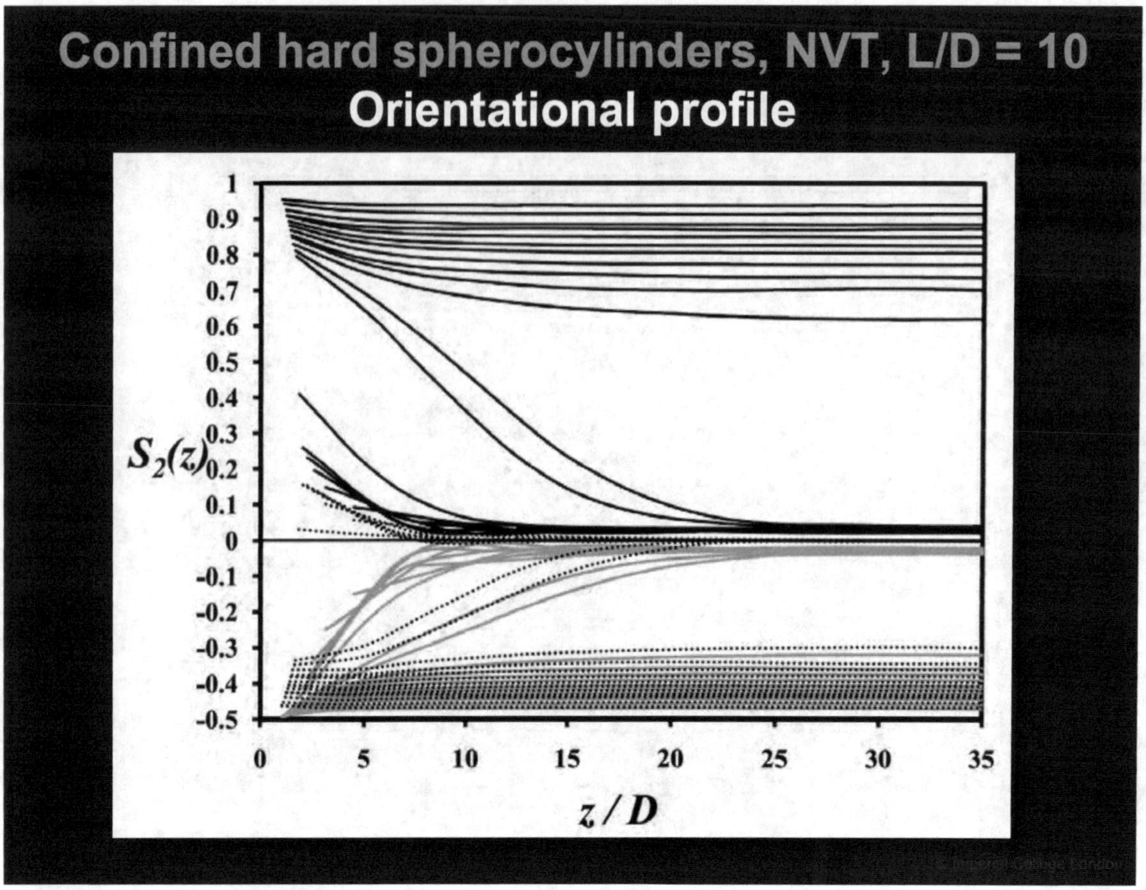
Confined hard spherocylinders, NVT, L/D = 10
Orientational profile
$S_2(z)$
z / D
© Imperial College London

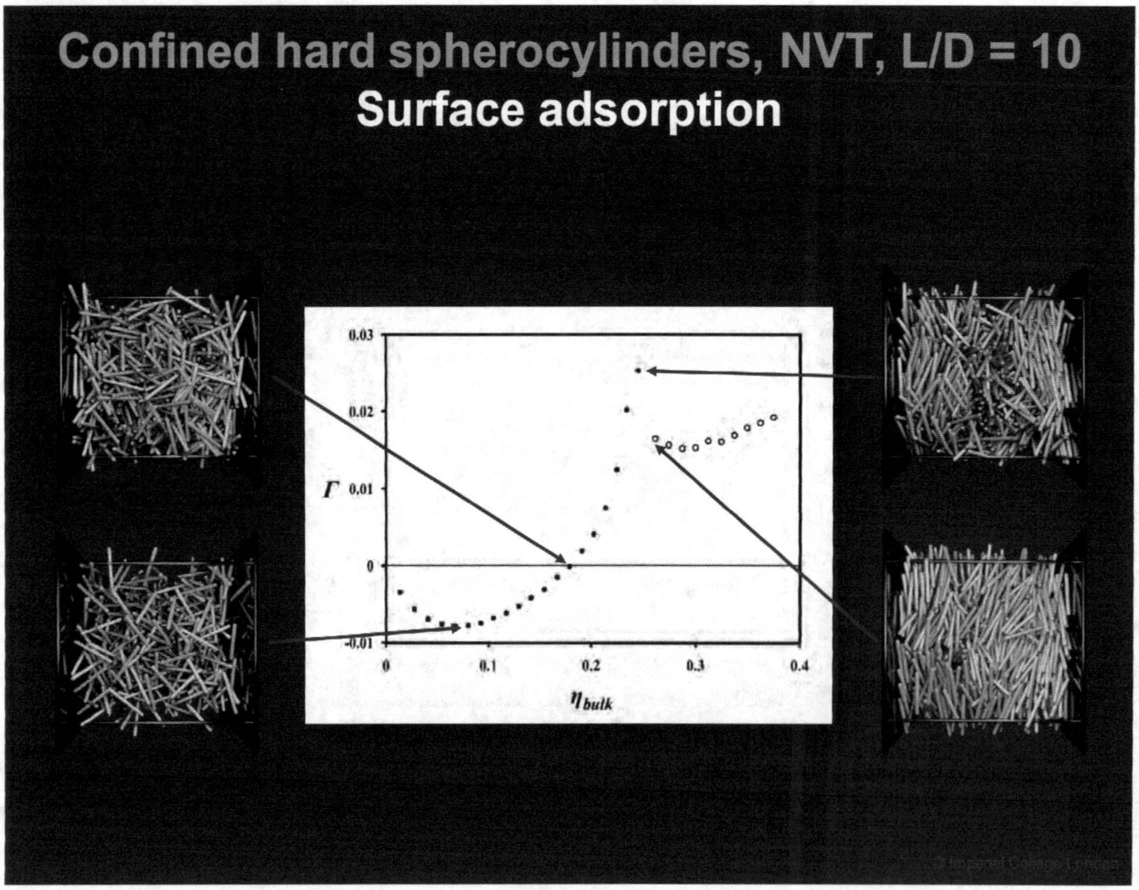
Confined hard spherocylinders, NVT, L/D = 10
Surface adsorption
Γ
η_{bulk}
© Imperial College London

Confined HSC system

Normal pressure

- *HSC   L/D=10*
- *Hard-wall confinement*
- *Surface wetting*
- *Nematic ordering*
- *Surface Adsorption*

$$\Gamma = \int_0^{Lz} \left[\rho(z) - \rho_\text{b} \right] \mathrm{d}z$$

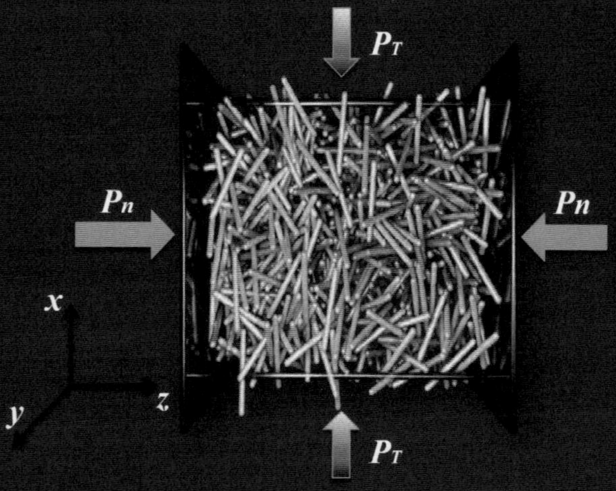

Tangential pressure

interfacial tension γ

$$\gamma = L_z \left(P_\text{n} - P_\text{T} \right)$$

Paul Brumby, PhD thesis, Imperial College London,2010

Calculation of the components of the pressure tensor and surface tension

Thermodynamic route

$$p = -\left(\frac{\partial A}{\partial V}\right)_{N,T} = -\lim_{[\Delta V \to 0]} \frac{\Delta A}{\Delta V}$$

$$p = \frac{N k_B T}{V} + \lim_{[\Delta V \to 0]} \frac{k_B T \ln \mathcal{P}_{nov}}{\Delta V}$$

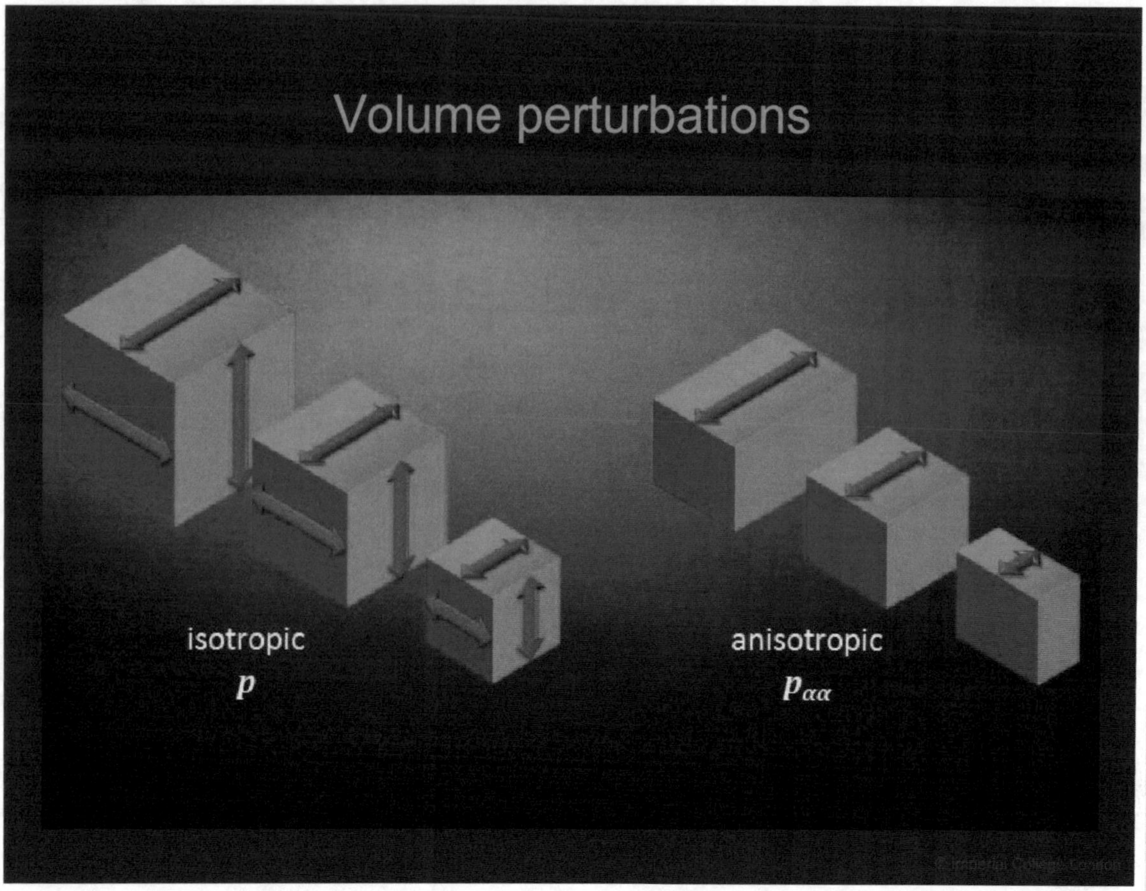
Volume perturbations
isotropic
p
anisotropic
$p_{\alpha\alpha}$
© Imperial College London

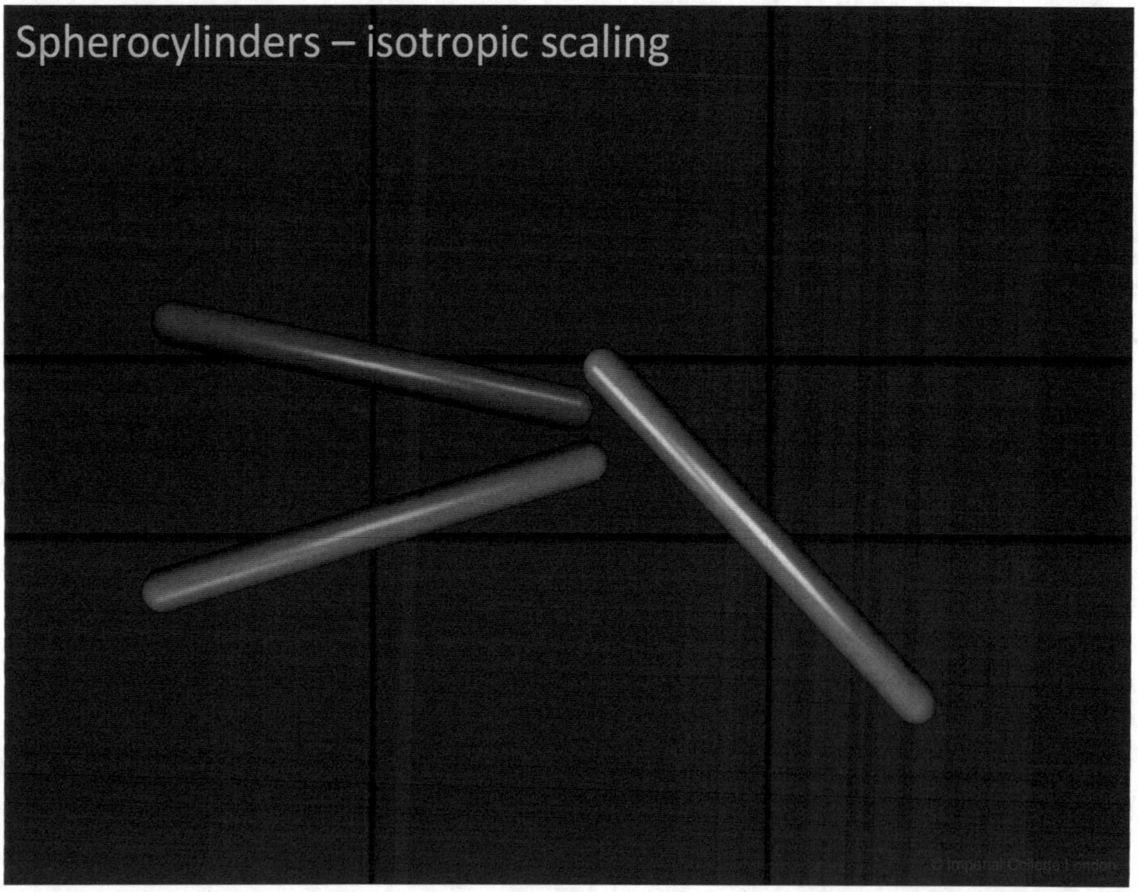
Spherocylinders – isotropic scaling
© Imperial College London

Spherocylinders – anisotropic scaling

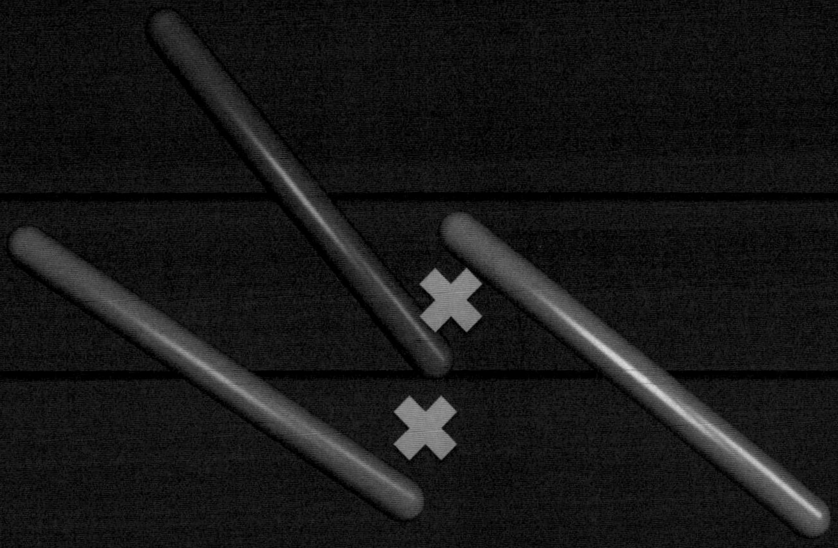

For non-spherical particles both the positive (expansive) and negative (compressive) volume changes have to be considered in evaluating the pressure tensor

$$p_{\alpha\alpha} = \frac{Nk_BT}{V_i} + \lim_{[\Delta V_{i\to j+}\to 0]} \frac{k_BT}{\Delta V_{i\to j+}} \ln P_{nov+} + \lim_{[\Delta V_{i\to j-}\to 0]} \frac{k_BT}{\Delta V_{i\to j-}} \ln P_{nov-}$$

Allows for the calculation of the wall-fluid surface tension as

$$\gamma = L_z\left(p_N - p_T\right)$$

Confined hard spherocylinders, NVT, L/D = 10

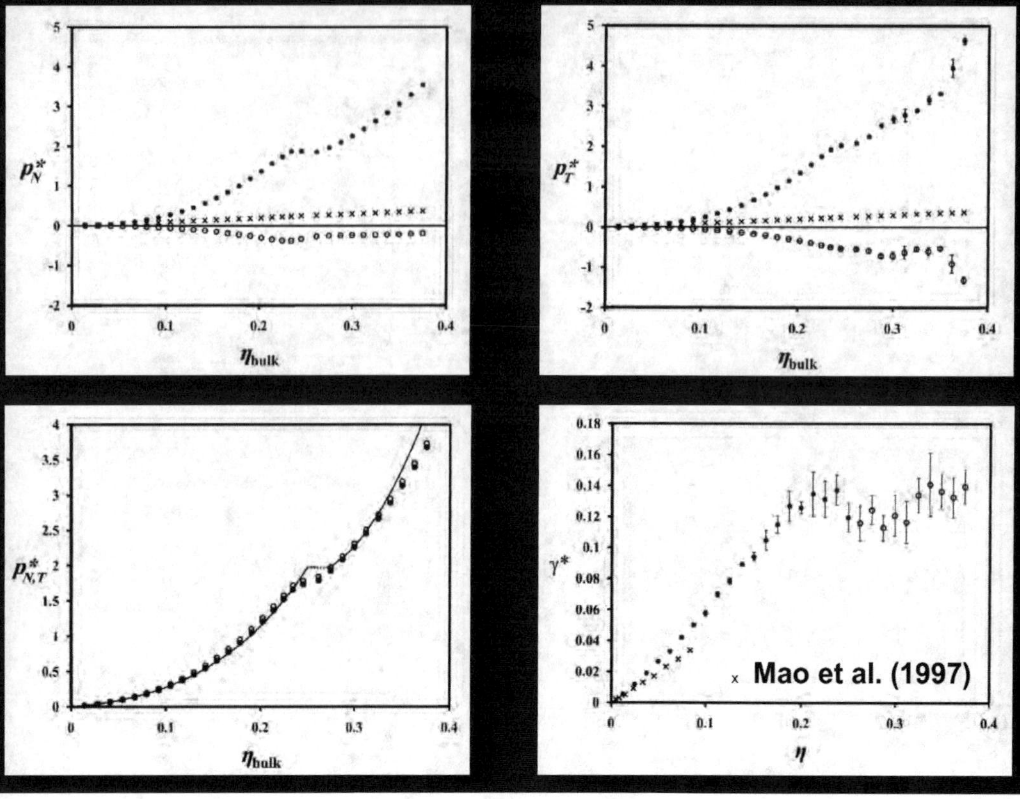

Conclusions

- First-order mechanical (virial) contribution characterises the planar interfacial tension. Henderson (1977)
- Large energetic fluctuations (second order) contribute to the interfacial tension of nanoscale drops in addition to mechanical contribution.
- The first-order mechanical (virial) route to the tension of drops is inappropriate.
- The components of the pressure tensor of non-spherical particles contain both compressive and tensile contributions.
- The two contributions are highly non-symmetrical.
- This provides a convenient route to the calculation of the fluid-wall surface tension of anisometric partciles.

371810 Nonlinear Optimization-Based Approaches for Modeling and Design of Heterogeneous Catalytic Systems

Abstract:

Microkinetic modeling, with input from computational chemistry calculations, is a key step in parameter estimation, experimental design, and rational catalyst design of heterogeneous catalytic systems. All these applications are, however, nonlinear optimization problems with an embedded microkinetic model that captures the dynamics of the reaction system. We, therefore, propose a generic methodology that combines nonlinear programming and differential algebraic systems solution for modeling and first-principles based design of heterogeneous catalytic systems.

For any given system, a comprehensive reaction network and DFT-derived thermochemical parameters of all species and transition states are taken as inputs. Subsequently, three classes of nonlinear optimization problems can be formulated and solved: (i) parameter estimation based on a simultaneous or sequential approach using activation barriers of reactions and binding energies of species as parameters [1, 2] to elucidate the detailed reaction mechanism, (ii) identification of optimal experimental conditions that maximize the coverage of or flux through an experimentally observable and mechanistically relevant surface intermediate, and (iii) determination of "Sabatier-optimal" binding energies of surface species and transition state energies of reactions that maximize turnover frequencies and/or selectivity of desired products.

The formulation of the optimization problems, selection of a relevant set of parameters, solution approaches, and multi-start schemes to identify a near-global minimum will be presented and discussed in the context of methanol synthesis from CO and CO_2 on transition metals.

References:

1. P. Rubert-Nason, M. Mavrikakis, C. T. Maravelias, L. C. Grabow, and L. T. Biegler., AIChE J 2014, 60, 1336 – 1346
2. L. T. Biegler, Nonlinear Programming: Concepts, algorithms, and applications to chemical processes. SIAM 2010

374764 Estimating Reaction Rates from Coverage Dependent Binding Energies

Abstract:

Creating accurate, compact rate models from first principles simulations remains a challenge in computational catalysis. Given a microkinetic description of all the steps in a catalytic mechanism and their sensitivity to their local surface environment, macroscopically observable rates can be rigorously estimated by suitable averaging over the microscopic rates, for instance using kinetic Monte Carlo. This approach can be further simplified by using the variation in apparent surface energetics as an estimator of microscale behavior, allowing for good fidelity using only an excess energy representation for surface species and a Brønsted–Evans–Polanyi (BEP) relationship to correlate overall reaction energies to energies of activation. In this work, density functional theory calculations are used to parameterize excess energy descriptions of adsorbates on a Pt (111) surfaces. These descriptions are used in conjunction with BEP relations to provide a steady-state description of catalytic oxidation networks. Results are compared with equivalent descriptions of the oxidation networks derived from Monte Carlo simulations and weighted distribution sampling.

Predicting Adsorption Energies for Use in Catalyst Design

One of the most successful paradigms for catalyst design is the use of adsorption energies to predict catalytic performance. Hence, understanding and predicting trends in adsorption energies can provide significant insight into catalyst design. We have developed general models for predicting adsorption energies, which can be used in rational design and high-throughput screening.

We have performed density functional theory calculations on adsorbates that bind to transition metal surfaces through various atoms, including C, O, H, N, S, and F. These calculations reveal that the dependence of the adsorption energy on the metal's d band and p band is independent of the identity of the adsorbing atom. However, the dependence on the adsorbate-surface coupling depends on the energy of the highest occupied molecular orbital (HOMO). Adsorbates with higher HOMO energies experience less Pauli repulsion. Combining these behaviors, the adsorption energy of any monovalent radical on any transition metal can be predicted using a single equation. This equation has single adsorbate-dependent fitting parameter that correlates with the HOMO energy, as well as a few universal parameters.

Because the adsorbate's behavior depends on the HOMO energy, adsorbates with similar HOMO energies correlate. Further, for adsorbates with very different HOMO energies, the coupling and d-band filling can be used to predict differences in the adsorption energy. Hence, adsorption energies of all adsorbates can be predicted from the adsorption energy of a single adsorbate.

These findings are useful for both rational design and high-throughput screening. This is particularly true because the surface electronic structure parameters can be related to the alloy structure. Understanding the factors controlling adsorption energies leads to an understanding of the factors controlling catalytic performance. Furthermore, the surface properties can be efficiently predicted, improving prospects for high-throughput screening. We apply our methods to several reactions, and experimentally test a novel catalyst for ethylene epoxidation with higher selectivity than Ag.

Metal nanoparticles find tremendous applications in nanotechnology and especially in modern catalytic technologies for energy generation and conversion. The interactions of the reacting molecules with the nanoparticle surfaces are key in the catalytic performance of the nanoparticles[1]. According to the Sabatier's principle, the metal-adsorbate interactions should be "just right". As an example, strong adsorption of one reactant over another will result to poisoning the catalyst by the preferential adsorption of the strongly bound adsorbate. On the other hand, weak adsorption results to desorption of the adsorbates at elevated temperatures and no reactions occur on the nanoparticle surface. One can fine tune the interactions of adsorbates with the nanoparticles by changing the nanoparticle morphologies[2], such us, their size, shape and metal composition. In this work[3], we use Density Functional Theory calculations to unravel the binding behavior of adsorbates (e.g. CO) as a function of nanoparticle morphologies, such as their size and shape. This work introduces novel adsorption concepts in catalysis and unravels very complex adsorption mechanisms on sub-nanoscale catalysts. At catalyst sizes of less than 1nm diameter, strong electronic effects dominate and govern i) adsorption and ii) catalyst restructuring as a function of surface coverage of the adsorbates.

1. Mpourmpakis, G.; Vlachos, D. G. *MRS Bulletin* **2011,** 36, (3), 211-215.
2. Mpourmpakis, G.; Andriotis, A. N.; Vlachos, D. G. *Nano Letters* **2010,** 10, (3), 1041-1045.
3. Austin, N.; Mpourmpakis, G. *submitted* **2014**.

375019 Uncertainty Quantification of a Computational Study of the Water-Gas Shift Reaction Catalyzed By Platinum Clusters Supported on Titanium Oxide (Pt/TiO2)

Abstract:

The water-gas shift reaction is the primary reaction for generating hydrogen for the chemical and petrochemical industries. In this paper a novel uncertainty quantification study is conducted of a computational model of the water-gas shift reaction occurring at the three-phase boundary of a Pt/TiO_2 catalyst by taking into account the inexact nature of density functional theory and constraining the results to thermodynamic experiments. This uncertainty quantification study provides informative model results in the presence of errors from DFT. The first objective of uncertainty quantification is to make reliable model results given such possible errors. A second objective is constraining density functional theory results to thermodynamic experiments. The thermodynamic constraints are accomplished by causing gas molecule corrections to sum to a constant amount. CO is given a large adjustment compared to other gas molecules due to prior knowledge about the PBE functional being unable to correctly describe the electronic structure of CO. Uncertainty in model results is represented by probability density plots and the dominant catalytic cycle is identified.

360463 Ab Initio Framework for Uncovering Structure-Property Relationships of Amorphous Catalysts

Abstract:

Approaches for modeling catalytic sites on amorphous supports lag far behind approaches for modeling catalytic sites on crystalline materials, e.g., zeolites and metal surfaces. A typical approach to model amorphous supports is to use cluster models with arbitrarily chosen constraints to represent the rigid amorphous support, but these constraints arbitrarily influence catalyst site activity. An alternative strategy is to use no constraints, but this results in catalytic sites with unrealistic flexibility. We present a systematic ab initio method to model isolated active sites on insulating amorphous supports using small cluster models. A sequential quadratic programming framework helps us relate chemical properties, such as the activation energy, to active site structure. The algorithm is first illustrated on an empirical valence bond model energy landscape. We then use the algorithm to model an off-pathway kinetic trap in olefin metathesis by isolated Mo sites on amorphous SiO_2.

390713 Computational Database of Metal-Oxide Surface Reactivities for Catalyst Design

Abstract:

We study surface reactivity of low index facets of rocksalt type Alkaline-Earth and first row transition metal oxide groups using small atoms and molecules (O, OH, CO, NO) .The computed database of adsorption and activation energies will be used to identify possible correlations with other quantities such as surface energies or electronic structure in order to establish scaling relations for future high-throughput screening efforts. A comparison will be made between DFT functionals of various levels of accuracy, e.g., GGA, GGA+U, GGA+vdW and GGA-hybrid, meta-GGA and hybrid meta-GGA, and compared to available experiments. This effort is part of the ``Predictive Theory of Transition Metal Oxide Catalysis" funded through the DOE Materials Genome Project.

376124 Development and Application of the SAFT-FMT-DFT Approach for Adsorption Equilibrium

Abstract:

Density functional theory (DFT) is commonly used to treat the adsorption of molecules in carbon and silica pores of various geometries. We have developed a DFT with an accurate molecular-based equation of state to calculate thermodynamic properties using fundamental measure theory (FMT), which is a rigorous approach for the treatment of homogeneous and non-homogeneous hard-sphere fluids. A theoretical framework results with adsorbing molecules treated as hard-sphere chains with square-well attractive interactions. The Mansoori-Carnahan-Starling-Leland and Carnahan-Starling-Boublik equations of state are used for the hard sphere interactions, and a version of the statistical associating fluid theory for potentials of variable range (SAFT-VR) is used to describe the square-well fluid. First and second order perturbative attractive terms are included in the theory. The theoretical predictions are in good agreement with published results for Monte Carlo simulations of the adsorption of chain molecules.

In this talk, the SAFT-FMT-DFT approach is used to develop a pore size distribution for BPL activated carbon based on nitrogen adsorption, and that pore size distribution is used to predict the adsorption of n-pentane on the activated carbon. Interaction parameters are obtained by simulating data for adsorption of nitrogen and n-pentane on the planar wall of a non-porous graphitized carbon. The 10-4-3 fluid-wall potential is used to model adsorption in the slit-shaped pores of the activated carbon. The single pore nitrogen isotherms simulated with the DFT theory show monolayer transitions, pore filling, and pore condensation. The predicted excess n-pentane isotherm is compared with a unique experimental n-pentane isotherm that extends into the Henry's law region. The predicted isotherm agrees well with the experimental results.

376012 Wetting at the Nano-Scale

Abstract:

At the macro-scale the extent to which a liquid wets a solid substrate is usually described in terms of the contact angle, θ_c, and the surface tensions involved. Depending on the liquid and substrate, the system is described as amphiphilic ('wetting', θ_c<90°) or amphiphobic ('non-wetting', θ_c>90°). Such a description has a number of limitations; in particular, it breaks down for sufficiently small nano-scale systems, and is limited to describing liquid, as opposed to gaseous or solid, adsorbed films. At a more fundamental level, wetting is determined by the competition between the adsorbate-substrate intermolecular forces and the adsorbate-adsorbate forces. Through a corresponding states analysis of the statistical mechanical description of such wetting systems it is possible to define a *microscopic wetting parameter*, $\alpha_{w,}$, that is a measure of wetting that applies at all scales and for any kind of adsorbed film (gas, liquid or solid).[1,2]

We illustrate the usefulness of this wetting parameter by considering the properties of a nano-phase confined within a porous material. In this case the dimensionless pore width, pore shape and wetting characteristics of the confined phase are of particular importance. Examples drawn from both experiment and molecular simulation studies will be presented for phase separations,[2,3] selective adsorption in the case of mixtures,[2] and pressure enhancement,[4,5] with emphasis on simple pore geometries. These examples illustrate the central role played by wetting, and also the breakdown of some concepts and macroscopic laws, such as Gibbs' surface thermodynamics for nano-phases confined within small pores. In particular, we show the breakdown of the Kelvin equation for vapor-liquid condensation, and the Gibbs-Thomson equation for freezing in nano-pores.

1. R .Radhakrishnan, K.E. Gubbins and M. Śliwinska-Bartkowiak, "Global Phase Diagrams for Freezing in Porous Media", *Journal of Chemical Physics*, **116**, 1147-1155 (2002).
2. Keith E. Gubbins, Yun Long and Malgorzata Sliwinska-Bartkowiak, "Thermodynamics of Confined Nano-Phases", *Journal of Chemical Thermodynamics*, **74**, 169-183 (2014).
3. L.D. Gelb, K.E. Gubbins, R. Radhakrishnan and M. Sliwinska-Bartkowiak, "Phase Separation in Confined Systems", *Reports on Progress in Physics*, **62**, 1573-1659 (1999).
4. Yun Long, Jeremy C. Palmer, Benoit Coasne, Małgorzata Śliwinska-Bartkowiak and Keith E. Gubbins, "Pressure enhancement in carbon nanopores: A major confinement effect", *Physical Chemistry Chemical Physics*, **13**, 17163-17170 (2011)
5. Yun Long, Jeremy C. Palmer, Benoit Coasne, Małgorzata Śliwinska-Bartkowiak, George Jackson, Erich A. Müller and Keith E. Gubbins, "On the Molecular Origin of High Pressure Effects in Nanoconfinement: Effects of Surface Chemistry and Roughness", *Journal of Chemical Physics*, **139**, 144701 (2013)

368222 Some Recent Developments in Modeling the Dynamic Behavior of Confined Fluids

Abstract:

Recently we have developed a dynamic mean field theory (DMFT) that describes the dynamics of adsorption and desorption of fluids in mesoporous materials. The theory predicts the evolution of the density distribution of the fluid in the porous material after a step change in the bulk pressure or chemical potential. An important feature of DMFT is that it is fully consistent with the mean field density functional theory (DFT) description of the thermodynamics. In addition the nucleation mechanisms in phase transitions of confined fluids are emergent features of the calculations. For instance the theory describes the nucleation of capillary condensation by liquid bridging. In this presentation we describe applications of DMFT to several systems. For pore networks we study the influence of pore connectivity upon uptake and desorption dynamics. Studies of adsorption/desorption dynamics of mixtures show the nature of the equilibration of the composition distribution in the system.

376177 Structure and Dynamics of Energy-Relevant Fluid-Solid Interfaces: A Long Journey with Peter Cummings through Very Small Times and Places!

Abstract:

My interactions with Peter Cummings began in 1994 when he became a joint ORNL/University of Tennessee (Knoxville) Distinguished Scientist, and later joint Chief Scientist of ORNL's Center for Nanophase Materials Science and Professor of Chemical Engineering at Vanderbilt University. In this talk I will review the development and ongoing application of our integrated molecular dynamics simulations and neutron/X-ray scattering studies of bulk aqueous and room temperature ionic liquid electrolytes and the changes in their structure and dynamics in the nanoscale region at neutral and charged surfaces (minerals, electrodes, etc.), including the effects of nanoconfinement.

Peter's collaboration with our Aqueous Chemistry and Geochemistry program began even before he joined ORNL, with quantitative experiments and modeling of the thermodynamic properties and phase equilibria of high temperature water and aqueous solutions into the supercritical regime[1]. This work applied the SPC water model, which could be readily adapted to simulate dissolved inorganic salts. This collaboration lead to the development of the Gaussian Charge Polarizable (GPC) force field for water[2], which is shown to be more accurate in describing the structure and phase envelope of pure water over very wide ranges of temperature and density[3]. My research group is now adapting GPC for aqueous electrolytes, and further demonstrates the excellent agreement between GPC and the most recent X-ray scattering data[4,5].

Over the last 14 years Peter has been a co-PI in our project funded by DOE's Office of Basic Energy Sciences, Geoscience Research Program, with the current name "Geochemical Equilibria and Reaction Dynamics", in which we seek to quantify the atomic-molecular origins of geochemical processes associated with fluid-rock interactions that control contaminant transport, nuclear waste isolation, fossil and geothermal energy exploitation, and the evolution of porosity/permeability in the subsurface. This work has broad applications in other areas of science, such as catalyst and nanomaterials synthesis and stability[6,7]. Since 2009, Peter has been a Thrust Leader, and is now the Deputy Director of the Fluid Interface Reactions, Structures and Transport (FIRST Center), of which I am the Director, which is supported by DOE's Office of Science, Office of Basic Energy Sciences. In this project, we have applied molecular modeling and X-ray/neutron scattering approaches to room temperature ionic liquid interactions with carbon electrode materials in planar, curved and nanopore-confined geometries[8,9].

[1]Cummings, P.T.; Cochran, H.D., Simonson, J.M., Mesmer, R.E., Karaboni, S. *J. Chem. Phys.*, 94:5606 (1991).

[2]Chialvo, A.A.; Cummings, P.T. *Fluid Phase Equil.*, 150:73 (1998).

[3]Paricaud, P., Predota, M., Chialvo, A.A., Cummings, P.T. *J. Chem. Phys.*, 122:244511 (2005).

[4]Chialvo, A.A., Vlcek, L. *J. Phys. Chem. B* (in review).

[5]Skinner, L.B.; Huang, C.; Schlesinger, D.; Pettersson, L.G.M.; Nilsson, A.; Benmore, C.J. *J. Chem. Phys.*, 138:074506 (2013).

[6]Wesolowski, D.J., et al., *Phys. Rev. B* 85:167401 (2012).

[7]Wang, H.-W.; Wesolowski, D.J., et al. *J. Amer. Chem. Soc.* 135:6885 (2013).

[8]Feng, G.; Li, S.; Presser, V.; Cummings, P.T. *J. Phys. Chem. Lett.*, 4:3367 (2013).

[9]Banuelos, J.L., et al. *Chem. Mater.*, 26:1144 (2014).

Integrated Modeling-Experimental Study of Structure and Dynamics of Room-Temperature Ionic Liquids under Hierarchical Micropore Confinement

Guang Feng,[1] Peter Cummings[2], José Leobardo Bañuelos[3], Gernot Rother[3], and David J. Wesolowski[3]

[1]*State Key Laboratory of Coal Combustion, School of Energy and Power Engineering, Huazhong University of Science and Technology, Wuhan 430074, China*

[2]*Department of Chemical and Biomolecular Engineering, Vanderbilt University, Nashville, TN 37235, USA*

[3]*Chemical Sciences Division, Oak Ridge National Laboratory, Oak Ridge, Tennessee 37831, USA*

ABSTRACT

Electrical double layer capacitors (EDLCs), also called supercapacitors, have attracted a great deal of attention in present electrical energy community, owing to their advantageous properties, such as high-power density and exceptional cycle life. Porous carbons have become the most widely used electrode materials in supercapacitors, in view of their high specific surface area, good electrical conductivity, relatively low cost, and so on; while room-temperature ionic liquids (RTILs) have emerged as promising electrolytes for supercapacitors, because of their exceptionally wide electrochemical stability window, excellent thermal stability, non-volatility and relatively inert nature. Despite considerable work on carbon supercapacitors, the details of what happens under porous carbon confinement still require in-depth exploration, even under zero bias.

In this talk we present structural and dynamic properties of 1-butyl-3-methylimidazolium bis((trifluoromethane)sulfonyl) amide ([Bmim][TFSI]) confined in ordered meso-/micro-porous carbon, using an integrated approach of modeling and experiment techniques, i.e., classical molecular dynamics (cMD), small angle X-ray scattering and neutron spin echo (NSE). High RTIL densities, obtained from cMD and from analysis of scattering data from samples with varying degrees of RTIL total pore volume filling, indicate strong RTIL/pore wall interactions mediated by the intrinsic affinity of liquids on the carbon surface and by the rough topology of surface. Consequently, cMD and NSE show a dramatic slowing of the overall collective dynamics, including an effectively immobilized fraction of RTIL at the pore wall. Increasing temperature from 296 K to 353 K reduces the amount of surface-immobilized RTIL molecules and lowers the overall RTIL density, allowing for an order of magnitude increase in the observed dynamics.

Keywords: Supercapacitor, Porous Carbon, Room-Temperature Ionic Liquids, Structure, Dynamics

376143 Transport Phenomena of Chemotactic Bacteria: Diffusion and Dispersion in Porous Media

Abstract:

Chemotaxis describes the ability of motile bacteria to sense chemical gradients in their surroundings and swim toward higher concentrations of chemicals that are beneficial to their survival. The chemotactic response in *Escherichia coli* bacteria is a well-characterized signal transduction mechanism that controls the run-and-tumble swimming behavior of individual cells. The motivation behind our experimental work has been to build on this fundamental knowledge of the underlying mechanisms to develop predictive models for bacterial migration in complex natural systems. Our focus has been on migration of chemotactic bacteria in porous media with application to bioremediation of polluted groundwater systems where chemical and structural heterogeneity influence their transport phenomena. I will present a series of experimental approaches that range from imaging chemotactic bands in microfluidic devices to monitoring dispersion within bench-scale microcosms to tracking the migration of bacteria introduced into a natural groundwater aquifer. Apparent diffusion and dispersion coefficients determined from the experimental observations are used in mathematical models to predict macroscopic-scale transport of bacterial populations. A dimensionless chemotaxis number is proposed to ascertain *a priori* the conditions under which a chemotactic response will impact bacterial transport relative to other processes such as advection and dispersion.

355295 De-Swelling Mechanisms of a Surface-Grafted Poly(NIPAAm) Brush

Abstract:

In this study, we investigated the deswelling mechanisms of hydrated surface-grafted poly(N-isopropylamide) (P(NIPAAm)) brush using full-atomistic molecular dynamics (MD) simulations at various temperatures such as 275, 290, 320, 345, and 370K. We clearly observed the deswelling of the P(NIPAAm) brush above the lower critical solution temperature (305K) below which the P(NIPAAm) brush is associated with water molecules stably. By analyzing the density profiles, the pair correlation functions & the coordination number, hydrogen bonding, solvent accessible surface area, and the free energy change of system, we explain that the detail mechanisms of deswelling of P(NIPAAm) brush above LCST.

Molecules to Engines: Combustion Chemistry of Butanol and their Application to Advanced Engines

Shamel Merchant,[1*] Dan Delvescovo[3], Hu Wang[3], Kevin M. Van Geem[2], Rolf D. Reitz[3], and William H. Green[1]

[1] Department of Chemical Engineering, Massachusetts Institute of Technology, MA, USA
[2] Laboratory for Chemical Technology, University of Ghent, Belgium
[3] Engine Research Center, University of Wisconsin-Madison
[*]Corresponding author: shamel@mit.edu

A major challenge in energy is the identification of viable liquid fuels as alternatives to petroleum-based fuels. There are a wide variety of candidate fuels to select from and assessing each new fuel is far from trivial, since rather small variations in chemical structure can cause large changes in fuel performance. Also, engine designs are changing rapidly; simulations which accurately predict how future fuels will perform in future engines are in many ways more valuable than knowing which fuels perform well in today's engines. A new combined computational-experimental approach is presented which makes it practical to rapidly construct accurate combustion chemistry simulations for proposed new fuels. In the new approach experiments and quantum chemical calculations are done in parallel, informing an evolving chemical kinetic model. This approach is used to understand and predict the combustion chemistry of the isomers of butanol.

A detailed kinetic model for butanol isomers is presented which is validated against a large number of datasets spanning the entire range of operating conditions seen during real engine operation. Overall we see that for many performance parameters the model predictions are about as accurate as the experiments and mechanistic insights into differing reactivity of the isomers can be made. Lastly we show how this detailed kinetic model can be applied in engine CFD simulations of a new type of engine, i.e. the reactivity controlled compression ignition engine (RCCI), in order to make predictions of how butanol will affect the engine efficiency and emissions.

384948 A Molecular Dynamics Study Assessing the Accuracy of the Generalized Amber Force Field to Predict the Thermophysical Properties of 20 Ionic Liquids

Abstract:

Ionic liquids (ILs) are organic salts that typically have melting points below 100°C. These salts are known as "designer" solvents, because small changes can be made to the structure (i.e. cation or anion) or chemical composition of the mixture that lead to large changes in the physical properties of the liquid; properties such as melting point, solvation power, and viscosity can all vary greatly with such subtle changes. Due to their low viscosities and negligible vapor pressures, as well as exceptional thermal and chemical stabilities, ILs have come to play important roles in many fields, such as electrochemistry, separation science, catalysis, organic synthesis, and biocatalysis and biomass conversion.[1] Despite the many advances ILs have led to in these areas of research, there is still a great need for novel quantitative approaches that can easily, quickly, and accurately predict the transport and thermodynamic properties of these unique solvents, with knowledge solely of their chemical structures. If this can be done, it would allow for many more ILs to be explored for various applications by decreasing the time to sift through and test the multitudes of possible cation/anion combinations.

This work critically assesses the ability of the Generalized Amber Force Field, or GAFF, to predict various properties of ILs. An entirely automated method has been created that allows for quick parameterization of these nonaqueous solvents. Specifically, thermodynamic properties of density and heat capacity and transport properties of self-diffusivity and viscosity have been computed, using classical Molecular Dynamics, for twenty industrially-relevant ILs. Results show that this simple force field is, in nearly every case, very adept at predicting all the computed properties of the ILs studied. Parallel simulations with Tip3p water have been done for comparison. This poster also presents a simple extension of the automated method in order to trivially enable simulation of biomass or biomolecules in the presence of ILs.

References

[1] Sambasivarao and Acevedo. *J. Chem. Theory Comput.* **5**, 1038–1050 (2009).

MULTISCALE MODELING OF THE ELECTRODE/ELECTROLYTE INTERFACE USING MODIFIED CHARGE OPTIMIZED MANY BODY (ECOMB3) POTENTIALS

Sneha A. Akhade[1]*, Andrew Antony[2], Tao Liang[2], Michael J. Janik[1], Janna M. Maranas[1] and Susan B. Sinnott[2]

[1]Department of Chemical Engineering, Fenske Laboratory, Pennsylvania State University, State College, PA 16802
[2]Department of Materials Science & Engineering, 549 Gale Lemerand Drive, Gainesville, FL 32611, University of Florida
*Email: saa243@psu.edu

Present day advances in computer simulation have greatly accelerated material discovery and investigation of bulk and surface properties that are relevant to the reactivity and performance of the material. However, all processes involving heterogeneous catalysis within the electrochemical device occur at the interface. Our main objective is to develop robust methods that can provide a reasonable description of the electrode/electrolyte interface and can correctly mimic the complex underlying interfacial chemistry. We employ a combined Density Functional Theory (DFT) with a force field (FF) based classical approach to accurately treat the electrochemical interface. Towards constructing a realistic interface, two key features are addressed - metal polarizability and the interfacial electrolyte chemistry in the presence of a potential-controlled field. In our previous study [1], we employ a reactive modified central force field (mCFF) that uses simple point charges with pairwise interactions to model reactive water that can dissociate under the influence of potential controlled electrodes. The Pt (111) electrode is held at constant potential by using a charge fluctuation approach called Electrode Charge Dynamics (ECD) [2]. In our present study, we are developing third-generation Charge Optimized Many Body potentials (COMB3) [3] in conjunction with an electrode charge equilibration scheme (ECOMB3). Our results compare and test the efficacy of the two models using useful tools that are developed to probe and measure relevant electrochemical properties that can provide a reliable atomistic description of the complex electrode/electrolyte interface.

1. Yeh, K.-Y., M.J. Janik, and J.K. Maranas, *Molecular dynamics simulations of an electrified water/Pt(111) interface using point charge dissociative water.* Electrochimica Acta, 2013. **101**(0): p. 308-325.
2. Guymon, C., et al., *Simulating an electrochemical interface using charge dynamics.* Condens. Matter Phys, 2005. **8**: p. 335-356.
3. Liang, T., et al., *Classical atomistic simulations of surfaces and heterogeneous interfaces with the charge-optimized many body (COMB) potentials.* Materials Science and Engineering: R: Reports, 2013. **74**(9): p. 255-279

385792 *In silico* Design of Nanostructured Porous Materials for Environmental Applications

Abstract:

Computer simulations and molecular modeling techniques have played an important role in elucidating and providing understanding of observed phenomena in a multitude of fields. Particularly for porous materials, computer simulations have provided useful insights about the role of pore size, pore size distribution, chemical functionalities, and other physical properties in gas storage, separations, catalysis, and other applications. New computational techniques are starting to emerge such as large-scale *in silico* generation of porous materials and high-throughput screening that hold particular promise to help identify promising materials for particular applications, discover structure-property relationships, and determine the potential performance limits of these materials.

Our recent work has focused on the role of water in adsorption phenomena in porous materials. In particular, it has focused on the capture of carbon dioxide under humid conditions. Many materials excel in carbon dioxide adsorption under dry conditions, but they can decompose or their adsorption performance is severely affected by water due to the competition for adsorption sites within these materials. Hence, research efforts have taken place to develop materials that are hydrophobic but still can adsorb carbon dioxide. An aspect that has been somewhat overlooked is the role that defects can have in the adsorption of both water and carbon dioxide. We have used grand canonical Monte Carlo (GCMC) simulations to assess the role of defects in water adsorption in a class of zirconium metal-organic frameworks (MOFs). We show how a relatively small number of defects can shift the water adsorption behavior, going from hydrophobic to hydrophilic. Additionally, we discovered that related metal-organic materials contain both hydrophobic and hydrophilic cavities.

Another important potential application of porous materials is hydrogen storage. MOFs hold particular promise in this area due to their large pore volumes and surface areas. However, these materials falter near room temperature due to their low heats of adsorption. To address this problem, we have investigated MOFs containing open metal sites to increase the heats of adsorption. Previous work had used quantum mechanical calculations to quantify the interaction energies between hydrogen and the open metal sites, and the results were fit to a Morse potential. We used this potential in GCMC simulations to assess the performance of these materials for hydrogen storage. Mg alkoxide functionalization turned out to be a promising candidate. Then, using recently developed methods for large-scale generation of MOFs, we generated over 18,000 structures containing various numbers of Mg alkoxides and a wide range of physical properties. In a high-throughput manner, the hydrogen storage capability of these structures was assessed using GCMC simulations, and we identified promising candidates, obtained novel structure-property relationships, and quantified what could be the performance limits of these materials. Currently, we are working on novel ways to generate new MOF structures with various topologies. This could lead to new, undiscovered porous materials that can hold promise for various environmentally relevant applications.

Title: Identification and Allosteric Regulation of Oxygen Pathways in Monomeric Sarcosine Oxidase via Single-Sweep Free Energy Reconstruction

Introduction:

Monomeric sarcosine oxidase (MSOX) is a flavoprotein D-amino acid oxidase that catalyzes the oxidation of sarcosine (N-methylglycine) to produce formaldehyde, glycine, and hydrogen peroxide. It is a 44 kDa two-domain protein consisting of 385 residues and a single covalently bound flavin adenine dinucleotide (8α-S-cysteinyl-FAD) whose flavin ring sits at the interface between the so-called catalytic and flavin domains. MSOX is one of the most studied flavoprotein oxidases, and has both wild type and mutant crystal structures available, the latter aiding in identification of catalytically important residues. In spite of the fact that MSOX is structurally well-characterized, little is known about the precise location of oxygen activation at the flavin site and the pathways by which oxygen accesses this site from the solvent, especially when the substrate channel is blocked with bound substrate. The purpose of this work is to advance our understanding of MSOX function by using all-atom simulations to map the diffusion pathways and localization sites for oxygen in MSOX and to characterize them thermodynamically [1].

Methods:

Traditional all-atom molecular dynamics simulations are insufficient for these purposes due to inadequate sampling. Therefore, enhanced sampling with temperature-accelerated MD (TAMD) coupled with the single-sweep method of free-energy construction are employed to map the free energy of oxygen as a function of position within the accessible protein volume. String method is then applied to determine pathways of minimum free energy on this surface connecting surface portals to the protein active site. We perform this mapping on three states of MSOX: (i) MSOX bound with the substrate mimicking inhibitor 2-furoic acid (FOA), and MSOX with an (ii) open or (iii) closed flaving-admitting cleft. In our composite approach, for each MSOX state, we first applied TAMD to the O2 center of mass to sample as much of the accessible protein volume as possible. Applying single-sweep we then took a subset of these sampled locations and conducted restrained MD at each one to compute thermodynamic mean forces with respect to the O2 position. The free energy as a function of oxygen position was then calculated using a Gaussian radial basis function expansion. The location of the deepest minima on this surface were then identified using steepest decent, and used as the input for string method. String method then identified four unique pathways leading to the active site.

Results:

All pathways are observed to converge in the active-site basin on the re-face of the flavin ring. No free-energy minimum is observed on the si-face in any of the three states. Pathway I is characterized by entering above the bridge indicative of flavin cleft opening, passing between Lys265 and Cys315, and terminating after passing through the minima observed on the re-face. Pathway II enters through the bridge that opens to admit the flavin cofactor, and then travels along the sugar backbone of the flavin, entering on the re-face near Cys315, terminating by passing through both minima present there. Pathway III is the longest and most interesting, as it traverses the entire FAD binding domain, then converges with Pathway II. It contains the deepest free energy minimum of any of the four pathways. Pathway IV enters via the large sarcosine entryway, terminating at the minima on the re-face (Fig. 1A).

We observe that the pathway thermodynamics were surprisingly sensitive to the presence of the bound FOA ligand (Fig. 1B). When substrate is bound, O2 localization free-energy is generally low, indicating oxygen will localize within MSOX but separate from the active site. When substrate (or product) unbinds, these minima vanish and O2 access to the active site is made easier by the reduction of barriers. Since binding of the inhibitor/substrate is the only major difference, this trend is indicative of subtle allosteric modulation of oxygen within MSOX, even far from the active site. This trend can be explained by analysis of the side chains surrounding the strings. We hope to explain the effects of this allosteric modulation in future work through analysis with the milestoning technique, where we will study the rate of transport along these four pathways.

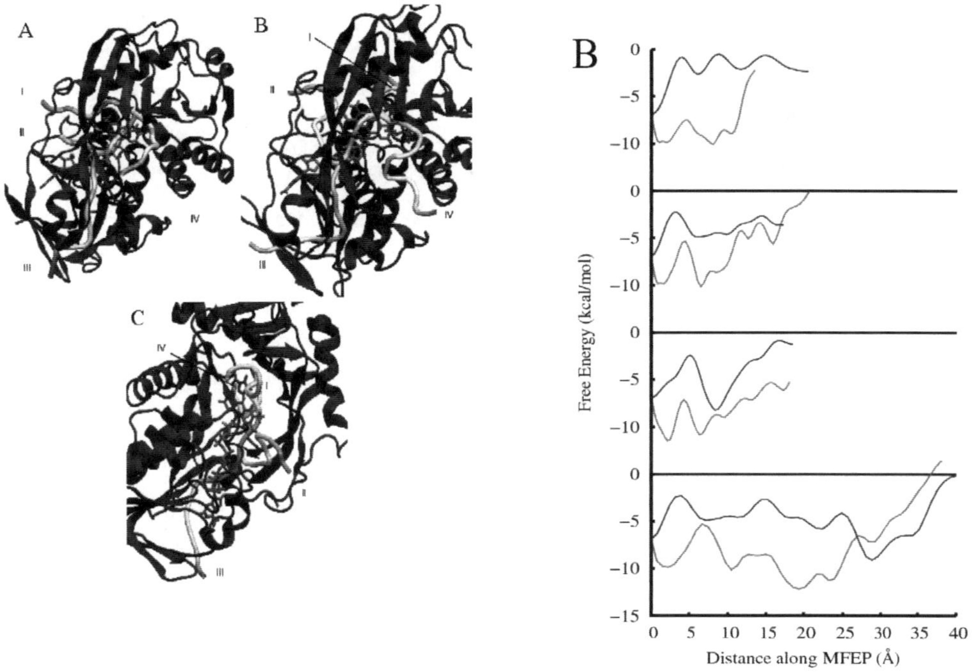

Figure 1. (A) MSOX structure showing pathways I, II, III, and IV. FADH cofactor in orange sticks. (B) Free energy profiles for O2 transport from active site (origin) to solvent in the case of the apo enzyme (blue) and inhibitor-bound (red).

Reference:

(1) Bucci, A., Abrams, C.F. (2014) Oxygen Pathways and Allostery in Monomeric Sarcosine Oxidase via Single-Sweep Free-Energy Reconstruction. *J. Chem. Theory Comput.* DOI: 10.1021/ct500088z

Computational Modeling of the Electrical-Sensing Properties of Single Wall Carbon Nanotubes

Shawn Bair, Pedro Cortes, Michael Radetic, and Donald Priour

Carbon nanotubes (CNTs) are an ever expanding field of research due to their unique physical and electrical properties since their discovery [1]. Based on how they are cut and rolled from a graphene, carbon nanotubes can exhibit semiconducting or metallic behavior. These electrical properties can change dramatically as molecules bond to their highly reactive surfaces. These property changes make CNTs very promising as detection structure in sensors. In this work, the electrical properties of single-wall carbon nanotubes in the presence of NO_2 has been modeled using density functional theory.

Carbon nanotube structure consists of a rolled sheet of graphene, which is itself a flat, one atom thick sheet of sp2 bonded carbon atoms. Single walled CNTs come in 3 types based on chirality, armchair, zigzag, and chiral. In this study, an (8,0) zigzag CNT was modeled. One of the most useful properties of CNTs is the reactivity of their exposed pi bonding surfaces. The curvature of these surfaces causes increased reactivity, as the electron cloud of the pi orbital is forced more to the outside of the tube. If the CNT is a semiconductor, the adsorption of molecules on the CNT surface causes a change in number of charge carriers, which alters the resistivity of the CNT. This reactivity, along with the change in electrical properties that results, is why electrons are being researched for use in sensors. [2]

To model the SWCNT, density functional theory (DFT) is used, first developed by Hohenburg and Kohn [3]. DFT simplifies the problem at hand by reducing the ground state electron wave function from a function of every electron position to a function of the electron density, lowering the number of variables in the function greatly. After the ground state electron wave function is found through DFT, non-equilibrium Green's functions (NEGF) are used to find a current through the CNT. [4] The program DFTB+[5] was used to carry out these calculations, using Slater-Koster orbital files from Eistner et al [6].

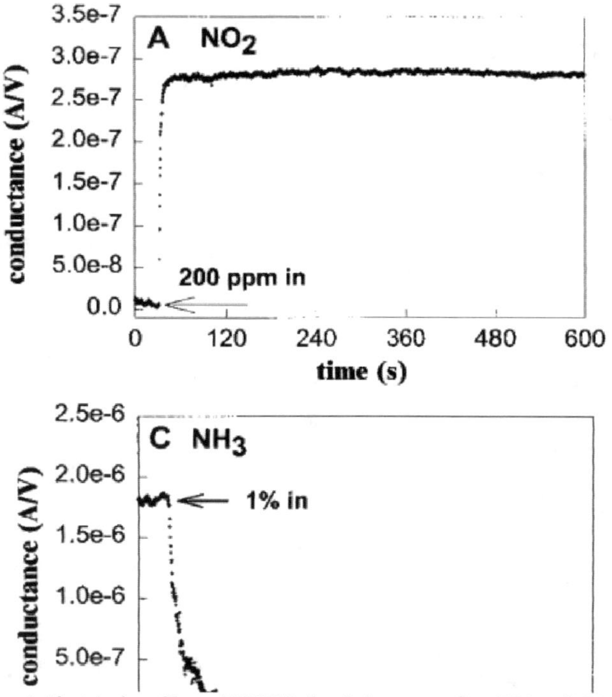

Figure 1. Electrical profile on SWCNTS after being exposed to NO2 and NH3

Kong et al [7] reported an experiment where a semiconducting CNT underwent changes in conductance on the scale of several orders of magnitude when it was exposed to NO_2 or NH_3 gas, as seen in Figure 1. The CNT in figure 1 is behaving as a p-type semiconductor, which has holes as the majority carrier. When the CNT is exposed to an electron withdrawing group such as NO_2, holes are added to the system, and vice versa when an electron donating group such as NH_3 is added. Indeed, depending on the interaction with either withdrawing or donating electron groups, the conductivity of the CNT can increase or decrease respectively.

In the present research work, the electrical current change on a single CNT as been modeled in the presence of NO2. Thus, to model the CNT, the first atom coordinates for the (8,0) SWCNT were obtained from Tubegen [8]. The structure of the CNT consisted of three parts, a source and a drain contact with constant potential, and a central area through which current flows. Here, the contacts were treated as semi-infinite in the direction of current flow, while the center region was based on 96 carbon atoms, along with the 3 atoms constituting the NO2 group. The fermi level for the contacts was found to be -4.3637 eV by using DFT on a single bare CNT in order to acquire the density of states (shown in Figure 2) and their occupations.

After obtaining the contact fermi levels, the current flow was calculated for various bias voltages both with and without the presence of the NO2 molecule. Here, the NO2 molecule was placed at 2.3 angstroms away from the CNT surface. The modeled results are shown in figure 3. As it can be observed from the figure, the CNT shows a higher current flow in presence of the NO2, particularly at lower bias voltages. However, the difference was smaller at higher bias voltages.

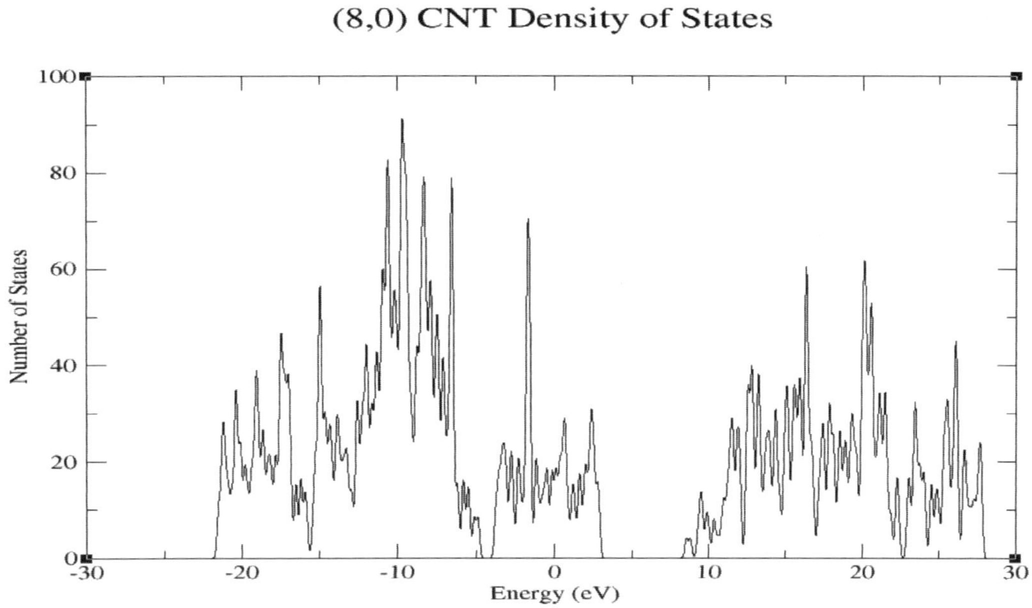

Figure 2. Density of States For 8,0 CNT. The Fermi level is at -4.3638 eV, where there is a gap in available states

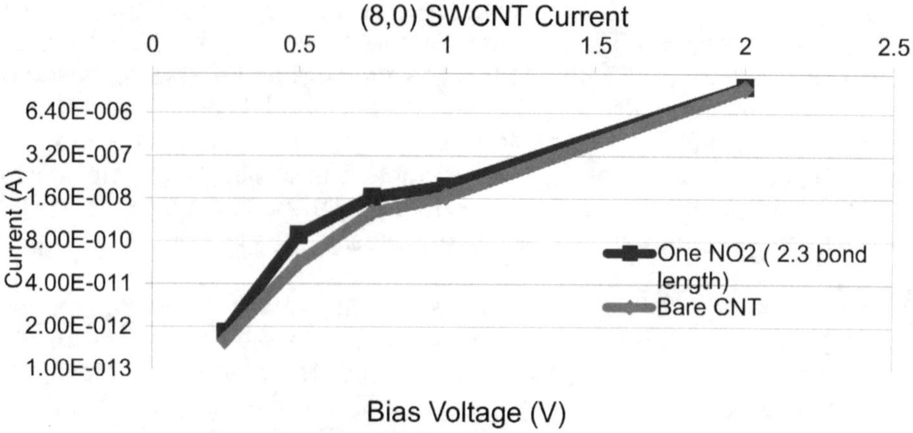

Figure 3. Current Vs Bias Voltage for 8,0 SWCNT.

The relation between the distances of the NO_2 was also investigated for the system. Simulations were done with a constant voltage while the NO_2 atom was placed at different distances from the CNT surface. The results are shown in Figure 4. The current ceases to be affected by the NO2 molecule at diatance larger than around 3.8 Angstroms.

8,0 CNT Current at 1 V Bias with Varying NO2 Distance

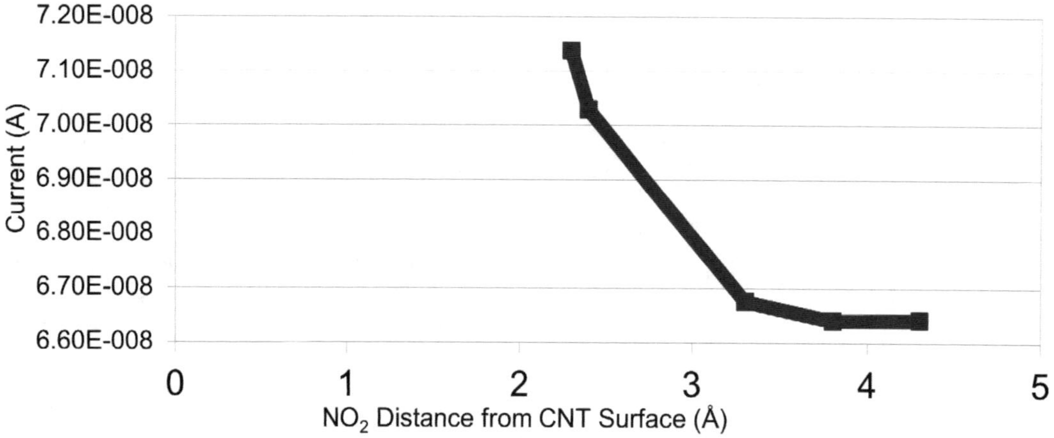

Figure 4. NO_2 distance versus current at 1 V Bias.

The general trends of these simulations match the experimental results observed by others such as Kong et al [7], though the absolute values are lower than expected. This is possibly due to the small CNT size in the simulation or the lack of metal contacts. Further ongoing work on this SWCNT modeling includes using metal contacts, larger simulations, as well as the incorporation of more and different molecules in the simulation.

References

[1] Iijima, Sumio (1991), "Helical microtubules of graphitic carbon", Nature 354: 56–58 (1991).

[2] Alkire, R.C. and Kolb, D.M. and Lipkowski, J. and Ross (2009), "Chemically Modified Electrodes", isbn: 9783527627066

[3] P. Hohenberg & W. Kohn, Phys. Rev. 136 (1964) B864

[4] Ferdows Zahid; Magnus Paulsson; Avik Ghosh; Supriyo Datta (2005), "Introduction to Molecular Conduction," https://nanohub.org/resources/8..)

[5] B. Aradi, B. Hourahine, and Th. Frauenheim. DFTB+," A Sparse Matrix-Based Implementation of the DFTB Method. J. Phys. Chem. A, 111(26):5678, 2007

[6] M. Elstner, D. Porezag, G. Jungnickel, J. Elsner, M. Haugk, T. Frauenheim, S. Suhai, and G. Seifert, Self-Consistent-Charge Density-Functional Tight-Binding Method for Simulations of Complex Materials Properties, Phys. Rev. B 58, 7260 (1998).

[7] J. Kong, N. Franklin, C. Zhou, S. Peng, K. Cho, H. Dai.] "Nanotube Molecular Wires as Chemical Sensors". Science 287, 622 (2000).

[8] TubeGen 3.4 (web-interface, http://turin.nss.udel.edu/research/tubegenonline.html), J. T. Frey and D. J. Doren, University of Delaware, Newark DE, 2011.

386890 A Comparison of Explicit Solvation Models for Density Functional Theory Simulations of Heterogeneously-Catalyzed Systems

Abstract:

Density Functional Theory (DFT) simulations have been used to measure catalytic activities for various homogenously- and heterogeneously-catalyzed systems. DFT is especially useful for calculations relating to fuel cell applications, which currently utilize expensive transition metal catalysts. DFT can be used to screen many possible catalyst material candidates, allowing for an analysis comparing cost and activity while saving time and money associated with experimental testing of many various candidate materials and compositions. DFT simulations have been used to successfully model gas-phase reactions on solid catalysts, such as those used in hydrogen fuel cells (HFCs), but modeling of liquid phase reactions on solid catalysts, such as those relevant to direct methanol fuel cells (DMFCs), is hindered by a lack of knowledge regarding the interactions between solvent molecules and surface-bound species. Thus, in order to effectively use DFT simulations to screen catalyst materials for fuel cell applications with liquid reactant streams, these interactions need to be understood more intimately. Our goal is to use DFT screening to reduce the use of transition metals such as platinum, which is heavily used as a catalyst and currently costs ~$1450 per ounce, to increase the cost-effectiveness, and therefore utilization, of fuel cells.

In this work, we aim to understand the effects of solvation on the oxidation of CO by OH on Pt(111), which has been shown to be the rate-limiting step in DMFCs [1]. Our long-term goals include using the knowledge gained from this work to effectively screen novel, cheaper catalyst materials for DFMCs. We compare various explicit solvation models at several adsorbate coverages of CO and OH to determine the optimal reaction environment for DMFC simulations. Solvation models considered include 2D and 3D ice crystal lattices, as well as 3D liquid-like and vapor-like models. We consider adsorbate coverages up to 1/3rdof a ML. Our results show that adsorption site, solvation model, water configuration, adsorbate coverage, and inclusion of dispersion forces all have important influences on the water-adsorbate interaction energies. Thus, it is essential to understand these effects in order to effectively screen potential catalyst candidates for heterogeneously catalyzed liquid phase systems such as those relevant for DMFCs.

References:

[1] H.A. Gasteiger, N. Markovic, J.P.N. Ross, E.J. Cairns, *J. Phys. Chem.*,1994, **98**:617.

387176 Oxygen Reduction Reaction on Pt-Free Catalyst Iron Phthalocyanine Functionalized Graphene for Fuel Cells

Abstract:

Metal phthalocyanines (MPc) have been widely employed for various applications including sensors, electronics and catalysis. In this work, we would present the oxidation reduction reactions (ORR) in a fuel cell in the cathode site catalyzed by an iron phthalocyanine functionalized graphene substrate. The desired Iron phthalocyanine is already a commercially available compound. The overall reaction has been modeled by using ab initio density functional theory (DFT) simulation. In this ORR reaction, we will estimate energetics, energy barriers and rate of reaction during the catalyzed reaction and determine to a degree the practicality of the system as a Pt-free catalyst in a fuel cell. We will also present the role of the central metal and several single bridging atoms such as N and C for efficient oxygen reduction reaction.

Acknowledgement:

DFT calculation work is supported from NSF TeraGrid (XSEDE) resources under allocation numbers [TGSEE090006] and [TG-DMR130127]. Use of the Center for Nanoscale Materials is supported by the US Department of Energy, Office of Science, Office of Basic Energy Sciences, under Contract No. DE-AC02-06CH11357.

387251 Molecular Models of Electrolyte Aqueous Solutions: Simulation Study on Limits of Their Applicability

Abstract:

Eleven of the most common non polarizable SPC/E-based aqueous NaCl solution force fields and two polarizable force fields based on SWM4-DP and BK3 models of water have systematically been examined with respect to the accuracy of their prediction of the solution density and chemical potential over the entire concentration range at ambient pressure and temperature. The density was obtained by Monte Carlo simulations at constant pressure, temperature, and number of molecules. Dependence of the electrolyte chemical potential on concentration was obtained using a recently developed Osmotic Ensemble Monte Carlo method [1] simulating the system at constant pressure, temperature, number of solvent molecules, and electrolyte chemical potential. Polarizable models were simulated using multiparticle move Monte Carlo method [2].

It is found that the results of the force fields considered are scattered over a wide range of values, and none is capable of producing quantitatively accurate results over the entire concentration range for the properties considered [3]. The polarizable models reproduce the slope of the electrolyte chemical potential vs. concentration at high concentrations better than the SPC/E-based models.

References

[1] F. Moučka, M. Lísal, J. Škvor, J. Jirsák, I. Nezbeda, W. R. Smith *J. Phys. Chem B* **115**, 7849(2011).

[2] F. Moučka, I. Nezbeda, W. R. Smith *Mol. Simul.* **39**, 1125(2013).

[3] F. Moučka, I. Nezbeda, W. R. Smith *J. Chem. Phys.* **138**, 154102(2013).

387901 Electrowetting of Electrolyte Aqueous Solutions: Simulation Study in an Open Statistical Ensemble, Confined Geometry, and Electric Field

Abstract:

Phenomena observed at liquid-solid interfaces in the presence of an external electric field have been studied at molecular scale. Of particular interest were density and concentration profiles, pressure tensor components, and wetting free energy. For this purpose, a novel efficient method for computer simulations in an open statistical ensemble, that allows studying aqueous solutions of electrolytes at these conditions, has been developed. The method simulates confined solutions at prescribed chemical potential values of its individual species. The particles are inserted/deleted to/from a nanopore through a sequence of intermediate states containing fractional particles that are partially coupled to the solution.

The method has been applied to aqueous solutions of NaCl confined in nanopores of various widths and considered to be in equilibrium with infinite bulk reservoirs of various concentrations at ambient pressure and temperature.

388200 Population Analysis of Sodium Chloride Crystals Having Unusual Stoichiometries

Abstract:

Recently, Zhang et al. (Science, Dec. 2013, Vol. 342, pp. 1502-1505) predicted and synthesized crystals of sodium chloride compounds that have novel stoichiometries under high pressure conditions (in excess of 1 million atmospheres of pressure): $NaCl_7$, Na_3Cl, Na_2Cl, Na_3Cl_2, NaCl, and $NaCl_3$. We used the Density Derived Electrostatic and Chemical (DDEC) and Bader atomic population analysis methods to study atomic charge states and bonding between atoms in Zhang et al.'s ten published crystal structures. The computed DDEC effective bond orders showed that in several of these structures the chlorine atoms bond to form chains along the crystals, which confirms Zhang et al.'s electron localization function (ELF) results. Bader and DDEC net atomic charges are also compared and discussed for these crystals.

388814 Molecular Scale Insights of the Structure Selective Growth of Bionanostructures: A Coarse-Grained and Metadynamics Based Study of Biosilica

Abstract:

In recent years there has been increasing interest in using computational tools to understand the bio/nano interface. Examples include protein adsorption, interfacial biocatalysis, and templated growth of bionanostructures such as biosilica. This poster will present a study of how various computational tools can be used to study molecular scale driving forces in peptide aggregation as well as how small peptide aggregates with identical composition but different structure template the growth of unique mesoscale bionano morphologies. The LK peptides are short amino acid sequences composed of leucine and lysine that adopt sequence-dependent secondary structure (e.g., alpha helix, beta, 3_{10} helix). These peptides have been well known to facilitate the growth of biosilica composites when placed in saturated solutions of silicon dioxide. Interestingly, it has recently been shown that the choice of LK peptide can dramatically change the resultant silicon dioxide nanoparticles. Guided by the underlying hypothesis that fine differences between small peptide aggregates are driving the unique changes in the morphology, we have sought to use molecular simulation to better understand this process. We have conducted coarse grained molecular dynamics simulations and additionally applied the metadynamics technique to study free energy landscapes of LK peptide aggregates.

We began our simulations by converting atomistic representations of our systems of randomly oriented LK peptides to MARTINI coarse grained (CG) models with explicit polarizable coarse grained water. By conducting molecular dynamics simulations with metadynamics, we were able to effectively explore and uncover the free energy of the aggregation of these peptides. Upon inspection of the free energy surface, it was clear that the systems have several metastable minima that would not be readily overcome in normal molecular dynamics systems and that enhanced sampling techniques are necessary to fully explore similar systems. Additional simulations used reverse coarse-graining to study the stability of the various aggregates at atomic resolution. Starting from a well-known scientific question about the preference of LK alpha-14 peptides to form tetrameric aggregates, we proceeded to address the question of how the different aggregates interact with amorphous silica nanoparticles. Finally, we used all of the studies at atomic and coarse resolution to quantitatively estimate the degree to which the simple CG models accurately capture the essential physical interactions in corresponding atomistic models.

389649 The Analysis of Temperature-Dependent Structural Properties of Mixed-Dispersed Model of 50 Percent Dipalmitoyl Phosphatidyl Choline and 50 Percent Lysolipid Using Molecular Dynamics Simulations

Abstract:

Drug delivery and its release depend greatly on the ability to control the permeability of the vesicle encapsulating the drug. Liposomes, which are spherical nanoparticles composed of one or more phospholipid bilayers, have emerged as a promising delivery system for potent chemotherapeutics. It is well known that the permeability of drug through lipid bilayers exhibits a maximum as the lipid bilayer undergoes the transition from gel to liquid-crystalline phase, which has been studied intensively in order to control the permeability. One of the ways for the controlled drug release is an addition of the lysolipid, which affects the transition properties such as temperature.

The study runs fully atomistic molecular dynamics (MD) simulation with flat liposome mixed-dispersed model. It is comprised of 50 percent DiPalmitoyl Phosphatidyl Choline (DPPC) and 50 percent lysolipids of MonoPalmitoyl Phoshatidyl Choline (MPPC). The changes in structure of mixed-dispersed model of DPPC and MPPC with the varying temperature are investigated. To do so, structural properties including the density profile, the interfacial tension, the area per lipid, and the free volume fraction are obtained.

390126 Graph-Based Evolutionary Algorithm for De Novo Molecular Design Under Multi-Dimensional Constraints

Abstract:

The area of computer-aided molecular design has greatly influenced the rate and cost at which novel chemicals with desired attributes have been identified. As such, great effort has been invested in new methodologies which allow for the solution of larger and more complex problems of this nature. The application of evolutionary algorithms is one such technique which has shown promise in the solution of large combinatorial, and highly non-linear molecular design problems. The typical approach begins with a population of randomly generated individuals and the more fit members are stochastically selected to undergo computational analogues of natural recombination and mutation. This process is iterative until the resultant population possesses the desired attributes, which could be evaluated with existing quantitative structure property (activity) relationships (QSPR's) or other types of property models.

It has been shown that many molecular properties or attributes are often best characterized by a combination of descriptors with varying dimensionality. This could, for example, include a combination of graph theoretical two-dimensional descriptors, such as the chi connectivity index, along with three-dimensional descriptors, which capture important spatial characteristics. The inverse solution to property models of this nature, which entails identifying candidate molecular structures with the desired properties as defined by the given model, is often highly non-linear in nature. In addition, the use of molecular fragments, as often practiced in the de novo design of novel structures, can lead to a combinatorially large search space which becomes intractable for exhaustive solution techniques. The application of an evolutionary algorithm provides a powerful method for the solution of these types of molecular design problems in which there are often multiple objective constraints with high computational complexity and a large search space.

This approach utilizes a fragment based descriptor known as the Signature descriptor, which is represented as a molecular graph, as building blocks to generate candidate solutions. The nature of signature descriptors allows for control over the desired chemical search space as well as convenient reconstruction of the resultant molecular graphs. This descriptor has proven useful in the solution of problems with topological constraints and has recently been extended to tackle problems of higher dimensionality, including three dimensional as well as four dimensional (conformational ensemble) constraints. The graph-based operators necessary for such an approach will be outlined as well as exemplified through a case study which will highlight the advantages of this algorithm.

390828 Quantification of DNA Cleavage Specificity in Hi-C Experiments

Abstract:

Hi-C experiments [1] produce large sets of DNA sequence read pairs, which are typically analyzed to deduce interactions across different loci within whole genomes. Here we show that the same read pairs can also be analyzed to characterize a key step in Hi-C experiments, the digestion of cross-linked chromatin with a restriction endonuclease. Specifically, we describe a computational method to estimate, directly from Hi-C reads, the fractions of cleavages resulting from random breakage and star activity. We have validated this method using Hi-C fragments from computer simulations, and its application to experimental Hi-C data sets reveals significantly different patterns of cleavage across different cell types. Knowledge of cleavage fractions will enable researchers to optimize Hi-C experimental conditions, develop accurate Hi-C simulations for testing computational algorithms, increase the number of interactions deducible from each Hi-C data set, and investigate biophysical properties that correlate with patterns of DNA cleavage.

[1] Lieberman-Aiden et al., *Science* 2009, 326, 289-293

IMPROVED APPROACHES FOR OPTIMIZING FORCE FIELDS OF COMPLEX MIXTURES USING MULTISTATE REWEIGHTING AND DIVERSE DATA SOURCES

Multistate reweighting methods, such as the multistate Bennett acceptance ration (MBAR) can predict free energies and the uncertainties of those free energies over poorly sampled or unsampled thermodynamic states. I have used multistate reweighting to efficiently optimize the nonbonded parameters of a molecular dynamics force field to fit a wide range of experimental data. This optimization scheme allows for the exploration of a large, highly dimensional parameter space. The developed optimization scheme optimizes the nonbonded parameters over a large number of atom types simultaneously, using experimental data over a range of concentrations to develop a force field that is accurate over a range of molecule types and concentrations. Increasing the scope of the optimization, both by increasing the range of concentrations as well as increasing the number of experimental measurements used in the optimization, yields a force field that is both accurate and transferable.

I have used the multistate reweighting optimization scheme to optimize the nonbonded parameters of the molecules found in fifteen amino acid side chain analogs, using the Generalized Amber Force Field (GAFF) nonbonded parameters as the initial parameters. The nonbonded parameters were optimized to accurately reproduce the free energies of solvation for the amino acid side chain analogs at infinite dilution in water, the free energies of hydration for water at infinite dilution in various organic molecules, and the thermodynamic properties of pure organic molecules.

356232 Gibbs Ebsemble Simulation of the Phase Diagrams of Simple Fluids: Ar, Kr, and CH4 Using Temperature Dependent Interaction Parameters

Abstract:

The use of temperature dependent interaction parameters, instead of using temperature independent interaction parameters, may lead to improvement in the prediction of the vapor-liquid coexistence curve. Published second virial coefficient data were used to fit a simple two parameter temperature dependent model for the collision diameter and well depth parameters of the Lennard-Jones potential energy model. Subsequently, vapor-liquid coexistence curves for argon, krypton, and methane were simulated using NVT Gibbs ensemble Monte Carlo (GEMC). The simulations were carried out using both temperature independent and temperature dependent parameters. The critical temperature and density were determined using the Ising-scaling model. The results using temperature dependent parameters produce, in general, more accurate phase diagrams compared to the diagrams generated using temperature independent interaction parameters. The root mean square deviation (RMSD) is reduced by 42.1% using temperature dependent interaction parameters for argon. Also, there was no significant difference between the results obtained using temperature dependent interaction parameters and the highly accurate and computationally demanding phase diagrams based on three body contributions.

355940 Molecular Simulations of the Critical Point for Molecules That Decompose Experimentally

Abstract:

An assumption that has been made by the BYU DIPPR 801 staff in estimating property values for large molecules is that all families eventually converge to the n-alkane family with increasing chain length. For this purpose, much effort has gone into examining the n-alkane family. Specifically, the properties of n-alkanes as the carbon number approaches infinity have been studied to establish asymptotic convergence to other families. Since the critical point is an essential value for many predictive methods of thermophysical and transport properties, it is particularly important to the DIPPR 801 database.

The difficulty with determining the critical point for large n-alkanes is that their decomposition temperature is much lower than the critical point for carbon chains longer than C10. For this reason, molecular simulations (specifically GEMC in this work) have been a popular means of estimating critical points for thermally unstable molecules. Prior to this work, however, simulations had only been performed to establish a trend for the intermediate sized compounds.

This research focuses on simulating molecules large enough to determine the infinite carbon length behavior. The objective is to determine whether or not the methylene contribution to the critical volume eventually becomes constant. This conclusion will dictate if the critical density approaches a non-zero constant value for an infinite carbon chain.

MOLECULAR DESCRIPTORS BASED RANDOM FOREST PREDICTORS OF ILs PROPERTIS

Zelimir F. Kurtanjek
University of Zagreb, Croatia

Abstract
Ionic liquids (ILs) have been considered as basis for numerous new green and sustainable chemical and bio-based technologies. Numerous possible combinations of cations and anions enable application of computer combinatorial design of ILs to tailor properties to specific technology constraints. Here are applied random forest models based on 797 molecular descriptors for prediction of physical and biological properties. The models are derived for combinations of the following cations: imidazole, pyridinium, quinolinium, ammonium, phosphonium; and anions: BF_4, Cl, PF_6, Br, CFNOS, NCN_2, $C_6F_{18}PBF_4$, Cl, PF_6, Br, CFNOS, $C_6F_{18}P$. The matrix of molecular descriptors is decomposed by singular value decomposition to extract maximum information contained in variance and also by maximization of t-statics for class separation. In particular, derived are the models for ILs biological impact on acetycholinesterase inhibition. Derived are models for toxicity classification and regression of E_{50} concentrations. Proposed are methodologies for RF post-modeling probabilistic global and local analysis of main molecular features. From component analysis of the principal targets the most important molecular descriptors for each of the properties are determined. The derived models are validated by ten fold cross validation resulting in 10% average errors by individual decision trees, and considerable improved prediction accuracy, when compared to single tree and standard chemometric models, were obtained by random forest predictors for toxicity classification at error level 2% and for E_{50} with Pearson regression coefficient R^2=0.992.

Key words: Molecular descriptors, ionic liquids, random forest model

Introduction

For last several decades application of ILs has been in focus for development of new green and sustainable technologies. Among many promising examples are productions of biofuels from waste cellulose and polymeric products from biomass feedstock.[1] Practically limitless combinatorial opportunities for selection of cations and anions enables tailoring ILs properties to specific technological needs. About 30 physical, chemical and biological properties are available in NIST and Merck data bases and literature.[2-4] The combinatorial opportunities have led to development of mathematical models for property predictions and ILs computer design. Various models can be found in literature, mostly based on assumption of functional continuity, linear or nonlinear, between preselected molecular identities and properties. This produced models with acceptable applicability, but constrained to specific ions pre-selections. In order to release constraints, in this work modelling is based on extensive list of molecular descriptors capturing complex chemical information. For each IL combination are calculated in total 663 descriptors (1-2 D), and 134 (3D).[5] The variable spaces of the descriptors for each ion, cation and anion) are decomposed by singular value decomposition to 10 principal components respectively, with 99,2 % of retained variance.

Model

Molecular descriptors of the selected cations and anions form the data matrix $\mathbf{X}(N_M x 2N_D) = (X_C, X_A)$, and the corresponding properties (EC_{50}, μ, T_m) form the output matrix **Y** (N_Mx3) with N_M=400 and N_D=797. Prior to modeling all the data are auto-scaled. Correlation matrices for cations and anions reveal high degree of interdependences with average R^2 =0.4 (presented in Fig.1). Each of the matrices are represented by the corresponding singular value decompositions (SVD). The decompositions of the variances are applied for extraction of molecular similarities (Eq. 1) and for maximization of *t*-statistics for class separation (Eq. 2).

$$\left(\mathbf{X}_C^T\mathbf{X}_C\right)\mathbf{v}_{C,i} = \lambda_{C,i}\,\mathbf{v}_{C,i} \qquad \left(\mathbf{X}_A^T\mathbf{X}_A\right)\mathbf{v}_{A,i} = \lambda_{A,i}\,\mathbf{v}_{A,i} \qquad j = 1,2\cdots M \tag{1}$$

$$\left(\mathbf{C}^{-1}\,\mathbf{V}\right)\mathbf{l}_i = \lambda_i\,\mathbf{l}_i \qquad \mathbf{LDA} = \mathbf{X}\cdot\left(\mathbf{l}_1\middle|\mathbf{l}_1\middle|\cdots\middle|\mathbf{l}_K\right) \tag{2}$$

Each ILs combination is projected to the low dimensional spaces spanned by the first 10 corresponding SVD eigenvectors (Eq. 3).

$$\mathbf{P}_C = \mathbf{X}\cdot\left(\mathbf{v}_{C,1}\middle|\mathbf{v}_{C,2}\middle|\cdots\mathbf{v}_{C,K}\right) \qquad \mathbf{P}_A = \mathbf{X}\cdot\left(\mathbf{v}_{A,1}\middle|\mathbf{v}_{A,2}\middle|\cdots\mathbf{v}_{A,K}\right) \tag{3}$$

Applied are decision tree (DT) with supervised trainings for classification and regression (Eq. 4) and for each model a consortium of 500 trees forming random forests (RF), Eq. 5.

$$\hat{\mathbf{Y}}_{DT}^{P} = DT\left(\mathbf{Y},\mathbf{P}_C,\mathbf{P}_A\right) \qquad \hat{\mathbf{Y}}_{DT}^{L} = DT\left(\mathbf{Y},\mathbf{LDA}\right) \tag{4}$$

$$\hat{\mathbf{Y}}_{RF}^{P} = RF\left(\mathbf{Y},\mathbf{P}_C,\mathbf{P}_A\right) \qquad \hat{\mathbf{Y}}_{RF}^{L} = RF\left(\mathbf{Y},\mathbf{LDA}\right) \tag{5}$$

All calculations are performed by the algorithms provided in *R* software.[6-8]

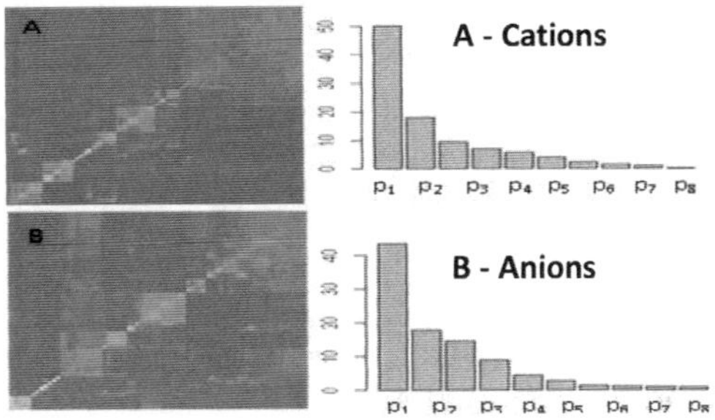

Fig. 1. „Heat maps" of Pearson R^2 matrices and SVD of ILs descriptors

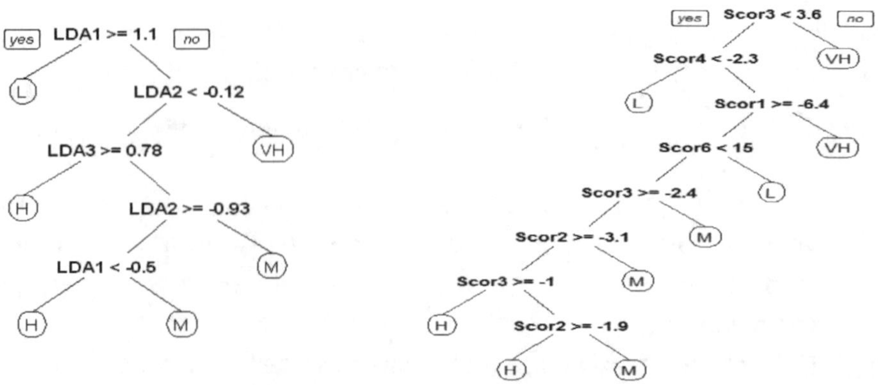

Fig.2. DT (decision tree) models for ILs classification (L low, M medium, H high, VH very high) of acetylcholinesterase inhibition. (A) LDA linear discrimination and (B) PCA decomposition.

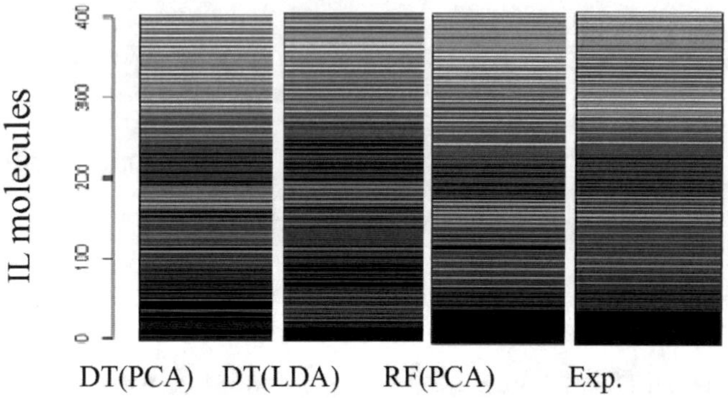

Fig. 3. Stacked bar charts for predictions ILs toxicity classification (L, M, H, VH) based on decision tree DT(PCA error 10%, LDA error 13%), random forest (RF, error 2%).

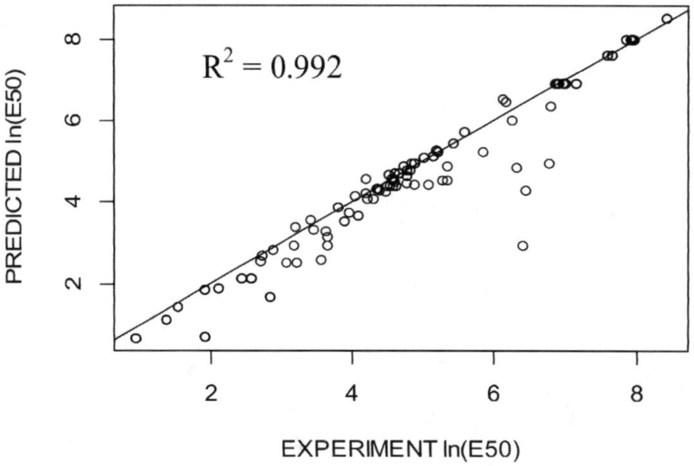

Fig. 4. Prediction of E_{50} *acetycholinesterase inhibition* concentrations by a random forest regression model.

Probabilistic analysis of RF models

Generally, RF models are considered to be the best choice for modeling complex systems, especially for the case with high dimension (n >> p) of modeling factors (n) compared to observed samples (p). Due to statistically weighted agglomeration of large number of randomized individual trees the main problem of model overfitting and biased inference is minimized. However, RF models are also considered untransparent since they do not have a closed mathematical form, they are nonparametric models (resulting in lack of confidence intervals), and can not be graphically presented in a canonical form. Although they are untransparent, due to RF accurate prediction and fast response (low computational demand), for example, they are very applicable for imbedding into on-line process control decision loop. Presently RF models are difficult for information extraction, i.e. understanding model underpinning.

Here are derived new methods for probabilistic analysis, i.e. "unscrambling" RF models. Proposed are MC (Monte Carlo) and FAST (Fourier Amplitude Sensitivity Test) for estimation of global parameter importance, local parameter gain factors, and nonparametric distribution (approximated by box-whiskers plots) of global gain factors. Simulations of RF predictions are applied for estimation of variances contributed by for each $\mathbf{P}_j$ cation and anion principal components, $\mathbf{P}_C$ and $\mathbf{P}_A$. To estimate "global" variance Q_j^2 of each principal component to RF response random samplings are drawn from assumed uniform distributions in their respective ranges:

$$Q_j^2 = \sigma^2\left(E\left(RF(\mathbf{P}_C, \mathbf{P}_A \middle| \mathbf{P}_j)\right)\right) \qquad \mathbf{P} \in \mathbf{U}\left[-\sigma, +\sigma\right] \tag{6}$$

To recover information on variance contribution S_i^2 for each molecular descriptor the projection coefficients $w_{i,j}$ (i=1,M, j=1,N) of the descriptors to principal components are accounted

$$S_i^2 = \sum_{j=1}^{j=N} w_{i,j}^2 \, Q_j^2 \qquad j = 1,..N \tag{7}$$

where N is the total number of principal vectors and M is the number of molecular descriptors.

For regression RF models local deterministic effects of molecular descriptors **x** (autoscaled) on response are evaluated as relative local gain factors g_i defined by

$$g_i = \frac{1}{RF} \frac{\partial RF}{\partial x_i}\bigg|_{\mathbf{X}} \qquad i = 1,2..M \tag{8}$$

For numerical evaluation of the descriptor gains applied is the "chain" rule of implicit derivation with respect to each principal component and the finite difference approximation:

$$g_i = \sum_{j=1}^{j=N} \left(\frac{\partial RF}{\partial P_j} \frac{\partial P_j}{\partial x_i} \right) = \sum_{j=1}^{j=N} w_{i,j} \frac{\Delta RF}{\Delta P_j} \qquad i = 1,2..M \tag{9}$$

The local descriptor gains provide useful information for inverse modeling in a sense of recursive corrections in small steps in a search space. Additional "global" information G_i can be obtained by MC simulation of the gains and presented in a form of probability distribution function (pdf)

$$G_i = \mathrm{pdf}\left(g_i\left(\mathbf{x}\right)\right) \qquad x_i \in U\left[-\sigma_i, +\sigma_i\right] \tag{10}$$

Obtained *G* pdf functions are in non-parametric forms and inference of each descriptor importance, mutual correlation and synergy effects can be easily observed by graphical inspection of the corresponding "box and whiskers" plots.

Results and Discussion

Results for correlation and SVD analysis presented in Fig. 1 reveal high degree of dependencies among the descriptors for the given set of ions yielding effective decomposition by the first ten components for each cation and anion data set. Results of supervised classification by individual decision trees are given in Fig. 2-3. LDA and PCA decomposition give similar classification errors of about 10 %. However, combinations of 500 trees into random forest models improved classification to error of 2 %. Applicantion PCA based random forest regression model yields Pearson regression coefficient R^2=0.992 (Fig. 4). Inspection of variable importance, i.e. individual molecular descriptor contribution to the key (top) tree branching reveal that for the enzyme level inhibition the most contributing descriptors, depicted in Fig. 5 are: 1) number of aromatic atoms and bonds; 2) sum of the absolute value of the difference between atomic polarizabilities of all bonded atoms in the molecule (including implicit hydrogens); 3) molecule complexity descriptor; 4) number of P atoms; 5) molecular distance edge between all secondary and tertiary carbons; 6) excessive molar refraction, 7) number of 6 membered rings; 8) number of 5 and 6 membered rings (includes counts from fused rings); 9) number of 6-membered rings (includes counts from fused rings) containing heteroatoms (N, O, P, S, or halogens).

Proposed are new methods for probabilistic post-modeling analysis of RF models. The methods are based on MC, or alternatively FAST, simulation of local gain factors and global properties of variance contribution and sensitivity gains. The methodology is aimed for extract of information from RF models and its application for inverse modeling objectives.

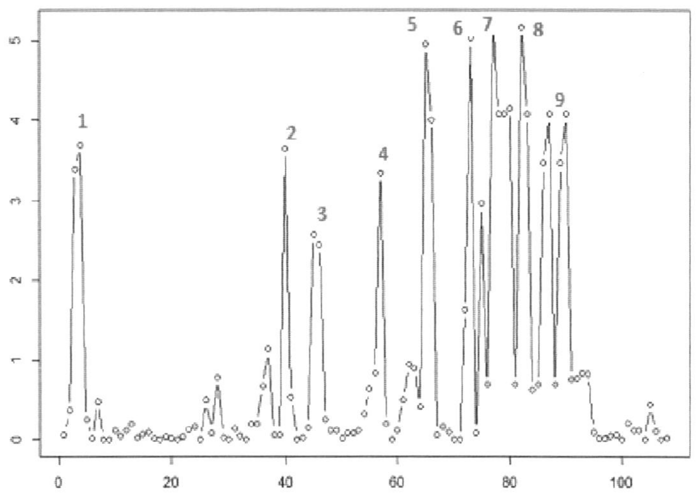

Fig. 5. Spectrum of the molecular descriptor contributions to "score 3" corresponding to the top of the PCA decision tree.

Conclusions

Applied are principal component and linear discrimination component based random forest models, based on extensive list of molecular descriptors, for prediction of ILs toxicity classification levels and E_{50} concentrations for *acetylcholinesterase* inhibition. Results obtained by forest models compared to standard (linear) chemometric models (results for PCR and PLS are not shown here) exhibit high improvement in model accuracy of prediction. The obtained error in classification prediction is on a level of 2% and for the regression model R^2 =0.992.

Obtained improved accuracy of the RF models makes them potentially suitable for ILs combinatorial search in design of green chemical and bio-based technologies.

Acknowledgement

This work was supported by Ministry of Science, Education and Sports of Republic of Croatia, project 058-1252086-0589.

References

1. Liu W., Zheng F., Li J., Cooper A., An Ionic Liquid Reaction ans Separation Process for Production of Hydroxymethylfurfural from Sugars, *AIChE J.,* 2014: 60: 300-314
2. NIST, Ionic Liquids Database Standard Reference Database #147, <ilthermo.bouder.-nist.gov /ILThermo/mainmenu.uix>.
3. The UFT/ Merck Ionic Liquids Biological Effects Database, <www.il-eco.uft.uni-bremen.de>, accessed on14/12/2013.
4. Suojiang Z., Lu X., Zhou Q.,. Li X, Zhang X., Li S., Ionic Liquids, Physichochemical Properties, Elsevier, 2009.
5. Chun Wei Yap, PaDEL-Descriptor: An Open Source Software to Calculate Moleclar Descriptors and Fingerprints, *J. Comp. Chem.*, 2010:32: 1466-1474
6. Svetnik V., Liaw A., Tong C., Culberson J.C., Sheridan R.P., Feuston B.P., Random forest: a classification and regression tool for compound classification and QSAR modeling, *J. Chem. Inf. Comput. Sci.*, 2003; 43: 1947–1958.
7. Breiman L., Cutler A.,< www.stat.berkeley.edu/~breiman/ RandomForests>.
8. R Development Core Team, , R: A language and environment for statistical computing. R, Vienna, Austria,www.R-project.org 2011.

   T. Therneau, B. Atkinson, B. Ripley, 2013, CRAN – Packagge rpart, <cran.rproject.org/web/ packages/rpart/index.html>.

364785 Mbuild: A Hierarchical, Component Based Molecule Builder

Abstract:

The biophysics simulation community has put considerable effort into creating tools and databases for building and parameterizing biological molecules with minimal effort, e.g. the Protein Data Bank[1], VMD[2], AmberTools[3]. Such toolchains allow researchers to generate input files for complex structures, such as proteins and DNA, that can run on most molecular dynamics simulation engines with little to no manual intervention. However, while the biophysics community's tools provide excellent functionality for biological system setup, they do not allow one to easily generate arbitrary structures and in many cases do not provide functionality when semi-infinite substrates are present in the system or chemical bonding between separate species is required (e.g. self-assembled monolayer systems). Unique aspects of non-biochemical systems, such as monolayers, heterogeneous polymer melts and surface bound polymer brushes, require a different approach. These systems may not be regular and thus defining a small unit cell that could be duplicated is not always possible.

mBuild is a hierarchical, component based molecule builder tool that relies on equivalence relations for component composition. Every component can recursively contain particles and other components to generate arbitrary, hierarchical structures where every particle represents a leaf in the hierarchy. Bottom-level components, such as an alkyl group or a monomer, can be hand-drawn using software like Avogadro[4] and are then connected using an equivalence operator which matches defined attachment sites between two components. mBuild supports parameterized structures through generative modeling. This allows for declaratively expressing repetitive structures, e.g. polymer chains, crystal structures, planar or spatial tiling, as well as for parameterizing the affine transformations applied to subcomponents of a composite component. Accompanying mBuild is a version-controlled library of components which promotes reuse, as well as collaborative component development and curation. Additionally, mBuild contains an interface to a molecular dynamics forcefield database. Upon generating a structure with mBuild, users can choose a forcefield with which to generate their system's topology, parameterize the system and ultimately produce a usable simulation input script.

As a case study, we generated an ensemble of poly(ethylene glycol) monolayers, where surface density and patterning of the monomers as well as their height is trivially tunable via the Python interface/component parameters thus allowing rapid generation of input files for parameter sweeping studies. We also demonstrate the scalability of mBuild through a more complex example: a system of poly(2-methacryloyloxyethyl phosphorylcholine) or pMPC brushes attached to a silica substrate via an atom transfer radical polymerization initiator. Similarly, the structure can be tuned in terms of surface coverage, patterning and brush height. Such systems, with atom numbers in the range of a hundred thousand to several million, can be generated and parameterized in seconds to minutes of computing time on a conventional workstation.

[1] F. C. Bernstein, T. F. Koetzle, G. J. Williams, E. F. Meyer, M. D. Brice, J. R. Rodgers, O. Kennard, T. Shimanouchi, and M. Tasumi, "The protein data bank: A computer-based archival file for macromolecular structures," Archives of Biochemistry and Biophysics 185, 584–591 (1978).

[2] W. Humphrey, A. Dalke, and K. Schulten, "VMD: Visual molecular dynamics," Journal of Molecular Graphics 14, 33–38 (1996).

[3] R. Salomon-Ferrer, D. A. Case, and R. C. Walker, "An overview of the Amber biomolecular simulation package," Wiley Interdisciplinary Reviews: Computational Molecular Science 3, 198–210 (2013).

[4] M. D. Hanwell, D. E. Curtis, D. C. Lonie, T. Vandermeersch, E. Zurek, and G. R. Hutchison, "Avogadro: an advanced semantic chemical editor, visualization, and analysis platform." Journal of Cheminformatics 4, 17 (2012).

Interfacial Properties of Realistic Molecular Fluids: A Combined Molecular Simulation and Classical Density Functional Theory Approach

Ahmadreza F. Ghobadi, J. Richard Elliott
Department of Chemical and Biomolecular Engineering
The University of Akron
AIChE Poster Session, #223f

1

Also this week:
(475d) Extrapolation to an Infinite-Order Thermodynamic Perturbation Theory and Implications in the Critical Region
Hilton Crystal AF, Wednesday, 9:42 AM

(749d) Transferable Intermolecular Potential Models for a Broad Range of Organic Compounds
Hilton Crystal AF, Friday, 9:42 AM

Ahmadreza F. Ghobadi, J. Richard Elliott
Department of Chemical and Biomolecular Engineering
The University of Akron

2

Motivation

- Phenomenological Equations, Cubic EOS and SAFT

$$\frac{A^{res}}{Nk_BT} = \frac{A^{rep}}{Nk_BT} + \frac{A^{att}}{Nk_BT} + \frac{A^{assoc/col}}{Nk_BT} + \ldots$$

Challenges for surface active molecules:
- Difficult to separate.
- Diverse molecular structure.
- Low vapor pressure.
- No saturated liquid density.
- Solid at ambient temperature.
- No critical properties.
- No bulk fluid properties.
- pH dependent chemistry.
- Forming micelles at high concentrations.

3

Thermodynamic Perturbation Theory (TPT)

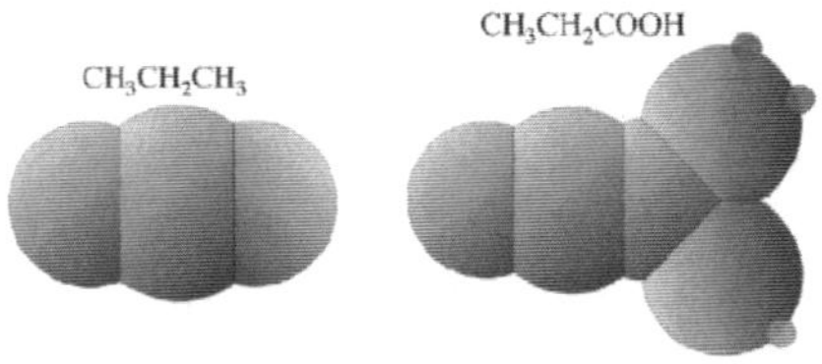

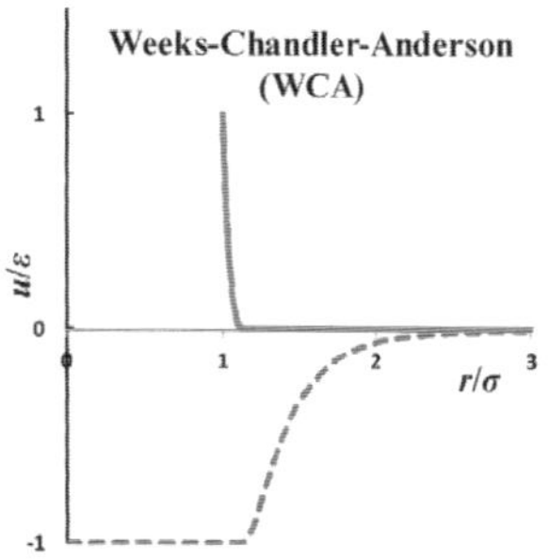

- SAFT-γ WCA EOS
- Site-based and Transferable

$$\frac{A^{res}}{Nk_BT} = \frac{A^{rep}}{Nk_BT} + \frac{A^{att}}{Nk_BT} \longrightarrow \frac{A^{att}}{Nk_BT} = \beta A_1 + \beta^2 A_2 + \beta^3 A_3$$

↓

Simulate the Reference (repulsive) part

$$A_1 = \frac{1}{N}\left\langle U^{att}\right\rangle_{rep}$$

$$A_2 = \frac{1}{2!N}\left[\left\langle (U^{att})^2\right\rangle_{rep} - \left\langle U^{att}\right\rangle_{rep}^2\right]$$

4

Thermodynamic Perturbation Theory (TPT) SAFT-γ WCA EOS

- Softness included
- Site-based and Transferable
- Extends directly to Interfaces via Fundamental Measure Theory

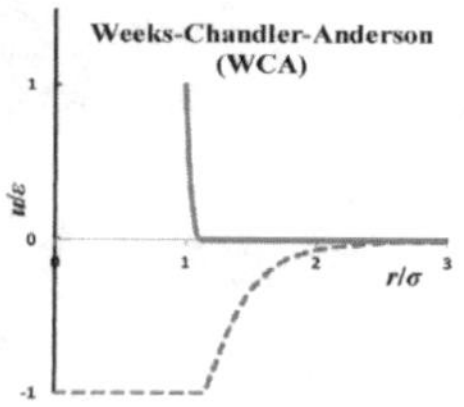

$$Z^{\mathrm{WCA}} = \frac{n_3}{1-n_3} + \frac{n_1 n_2 \gamma_1}{n_0 (1-n_3)^2} + \frac{n_2^3 \gamma_2}{12\pi (1-n_3)^3} - \frac{n_3 n_2^3 \gamma_3}{36\pi (1-n_3)^3}$$

$$n_\omega = \rho \sum_i^{NC} x_i \mathrm{M}_i^{(\omega)}, \quad \omega = 0-3$$

$$\mathrm{M}_i^{(0)} = \sum_k^{NS} \nu_{k,i} s_k, \quad \mathrm{M}_i^{(1)} = \frac{1}{2}\sum_k^{NS} \nu_{k,i} s_k d_{kk}, \quad \mathrm{M}_i^{(2)} = \pi \sum_k^{NS} \nu_{k,i} s_k d_{kk}^2, \quad \mathrm{M}_i^{(3)} = \frac{\pi}{6}\sum_k^{NS} \nu_{k,i} s_k d_{kk}^3$$

s_k = "shape factor," to correct from tangent sphere perspective to fused spheres. E.g. $M_i^{(3)} = (\pi/6)\,\Sigma\, \nu_k\, s_k\, d_k^3$ = molecular volume.

$$A_1\left[\{\rho(\mathbf{r})\}\right] = \frac{1}{2}\sum_{i=1}^{NC}\sum_{j=1}^{NC}\sum_{k=1}^{NS_i}\sum_{l=1}^{NS_j} \bar{g}_{0,kl}^{\mathrm{WCA}}\left[n_{3,kl}^{\mathrm{eff},1}\right] \qquad g_{0,kl}^{\mathrm{WCA}}\left[n_{3,kl}^{\mathrm{eff},1}(\mathbf{r})\right] = \frac{1}{1-n_{3,kl}^{\mathrm{eff},1}(\mathbf{r})} + \frac{3(\gamma_1+\gamma_2)n_{3,kl}^{\mathrm{eff},1}(\mathbf{r})}{4\left(1-n_{3,kl}^{\mathrm{eff},1}(\mathbf{r})\right)^2} + \frac{(3\gamma_2-\gamma_3)\left(n_{3,kl}^{\mathrm{eff},1}(\mathbf{r})\right)^2}{4\left(1-n_{3,kl}^{\mathrm{eff},1}(\mathbf{r})\right)^3}$$

$$\times \int d\mathbf{r}_1 n_{0;k,i}(\mathbf{r}_1) \int d\mathbf{r}_2 n_{0;l,j}(\mathbf{r}_2) u'_{CutShift,\,kl}\left(|\mathbf{r}_1 - \mathbf{r}_2|\right)$$

5

Part I:Homogeneous Molecules Without Partial Charges

JCP 139, 234104 (2013)

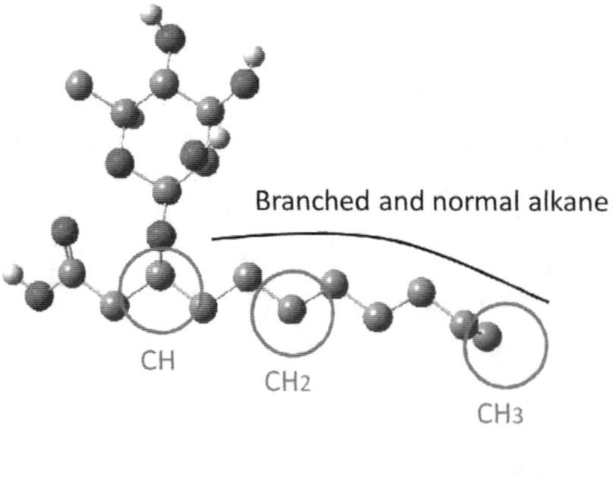

6

SAFT-γ WCA: n-Alkanes (TraPPE Force Field)

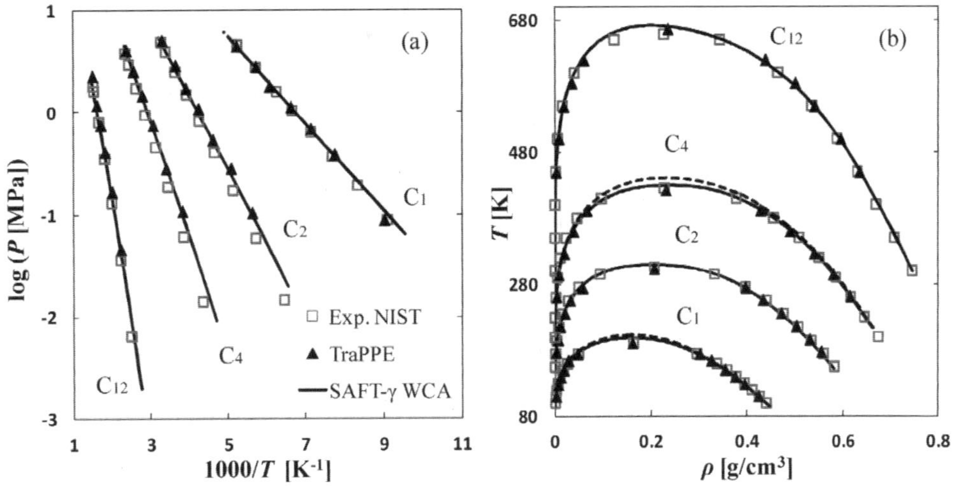

- Deviation from simulation: P^{sat} 4%, ρ^{liq} 1%
- Critical properties deviates around 1% from simulation

Simulate data for alkanes from Martin et al. JPCB 1998

7

SAFT-γ WCA: Branched Alkanes (TraPPE)

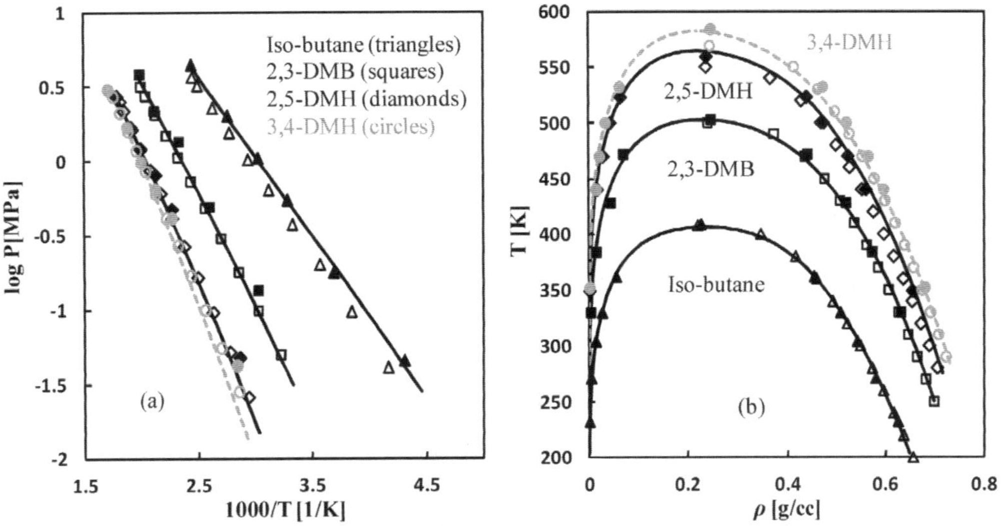

- SAFT-γ WCA EOS can distinguish between different isomers of the same chain length.

Simulate data for Branched alkanes from Martin et al. JPCB 1999

8

Part I:Homogeneous Molecules Without Partial Charges

Ghobadi and Elliott, *JCP* 139, 234104 (2013)

In this work, we develop a version of the Statistical Associating Fluid Theory (SAFT)-γ equation of state (EOS) that is compatible with united-atom force fields, rather than experimental data. We rely on the accuracy of the force fields to provide the relation to experimental data. Although, our objective is a transferable theory of interfacial properties for soft and fused heteronuclear chains, we first clarify the details of the SAFT-γ approach in terms of site-based simulations for homogeneous fluids. We show that a direct comparison of Helmholtz free energy to molecular simulation, in the framework of a third order Weeks-Chandler-Andersen perturbation theory, leads to an EOS that takes force field parameters as input and reproduces simulation results for Vapor-Liquid Equilibria (VLE) calculations. For example, saturated liquid density and vapor pressure of n-alkanes ranging from methane to dodecane deviate from those of the Transferable Potential for Phase Equilibria (TraPPE) force field by about 0.8% and 4%, respectively. Similar agreement between simulation and theory is obtained for critical properties and second virial coefficient. The EOS also reproduces simulation data of mixtures with about 5% deviation in bubble point pressure. Extension to inhomogeneous systems and united-atom site types beyond those used in description of n-alkanes will be addressed in succeeding papers.

9

Phase II Molecules with Partial Charges

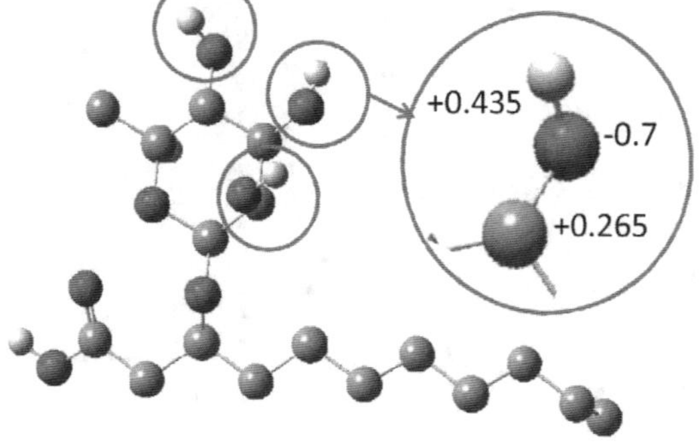

10

SAFT-γ WCA, CO_2 (TraPPE)

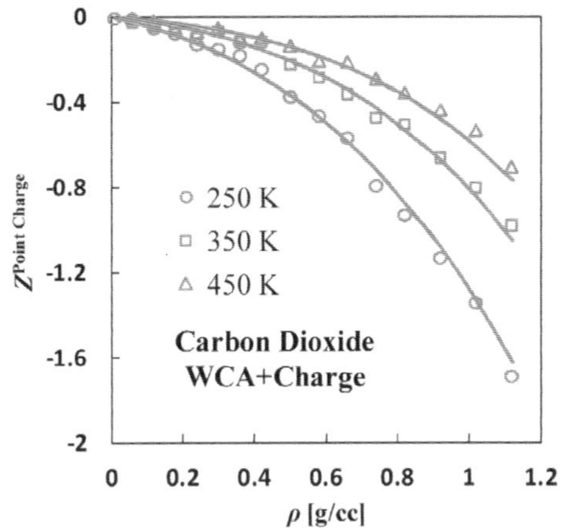

$$\frac{A^{ref}}{Nk_BT} = \frac{A^{sphere}}{Nk_BT} + \frac{A^{chain}}{Nk_BT} + \frac{A^{Assoc}}{Nk_BT}$$

$$\frac{A^{Assoc}}{Nk_BT} = \ln X - \frac{X}{2} + \frac{1}{2}$$

11

SAFT-γ WCA, CO_2

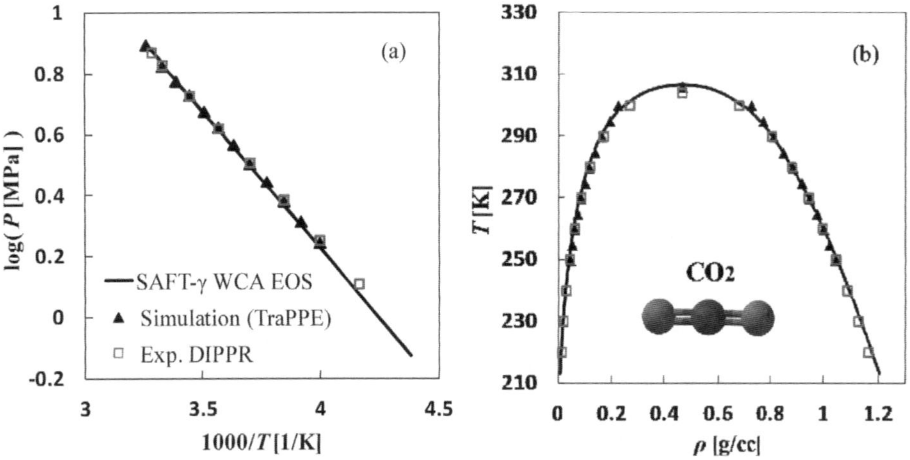

- Deviation from simulation: P^{sat} 3%, ρ^{liq} <1%
- The perturbation contributions are based on WCA+q reference

Simulate data for CO2 from Potoff et al. AICHE 2001

12

SAFT-γ WCA, Nitriles (TraPPE)

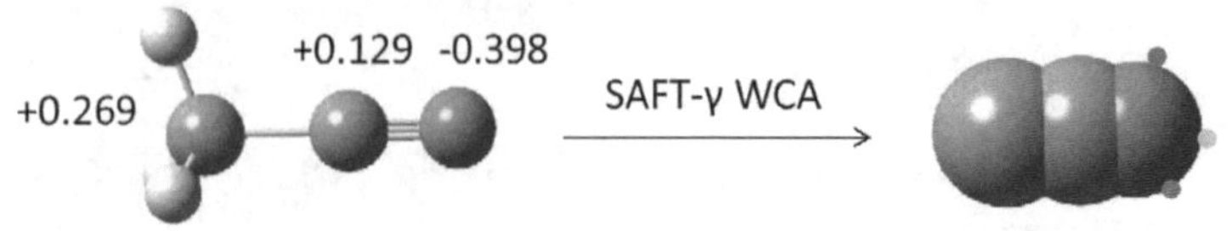

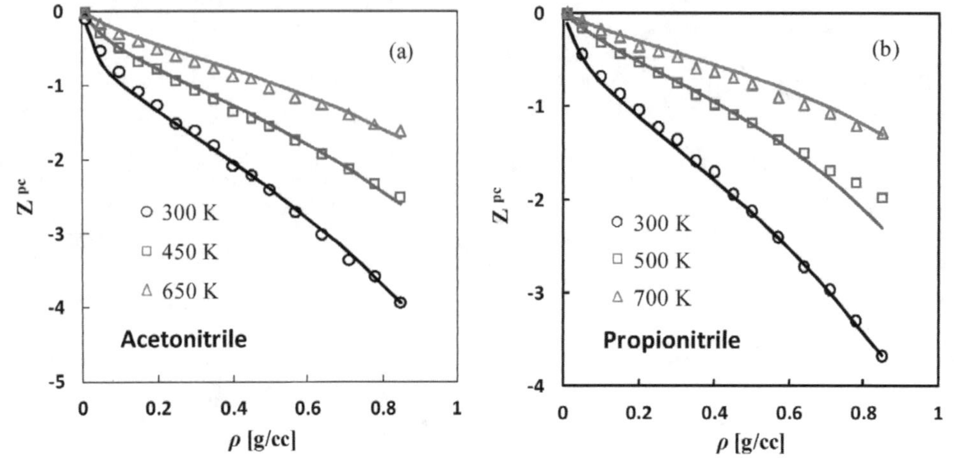

13

SAFT-γ WCA, Nitriles (TraPPE)

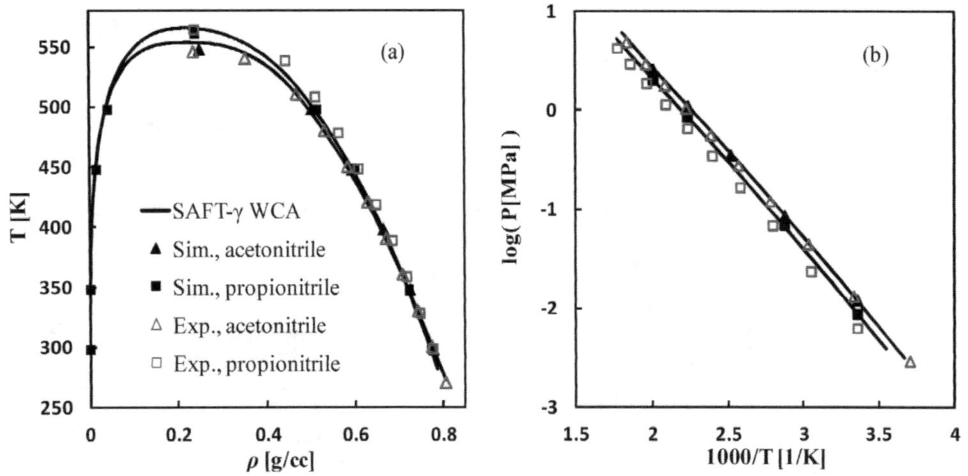

- Deviation from simulation: P^{sat} 3%, ρ^{liq} <1%
- Nitriles represent molecules with high charges density that don't form association

Simulate data for Nitriles from Wicks et al. JPCB 2005

14

SAFT-γ WCA, Ethers (TraPPE)

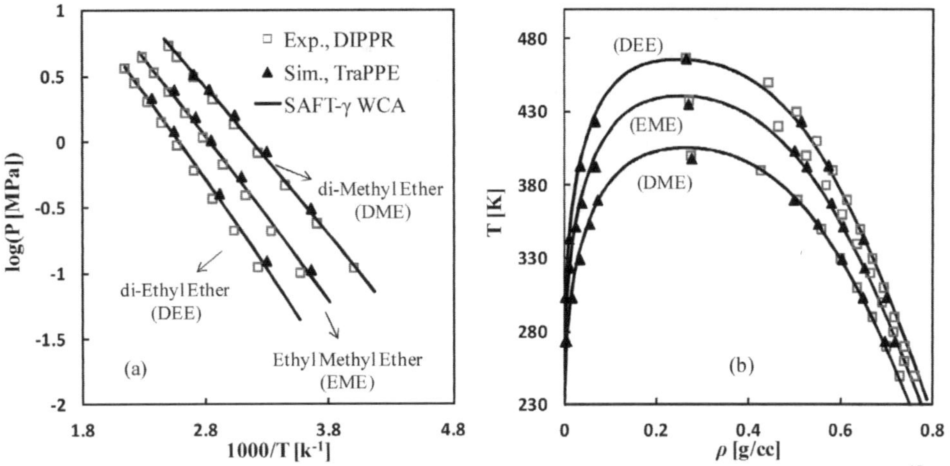

Simulate data for Ethers from Stubbs et al. JPCB 2004

15

SAFT-γ WCA, 1-alkanols (TraPPE)

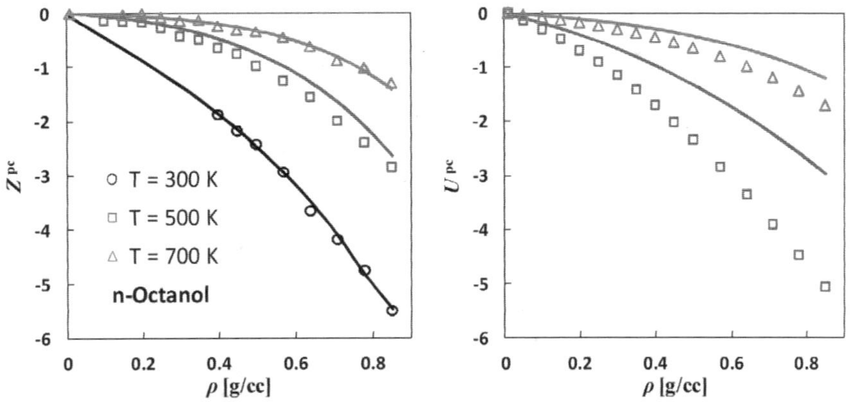

- The T-dependency of association term is not adequate for Associating compounds!

16

SAFT-γ WCA, 1-alkanols

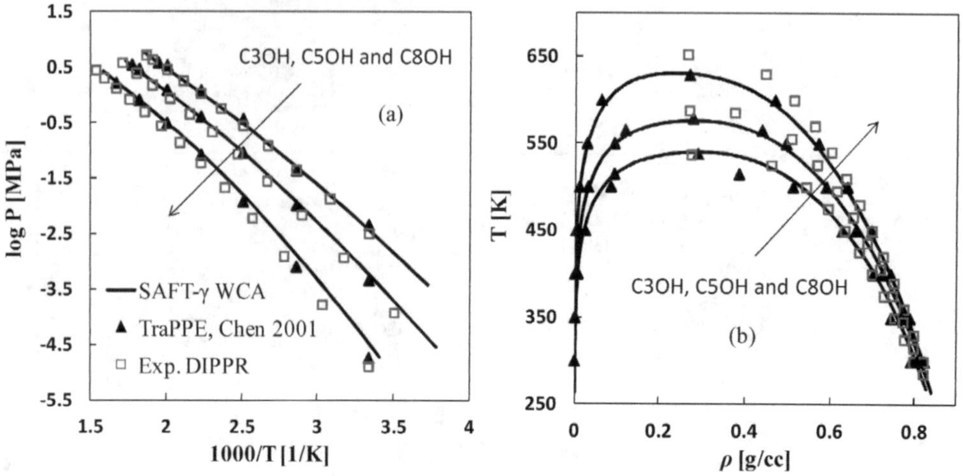

- Internal energy mainly affects vapor pressure, but not saturated densities.

17

SAFT-γ WCA, Water (TIP4P/2005)

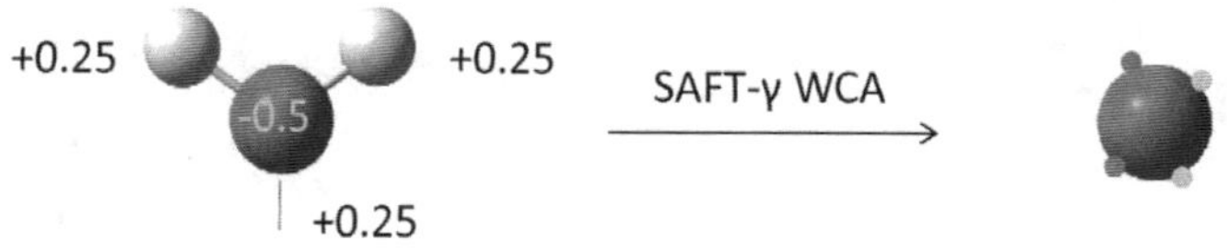

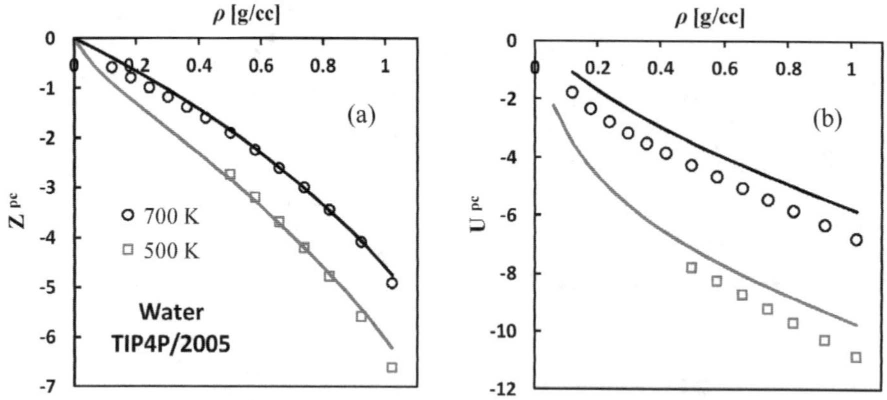

- Internal energy is underestimated, similar to 1-alkanols

18

SAFT-γ WCA, Water (TIP4P/2005)

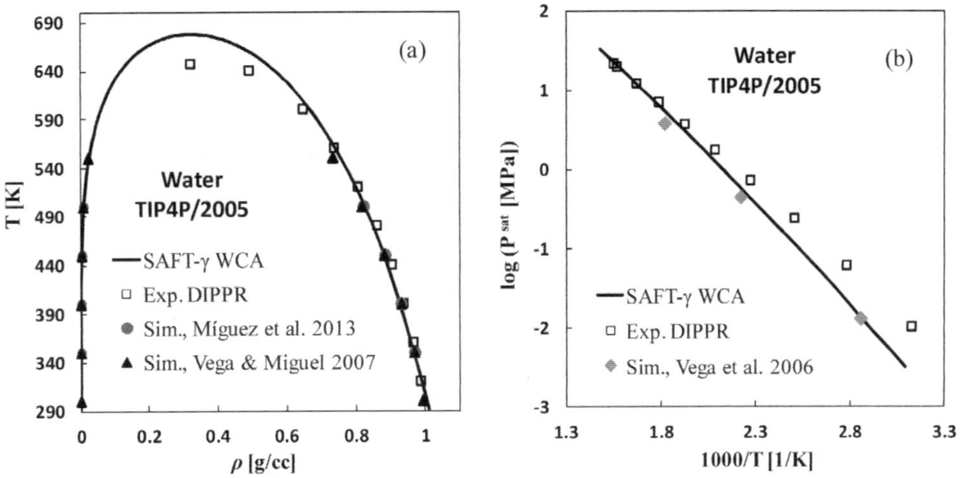

- TraPPE has no data for water. So, TIP4P/2005 model was used as it is proved to be the best force field for interfacial properties.

19

Phase II: Polar and Associating Systems

Ghobadi and Elliott, *JCP*, 141:024708 (2014)

In this work, a new classical density functional theory is developed for group-contribution equations of state (EOS). Details of implementation are demonstrated for the recently-developed SAFT-γ WCA EOS and selective applications are studied for confined fluids and vapor-liquid interfaces. The acronym WCA (Weeks-Chandler-Andersen) refers to the characterization of the reference part of the third-order thermodynamic perturbation theory applied in formulating the EOS. SAFT-γ refers to the particular form of "statistical associating fluid theory" that is applied to the fused-sphere, heteronuclear, united-atom molecular models of interest. For the monomer term, the modified fundamental measure theory is extended to WCA-spheres. A new chain functional is also introduced for fused and soft heteronuclear chains. The attractive interactions are taken into account by considering the structure of the fluid, thus elevating the theory beyond the mean field approximation. The fluctuations of energy are also included via a non-local third-order perturbation theory. The theory includes resolution of the density profiles of individual groups such as CH2 and CH3 and satisfies stoichiometric constraints for the density profiles. New molecular simulations are conducted to demonstrate the accuracy of each Helmholtz free energy contribution in reproducing the microstructure of inhomogeneous systems at the united-atom level of coarse graining. At each stage, comparisons are made to assess where the present theory stands relative to the current state of the art for studying inhomogeneous fluids. Overall, it is shown that the characteristic features of real molecular fluids are captured both qualitatively and quantitatively. For example, the average pore density deviates ~2% from simulation data for attractive pentadecane in a 2-nm slit pore. Another example is the surface tension of ethane/heptane mixture, which deviates~1% from simulation data while the theory reproduces the excess accumulation of ethane at the interface.

20

Phase III
Inhomogeneous Systems

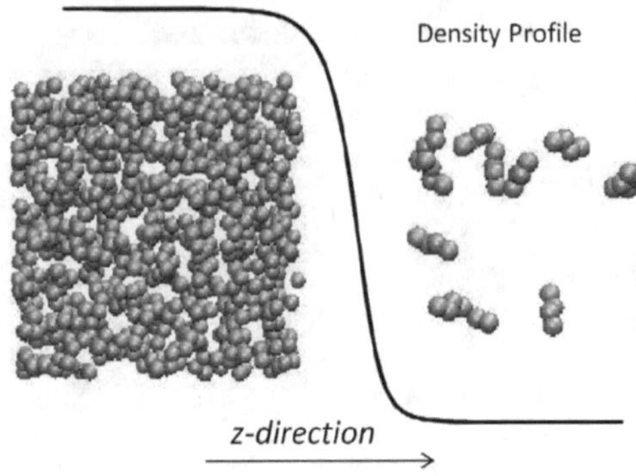

- Density Functional Theory:

$$\Omega\left[\rho(\mathbf{r})\right] = A\left[\rho(\mathbf{r})\right] - \int \rho(\mathbf{r}')\mu d\mathbf{r}' \qquad \frac{\delta A\left[\rho(\mathbf{r})\right]}{\delta\rho(\mathbf{r})} = \mu$$

21

Repulsive WCA Spheres Against a Soft-Wall

- Rosenfeld's Functional is generalized to account for softness.

$$\frac{A^{res}}{Nk_BT} = \frac{A^{sphere}}{Nk_BT} + \frac{A^{chain}}{Nk_BT} + \frac{A^{dispersion}}{Nk_BT} + \frac{A^{electrostatic}}{Nk_BT}$$

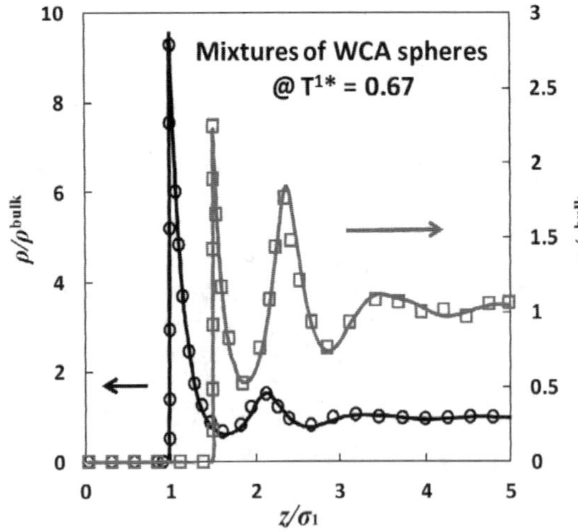

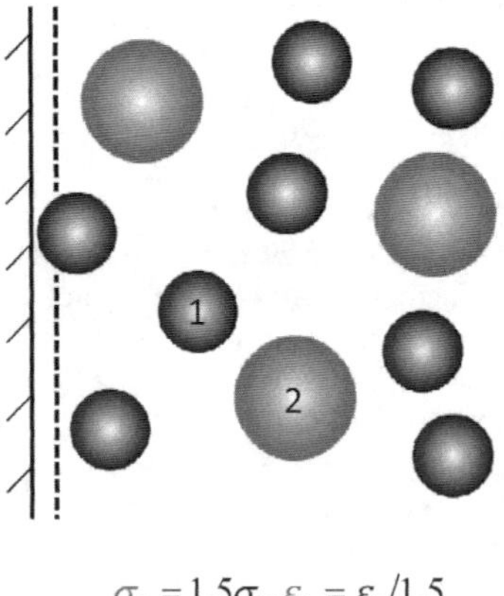

$\sigma_2 = 1.5\sigma_1, \varepsilon_2 = \varepsilon_1/1.5$

22

LJ Spheres Against a Hard-Wall

- Non-Local Attractive Functional. A2 and A3 are included.

$$\frac{A^{res}}{Nk_BT} = \frac{A^{sphere}}{Nk_BT} + \frac{A^{chain}}{Nk_BT} + \frac{A^{dispersion}}{Nk_BT} + \frac{A^{electrostatic}}{Nk_BT}$$

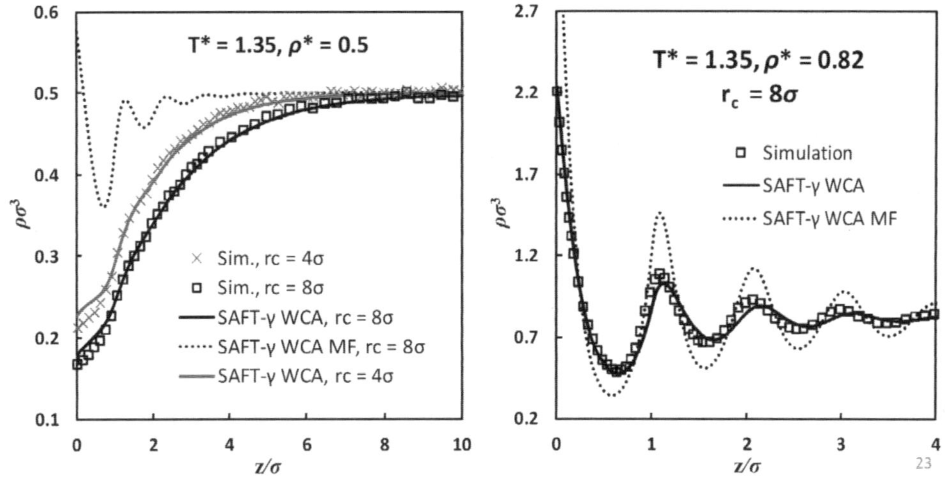

23

LJ Spheres in Graphite Slit-pore

- For attractive walls, the MF theory is not too bad!

$$\frac{A^{res}}{Nk_BT} = \frac{A^{sphere}}{Nk_BT} + \frac{A^{chain}}{Nk_BT} + \frac{A^{dispersion}}{Nk_BT} + \frac{A^{electrostatic}}{Nk_BT}$$

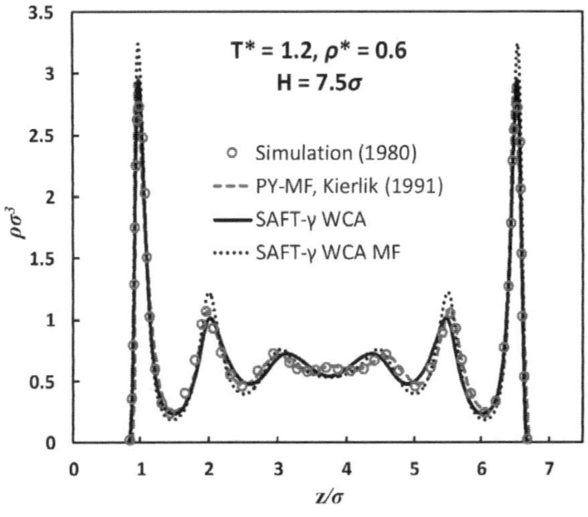

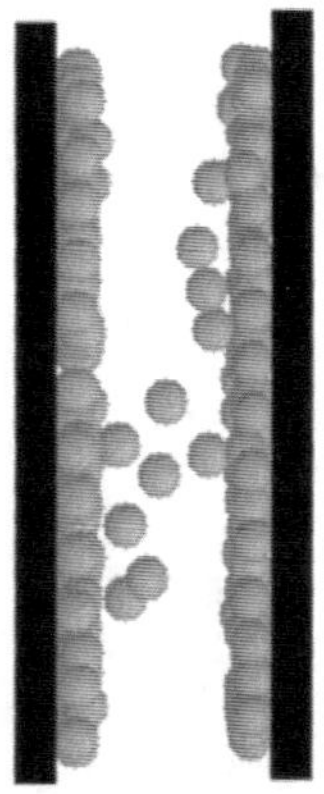

24

WCA Fused Soft Chains

$$\frac{A^{res}}{Nk_BT} = \frac{A^{sphere}}{Nk_BT} + \frac{A^{chain}}{Nk_BT} + \frac{A^{dispersion}}{Nk_BT} + \frac{A^{electrostatic}}{Nk_BT}$$

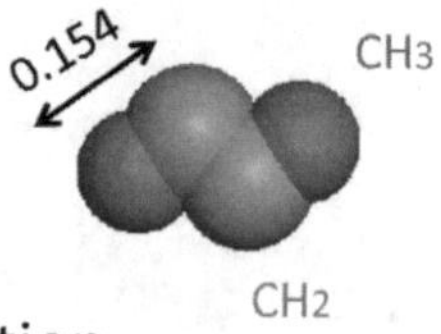

- Surface Excess deviates less than 7% from simulation

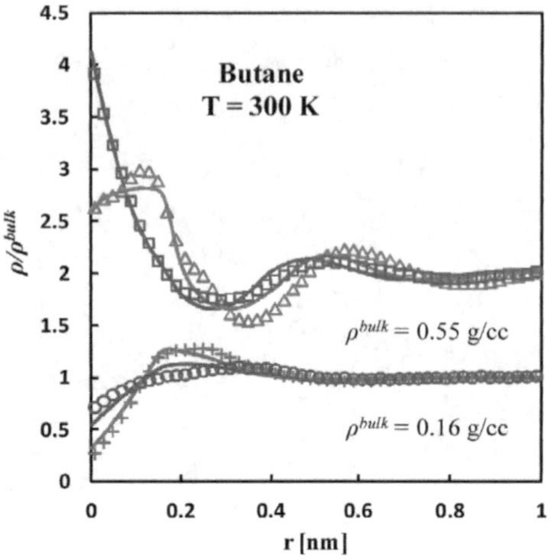

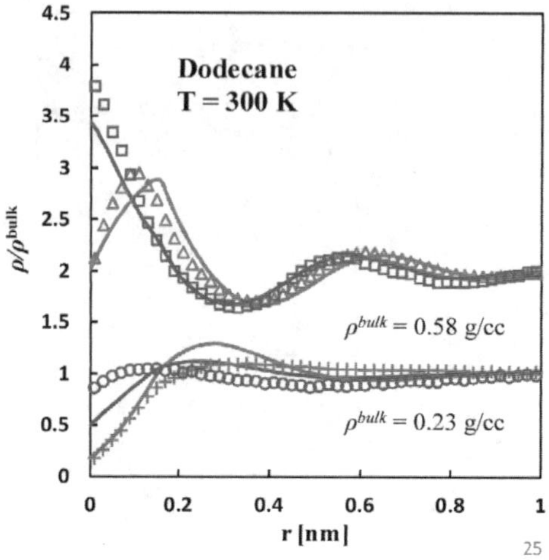

Attractive Fused Soft Chains

$$\frac{A^{res}}{Nk_BT} = \frac{A^{sphere}}{Nk_BT} + \frac{A^{chain}}{Nk_BT} + \frac{A^{dispersion}}{Nk_BT} + \frac{A^{electrostatic}}{Nk_BT}$$

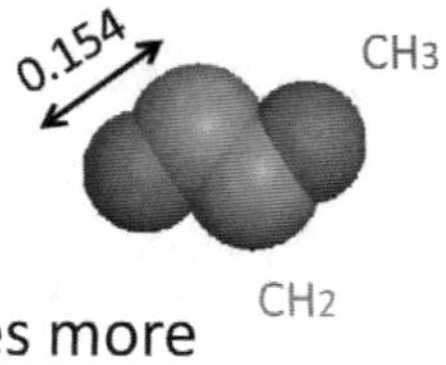

- For longer chains, dihedral angle change becomes more dominant

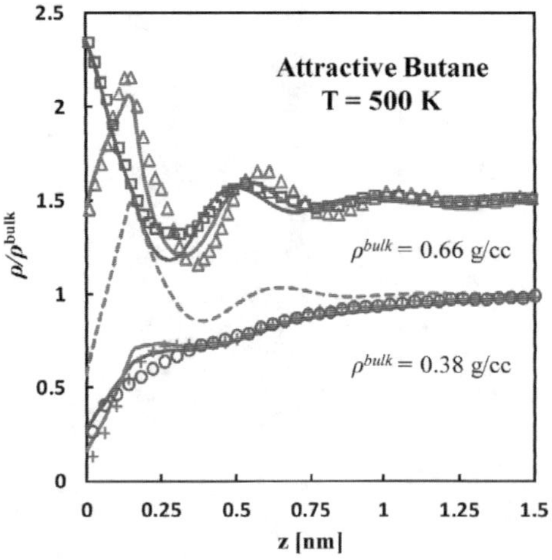

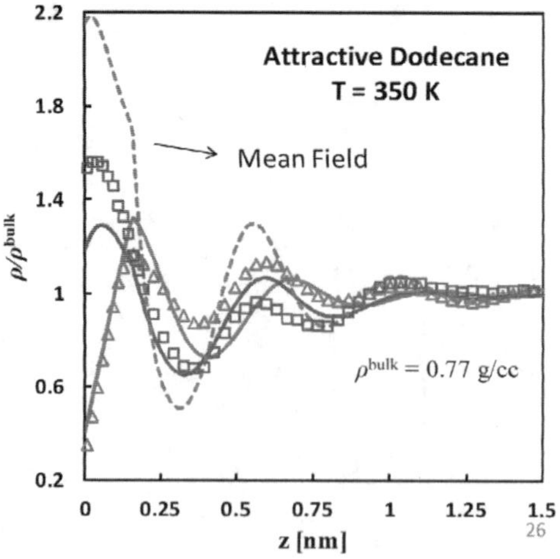

Fused Soft Chains with Partial Charges

- Adding electrostatic interactions lowers down the contact value

$$\frac{A^{res}}{Nk_BT} = \frac{A^{sphere}}{Nk_BT} + \frac{A^{chain}}{Nk_BT} + \frac{A^{dispersion}}{Nk_BT} + \frac{A^{electrostatic}}{Nk_BT}$$

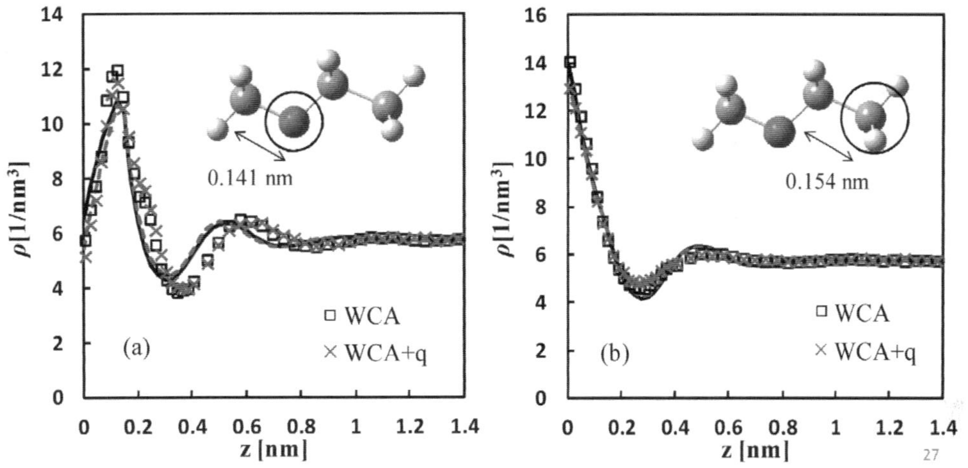

Surface Tension, Alkanes

- The deviation from simulation data ranges from 1% for Methane up to 4% for dodecane

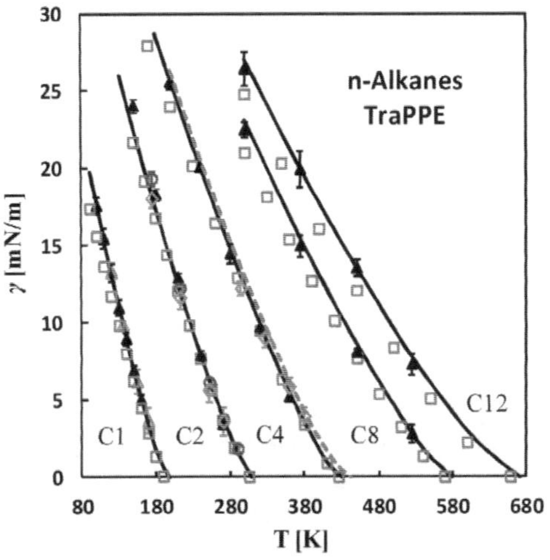

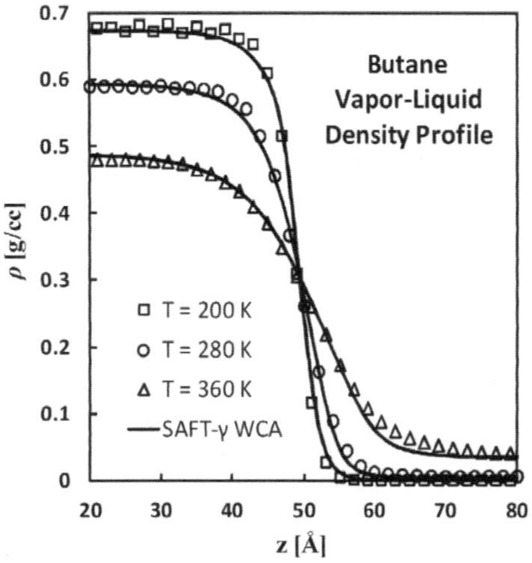

Surface Tension, CO2 and Nitriles

- The average deviation from simulation data is less than 4%
- United-atom force fields overestimate the surface tension at low T.

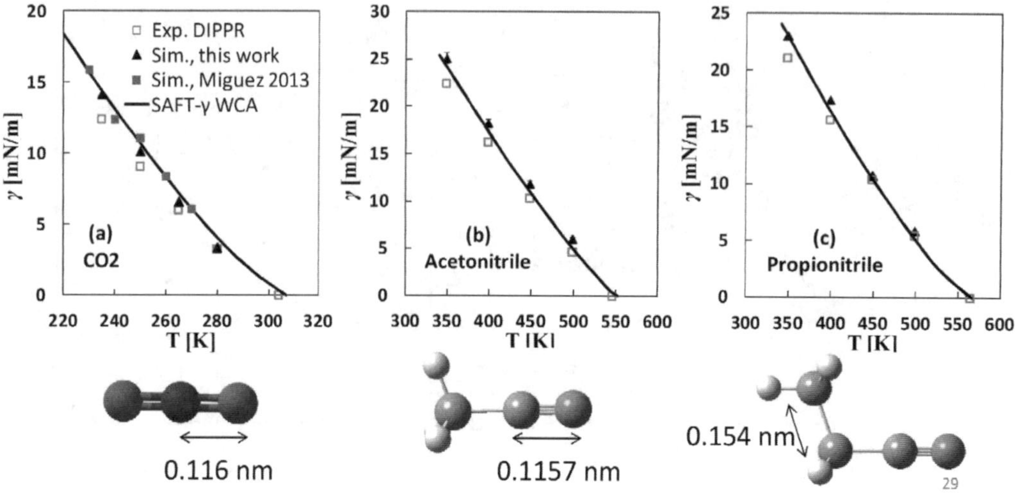

29

Surface Tension: Ethers, Alcohols and Water

- Surface tension of alcohols and water are overestimated and underestimated, respectively.

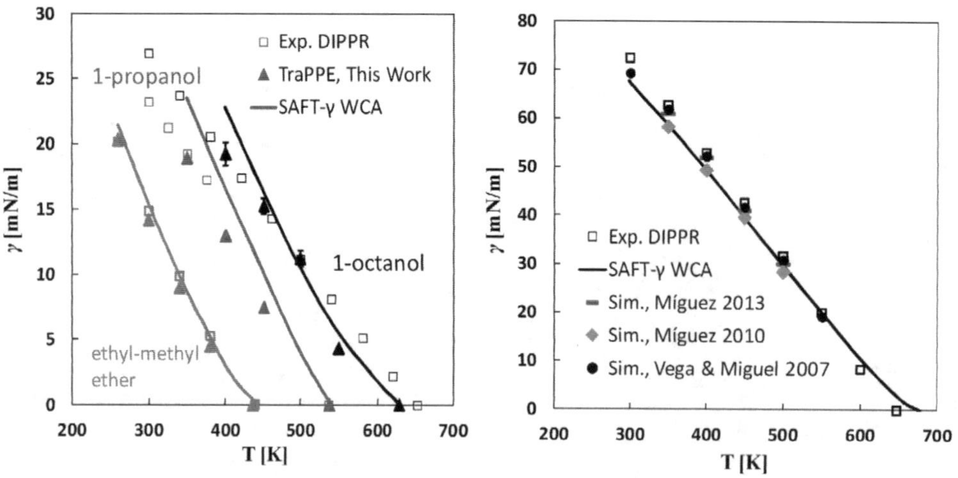

30

Density Profile of 1-hexanol and Water

- Spherical Symmetry
- Explicit Hydrogen

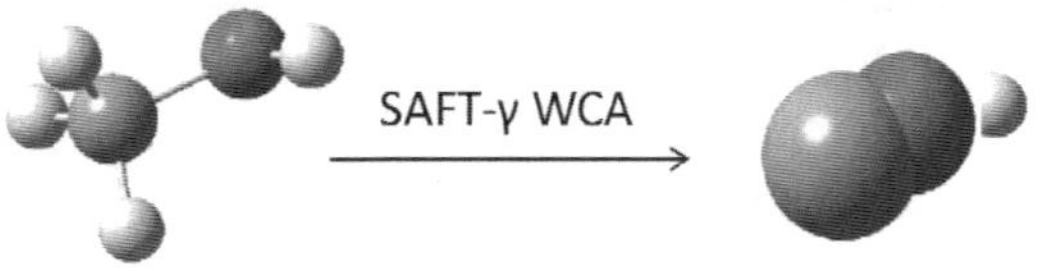

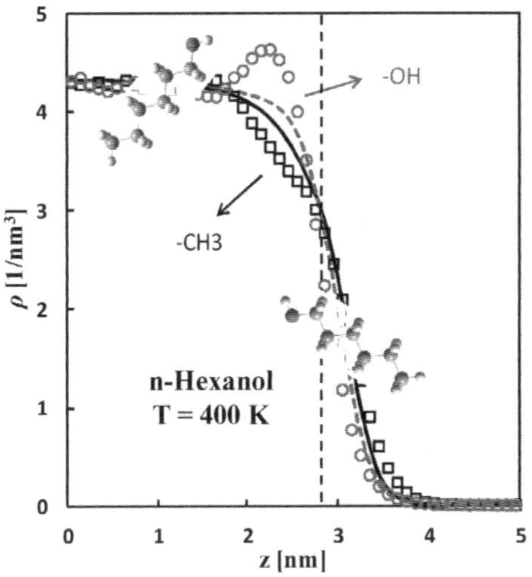

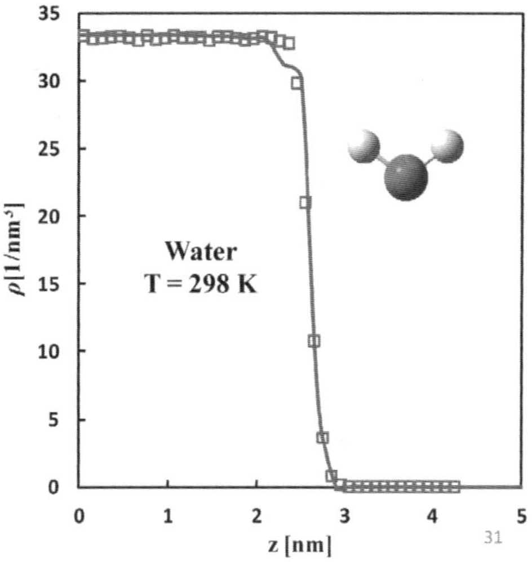

31

Surface Tension, CO_2/Ethanol

- The pure ethanol surface tension is overestimated.
- The pick at the interface is an indication of a chemical barrier.

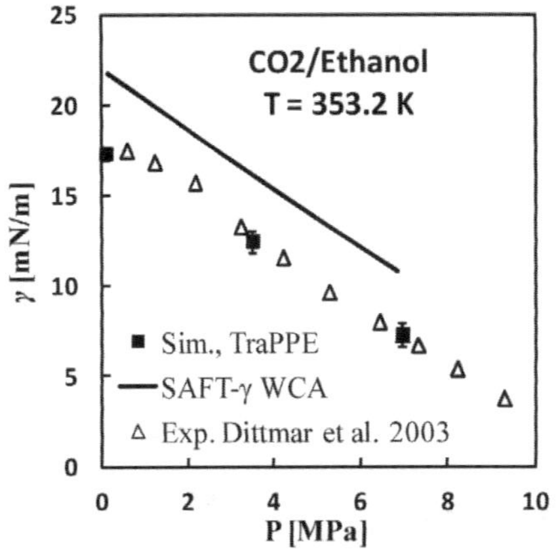

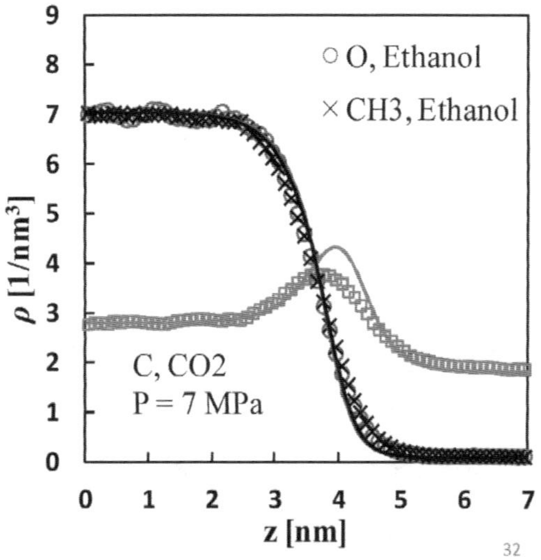

32

Surface Tension, Mixtures

- At low concentrations, CO2 behaves like a weak surfactant.

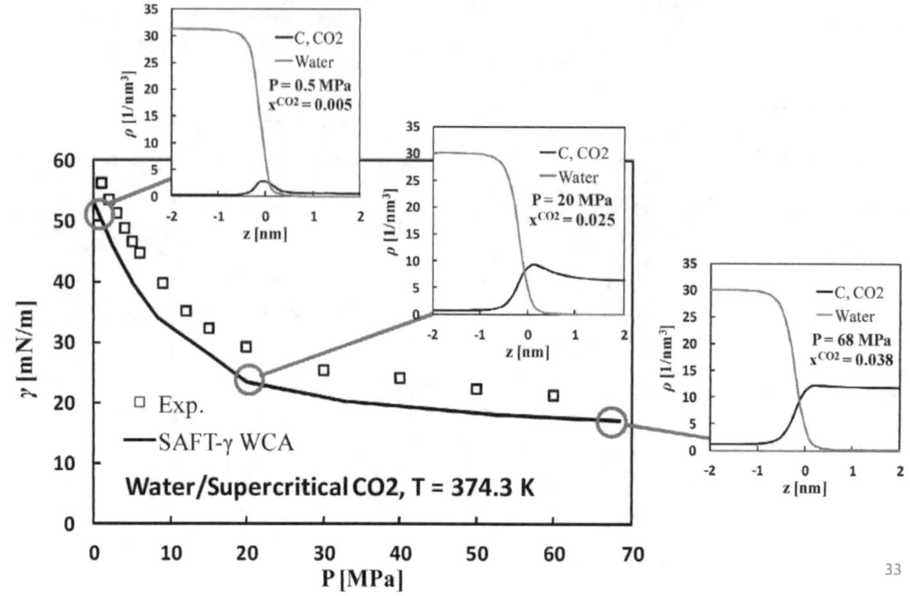

33

Microstructure of Surfactants

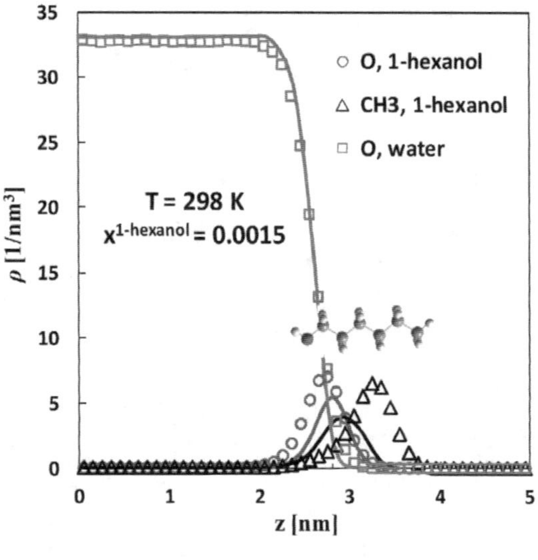

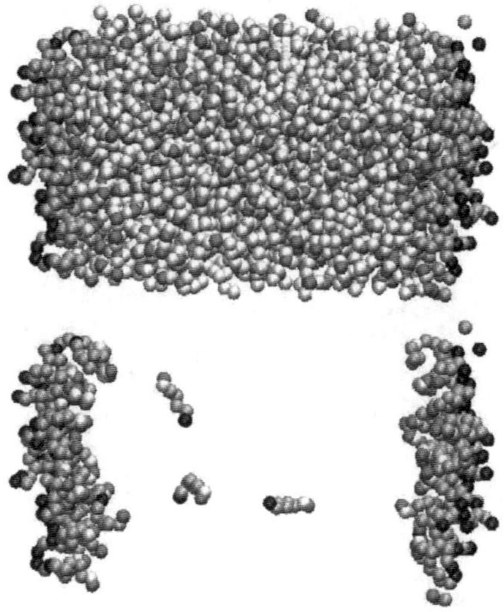

- Surface tension drops from 72 mN/m to 29±2.5.
- EOS predicts 41 mN/m for mixture and 35 mN/mfor pure 1-hexanol.

34

Phase III: Confined and Interfacial Systems
Ghobadi and Elliott, *JCP*, 141:024708 (2014)

In Paper I [A. F. Ghobadi and J. R. Elliott, J. Chem. Phys. 139(23), 234104 (2013)], we showed that how a third-order Weeks–Chandler–Anderson (WCA) Thermodynamic Perturbation Theory and molecular simulation can be integrated to characterize the repulsive and dispersive contributions to the Helmholtz free energy for realistic molecular conformations. To this end, we focused on n-alkanes to develop a theory for fused and soft chains. In Paper II [A. F. Ghobadi and J. R. Elliott, J. Chem. Phys.141(2), 024708 (2014)], we adapted the classical Density Functional Theory and studied the microstructure of the realistic molecular fluids in confined geometries and vapor-liquid interfaces. We demonstrated that a detailed consistency between molecular simulation and theory can be achieved for both bulk and inhomogeneous phases. In this paper, we extend the methodology to molecules with partial charges such as carbon dioxide, water, 1-alkanols, nitriles, and ethers. We show that the electrostatic interactions can be captured via an effective association potential in the framework of Statistical Associating Fluid Theory (SAFT). Implementation of the resulting association contribution in assessing the properties of these molecules at confined geometries and interfaces presents satisfactory agreement with molecular simulation and experimental data. For example, the predicted surface tension deviates less than 4% comparing to full potential simulations. Also, the theory, referred to as SAFT-γ WCA, is able to reproduce the specific orientation of hydrophilic head and hydrophobic tail of 1-alkanols at the vapor-liquid interface of water.

35

The calculation of potential dependent activation barriers for elementary reaction steps involving electron transfer is challenging using density functional theory (DFT) methods. This poster presents a simple DFT approach to estimating these barriers and calculating the symmetry factor for an elementary electroreduction reaction. We consider the reaction ($A^*+H^++e^- \rightarrow AH^*$) as consisting of an electrochemical step ($A^*+H^++e^- \rightarrow A^*+H^*$) and a non-electrochemical step ($A^*+H^* \rightarrow AH^*$), facilitating DFT analysis by providing a definable reference for the reactant $A^*+H^++e^-$ state and a tractable transition state search for a constant electron number reaction. We assume that the transition state for the electrochemical step and the non-electrochemical step are identical at the specific electrode potential (U^0) where $A^*+H^++e^-$ and A^*+H^* have the same free energy, therefore allowing us to find the free energy of the transition state for the original reaction. With the reactant, transition state, and product states and energy located, the formalism of Marcus Theory can be utilized to determine electrode potential dependent activation barriers. With some definition of the reaction coordinate, we fit two parabolas for the free energy surfaces for the reactant and product states. The intersection of the two parabolas is at the transition state. Incorporating the potential dependence of the elementary reaction free energy, the interaction can be determined as a function of potential, providing a potential dependent activation barrier. We explore utilized multiple definitions of the reaction coordinate, including the extent of charge transfer and a structural measure, emphasizing analysis of the initial step of CO_2 reduction on the Cu(111) surface.

371880 Mesoscale Simulations of Lipid Membranes Supported on Solid Surfaces

Abstract:

Using dissipative particle dynamics (DPD), we model lipid adsorption on solid supports. On metals and strongly hydrophobic substrates, lipids tend to adsorb in a lipid monolayer form. On hydrophilic and moderately hydrophobic surfaces, lipids adsorb in the bilayer form, with a water film of approximately 1-5 nm in width between the bilayer and the solid. Adhesion of silica to LB are to large extent determined by the free energy of the layer of water confined in the gap between the bilayer and the silica body. The very existence of the gap between silica and LB, as well as experimental measurements reported in the literature, testify that the disjoining pressure (a derivative of the characteristic thermodynamic potential by the distance between the bilayer and substrate) is non-monotonic and cannot be achieved with standard DPD potential. In order to enable non-monotonic disjoining pressure, we complement the DPD short-range repulsive potential with a long-range potential term between the support and the lipid tails. The parameters of the potential are fitted to the experimental and atomistic simulation results on disjoining pressure. With this model, we model lipid adhesion to hydrophobic and hydrophilic surfaces and consider the effect of surface curvature on lipid adhesion. This work was supported by NSF grant 1264702

Molecular modeling of homogeneous nucleation of [bmim$^+$][Cl$^-$] from the melt

Yan Shen[1], Xiaoxia He[1], Francisco R. Hung[1] and Erik E. Santiso[2]

[1] Cain Department of Chemical Engineering, Louisiana State University, Baton Rouge, LA 70803

2 Department of Chemical and Biomolecular Engineering, North Carolina State University, Raleigh, NC 27695

The study of solid phases of ionic liquids (ILs) confined inside nanoporous materials is relevant for the development of IL-based nanomaterials with unique properties (fluorescence, magnetic susceptibility). Although nucleation is a critical step in the process of crystallization, understanding of this behavior at the molecular level is still challenging, especially in the field of ionic liquids. Here we used molecular simulation to study the homogeneous nucleation of the IL 1-butyl-3-methylimidazolium chloride, [bmim+][Cl-], from its supercooled liquid phase to its orthorhombic crystalline structure. This IL is commonly used, has a relatively simple structure, and has been studied widely both in experimental and simulation work. The crystalline structure of [bmim+][Cl-] is described using local order parameters based on the pair distribution function for molecular crystals. The string method in collective variables (SMCV) is used to determine the minimum free energy path connecting the solid and liquid basins. The mechanism of nucleation and the free energy barriers associated with this process are determined and discussed.

375356 Green Leaf Volatiles on Atmospheric Air/Water Interfaces: A Combined Experimental/Molecular Simulation Study

Abstract:

Green leaf volatiles (GLVs) are a family of oxygenated hydrocarbons emitted by plants, especially under mechanical stress or damage. GLVs are believed to react with atmospheric oxidants both at the surface and within the bulk of water droplets (fog, mist, and rain), which can significantly contribute to the formation of secondary organic aerosols (SOAs). Both molecular simulation and experiments were conducted to investigate the properties of four selective GLVs, 2-methyl-3-buten-2-ol (MBO), methyl salicylate (MeSA), *cis*-3-hexen-1-ol (HxO), and *cis*-3-hexenylacetate (HxAc) in the air/water systems. Our simulations show that all the four GLVs have a thermodynamic preference to be adsorbed to the water droplets and stay at the air/water interface, as indicated by the deep free energy minima at those interfaces, and the stability of the GLV solutes at the air/water interface is mainly driven by energetic interactions between these solutes and water molecules. These results suggest that the air/water interface is the most likely reaction site for GLVs and atmospheric oxidants. Furthermore, the surface tension at the air/water interface is affected by adding those GLVs. These findings are relevant to the kinetics and mechanisms of the reactions between GLVs and atmospheric oxidants, which can be significantly different from those observed when the reactions take place in the gas phase or in the bulk of water phases.

375638 Molecular Dynamics Simulations of Hydrophobins Encapsulating Organic Molecules: Can Hydrophobins be Used As Oil Dispersants?

Abstract:

Hydrophobins are a class of proteins produced by filamentous fungi in soil. Preliminary experiments suggest that these proteins can encapsulate oil in *cylindrical* 'blobs', or gases in *cylindrical* bubbles, which implies a striking surface activity. These properties, as well as the abundance and ease of biosynthetic manufacture of hydrophobins, suggest that they could be used as 'natural' oil spill dispersants. Here we report molecular dynamics (MD) simulations where we probed the stability of benzene blobs encapsulated by hydrophobins. Our simulations were performed using the Martini coarse-grained model, and probed system sizes that are comparable to oil blobs realized experimentally by the Russo group. Our results suggest that our model hydrophobins do encapsulate benzene forming structures that are not spherical but rather elongated, as indicated by simulation snapshots and measurements of the characteristic dimensions (moments of inertia) of the structures. The stability of the bubbles, as indicated by measurements of the root mean squared deviations (RMSDs) from the initial sizes, is assessed and discussed.

375679 Direct Evaluation of Polypeptide Partial Molar Volumes in Water Using Molecular Dynamics Simulations

Abstract:

Molecular simulations of a series of tri- and tetrapeptides with the sequences gly-X-gly and gly-X-Y-gly, where X and Y are neutral amino acid side chains, have been performed to characterize their volumetric properties in water at 300 K. The chains were terminated with both zwitterionic and neutral N-acetyl/amide functionalities to examine end group effects. The simulations accurately capture the experimental effects of side chain mutations and increasing peptide length, providing an insight into the properties of oligopeptides and proteins in the idealized random coil reference state. For the zwitterionic terminated chains, however, simulations exhibit a constant, lower volume than experiment, suggesting simulations over predict electrostriction by the charged end groups. The simulation volumes are well described by a group contribution correlation, yielding excellent agreement between simulated side chain group volumes and experiment. The accuracy of our simulations and fidelity of the group additivity model to reproduce simulation volumes suggest it is possible to screen the potential effects of side chain interactions on group volumes when they are charged and more significantly influence neighboring hydration shells.

Multi-scale Modeling Study of poly(3-hexylthiophene) and [6,6]-phenyl-C_{61}-butyric acid methyl ester Towards Organic Photovoltaic Cell Application

Hanjong Yoo, Ki Chul Kim, Seung Soon Jang

School of Material Science and Engineering, Georgia Institute of Technology, 771 Ferst Drive NW, Atlanta, GA 30332-0245, USA

To date, organic photovoltaic cells have been extensively gained attention due to their promising potential in the industry. Despite their promising properties, there are many barriers to be overcome for practical applications. Particularly, its low efficiency needs to be resolved through constructing the optimistic morphology. There are three morphological quantities that relate to the efficiency. The domain size of the electron donor phase has to be small and the interface-to-volume ratio of the blend must be large. Also, the percolation ratio has to be high. In this study, poly(3-hexylthiophene) (P3HT) and [6,6]-phenyl-C_{61}-butyric acid methyl ester (PCBM) blends have been introduced as our active layer candidates. The state-of-the-art multi-scale modeling is employed to investigate the morphological properties of the active layer systems. We have developed our own forcefield parameters to accurately describe potential energy surfaces in the layer systems. Subsequently, coarse-grained models for P3HT and PCBM have been developed based on the improved atomistic force field parameters. Using the coarse-grained model, we can investigate much larger system during longer simulation time than using full atomistic model. We present the results from coarse grained model and its molecular dynamics simulation in comparison with those from the full atomistic model.

Title: The mechanical property of dental resin composite consisting of bisGMA and TEGDMA using molecular dynamics simulations

Author list: Yoon Hwan Edward Jeong, Jaeho Shin, Hean Gee Lee, Taewan John Kim, Hyea Hennim Hwang, Dr. Seung Soon Jang

Abstract: A resin-based dental composite has been studied for decades since its first appearance. Since such material has provided useful ways of treating dental patients, it has become popular in dentistry. In our study, the resin composite system consisting of bisphenol A glycidyl methacrylate (bisGMA) and triethylene glycol dimethacrylate (TEGDMA) with 1:1 ratio was simulated using molecular dynamics simulation to investigate the mechanical property by deformation and contraction. The simulation showed that the density of equilibrated system was approximately 0.993 ± 0.005 g/cm^3. During the extension simulation, the density was decreased to 0.888 ± 0.002 g/cm^3 due to the volume expansion. After deformation, the internal pressure and stress-strain of the structure was analyzed. The internal pressure and stress-strain were stabilized at 0.0085 Pa and 0.012 Pa. Under the compression, the simulation indicated that the density was reduced to 0.980 ± 0.002 g/cm^3. At first, the density increased until approximately 200 picoseconds as a result of decrease in the volume of the structure. After 200 picoseconds, the density started to increase and decrease continuously over the time interval of approximately 200 picoseconds. According to the internal pressure and stress-strain analysis, the structure was stabilized at 0.012 Pa and 0.014 Pa, respectively. From the two deformation modes such as extension and compression, we characterize the structural change at molecular level in conjunction of the stress through the system.

378373 Molecular Modeling of the Polymerization Behavior of Poly(vinylimidazole) in an Ionic Liquid Solution

Abstract:

Imidazole-based polymers, such as poly(vinylimidazole), may provide useful properties for membrane-based CO_2 capture applications, so it is becoming increasingly important to understand the formation mechanisms and underlying molecular structure of these materials. One of the challenges in the photopolymerization of 1-vinylimidazole (VIm) are slow rates and incomplete monomer conversion. We recently discovered that polymerization rate and overall conversion could be dramatically improved when VIm is in the presence of a bulky salt, lithium bis(trifluoromethylsulfonyl)imide $[Li^+][Tf_2N^-]$. In this work, molecular dynamics simulations are used to model the intermolecular conformational behavior and thermophysical properties of the VIm-$[Li^+][Tf_2N^-]$ complexes. Molecular conformations of different ratios of [VIm] to salt were characterized in order to identify the structural ordering induced by $[Li^+][Tf2N^-]$ on the polymerization of [VIm]. The same structural analyses were also performed on systems containing poly(vinylimidazole) [PVIm] in order to explore the conformations and thermophysical properties at different stages of the polymerization reaction. The calculated properties from the molecular dynamics simulations are compared against experimentally-derived structural features and found to be in agreement. Radial distribution functions indicate that $[Li^+]$ chelates with the "pyridine-like" nitrogen of the imidazole ring in both [VIm] and [PVIm], while the fluorine atoms of $[Tf_2N^-]$ have strong interactions with the vinyl groups. Also, as the relative concentration of the salt increases, the probability of vinyl groups aggregating in favorable conformations to polymerization tends to increase (at all temperatures studied). Furthermore, electronic structure calculations have also been performed to identify electrostatic interactions in the systems, such as partial charge assignments and potential dipole interactions with [VIm] and [PVIm].

Molecular Thermodynamic Analysis for an Abnormal Phase Behavior of PMMA solution in PEG200/alcohol solutions

Sang Min Lee, Young Chan Bae

Division of Chemical Engineering and Molecular Thermodynamics Laboratory,
Hanyang University, 133-791, Seoul, Korea

Keywords: Poly(methyl methacrylate), Cosolvency, Molecular thermodynamic
Corresponding author: Young Chan Bae, ycbae@hanyang.ac.kr

ABSTRACT

A new pair of non-solvent (PEG and alcohol) that enhances the miscibility of poly(methyl methacrylate) (PMMA) is first reported in this work. A possible mechanism for the enhanced solvation effect (cosolvency) is discussed based on experimental and simulation data. Some factors that affect the miscibility of PMMA such as temperature, molecular weight of polymer, chemical and structural characteristics of alcohols are investigated. A close relationship between the cosolvency of PMMA and the re-collapsing swelling behavior of PMMA particle gels is studied within the framework of molecular thermodynamic. Thermo-optical analysis (TOA) was employed to determine phase separation temperature and a dynamic light scattering (DLS) technique was used to measure the swelling transition of PMMA gel. A new lattice cluster associating fluid (LCAF) model is introduced for a quantitative description of the free energy of mixing, and the Flory-Rehner (F-R) chain model is employed for the free energy of elasticity. Guggenheim's approach for the frequency of occupation provides a basis for the LCAF model to take the probability of specific interactions between segments into account. Required interaction parameters are obtained from each binary system, and they are directly applied both to the ternary and swelling calculation. We apply the proposed method to PMMA solutions containing PEG200, PEG400, and a wide range of aliphatic alcohols. Our calculation results represent well with the experimental cosolvency and re-collapsing data, and show noticeable improvement in comparison with other models.

378962 Development of Reaxff Simulation Techniques for Assessing in Situ Phase Behavior of Pd-Based Catalysts

Abstract:

A detailed understanding of the structure of a catalyst's surface as a function of varying reaction conditions is essential for optimizing active and selective catalysts. Surface phase reconstruction under operating conditions is a particular concern for Pd-based catalysts, which are widely employed in industrial-scale oxidation and hydrogenation applications. Pd readily forms surface and bulk oxide phases when employed under oxidizing conditions, and can form complex hydride/carbide phases under a hydrocarbon-rich reactant phase. Density functional theory (DFT) can assess phase stability through the formalism of *ab initio* thermodynamics; however, the computational expense of DFT limits such investigations to idealized surfaced models. This has motivated the use of computationally inexpensive classical-force field methods that can reach larger size scales. This allows for more complex surface models that can better represent the catalytic behavior of the system. We have developed a Pd/O/C/H interaction potential for the ReaxFF reactive force field by fitting the interaction parameters of the potential to an extensive set of DFT data encompassing bulk and surface properties. Additionally, we developed a ReaxFF-based hybrid grand canonical Monte Carlo/molecular dynamics (GC-MC/MD) approach to assess the thermodynamic stability of oxide, hydride, and carbide phases in Pd clusters. Using this simulation method, we derive theoretical phase diagrams for Pd clusters under reactive gas phase environments, which demonstrate agreement with experimental results in the literature. This poster will present the detailed force-field development and the algorithmic approach using in the hybrid MC/MD method. The GC-MC/MD method is novel to ReaxFF, and is well suited to systems where phase restructuring influences catalytic activity, selectivity, and stability.

Association of Poly(2-oxzazoline) Chains with Water Molecules: Molecular Dynamics Simulation Approach

Byeong Jae Chun and Seung Soon Jang

Computational NanoBio Technology Laboratory, School of Materials Science and Engineering, Georgia Institute of Technology, 771 Ferst Drive NW, Atlanta, GA 30332-0245

We investigate the polymer-water interactions using full atomistic molecular dynamics (MD) simulation. The quantified data demonstrate the change of the interaction parameter, χ, as a function of water concentration. For this, first, the monomers of poly(2-oxazoline) series, the side chains of which are distinctive from each other, are geometry-optimized using B3LYP and 6-31G** and then Mulliken population analysis is performed for atomic charges. Second, the system of homogeneous polymer blend with various water contents is constructed. By performing MD simulation, the enthalpy of mixing is calculated from the system to estimate the interaction parameters and optimum amount of water content in each homopolymer blend.

$$\Delta H_{mixing} = \chi RT \phi_{polymer} \phi_{water}$$

The obtained data are compared to the experimental ones in order to validate our force field and simulation procedure. We believe this study provides a molecular-level understanding of association of polymer with water molecules for characterizing the macro-scale polymer complex in water phase.

Effect of the Modified Erythrosine B on Amyloid β 40 monomer as an inhibitor for the fibril using molecular modeling approach

Juho Lee[a], Woo Yaa Lee[a], Joy Kim[a], Sunju Kang[a], Hee Su Lee[a], Doyeon Koo[a], Hennim Hwang[a], Inchan Kwon[b], and Seung Soon Jang[a]

a. School of Materials Science and Engineering, Georgia institute of Technology, 771 Ferst Drive, Atlanta, GA 30332

b. School of Chemical Engineering, University of Virginia, 102 Engineer's Way, Charlottesville, VA 22904

Alzeheimer is most well known type of degenerative dementia. As a main cause of the disease, previous studies have focused on the formation of neuro-toxic plaque composed of 39 to 42 residue-long amyloid beta (Aβ) fibril. This has been found in the patient's brain with relatively high portion. Through a series of investigation about the mechanism of beta amyloid aggregation, it was found that initial Aβ monomer formed with high portion of alpha helical structure undergoes intermediate state for aggregation so-called aggregation-prone state (APS) which is partially folded. Erythrosine B (ER) is one of the prominent drug candidates to hinder the aggregation of the Aβ peptides. ER is a red food dye that has been approved by Food and Drug Administration (FDA). ER is a common xanthene dye approved not having any toxicity to humans under daily dose of 60 mg/kg. With a collaboration work, the effect of ER on the Amyloid beta monomer has been suggested as an inhibitor of the formation of aggregation-prone state structure and fibril elongation through *in vitro* experiment. Here, molecular modeling methods such as docking and molecular dynamics simulation was performed to study the effect of modifying functional group of ER and specific effect of modified ER on Aβ40 monomer.

379914 Why Pucker, Sugar? Computational Chemistry Investigations into Carbohydrate Ring Distortion for Enzymatic Action

Abstract:

Since the first crystal structure of a lysozyme was solved in the 1960s, researchers have noted that glycoside hydrolase (GH) enzymes consistently distort the conformation of the carbohydrate ring closest to the scissile bond away from the solution-stable $^4C_1$ chair conformation. A variety of theories have been proposed to account for this feature, the most promising of which focus on favorable stereoelectronic features of puckered rings, including accumulation of positive charge at the anomeric carbon, the site of nucleophilic attack. Quantum mechanical studies are an obvious choice for investigating such hypotheses. In this work, we present the most thorough electronic structure study to date of five biologically-paramount monosaccharides in vacuum, β-xylose, α-glucose, β-glucose, β-mannose, and β-N-acetylglucosamine. Exploiting the inherently parallel problem of comparing different monosaccharide conformations, we evaluated over 123,000 geometries for the set of molecules in the study, comprising all 38 IUPAC puckering geometries and exocyclic group orientations. We employed a step-wise screening approach of increasingly accurate methods to allow rigorous examination of the differences between their properties that lend particular puckers especially amenable to catalysis. We isolated both local minima and transition states to reveal the puckering interconversion landscape of each sugar. Comparing our findings to experimental structural studies of GH substrate distortion, we found that previously proposed correlations between enzymatically employed puckers and specific catalytically favorable electronic structure properties do not persist across all sugars in the study, mirroring important differences between the sugars. Additional factors, such as interactions between exocyclic groups, must be considered in a full picture of sugar puckering. These results demonstrate that exocyclic group conformations must be explicitly considered and reveal a complex optimization problem reflecting the complexity of glycobiology. It contributes a more nuanced understanding relationship between substrate conformation and reactivity.

382820 Molecular Dynamic Simulations of Cross-Linked Polyamide Reverse Osmosis Membrane

Abstract:

Molecular dynamics simulations are used to construct an atomistic model of an aromatic polyamide reverse osmosis membrane. An aromatic polyamide membrane is generated from a highly cross-linked polymerization between m-phenylenediamine (MPD) monomers and trimesoyl chloride (TMC) monomers. An atomistic model of polyamide membrane is successfully built using a heuristic approach. The unit cell length and the target density of polyamide membrane are reached, and they are in good agreement with the experimental values. A number of water molecules is added into a polyamide membrane in order to generate the hydrated polyamide membrane. Further molecular dynamics simulations are performed. From the solvated membrane, water permeability is yielded. Furthurmore, the effects of temperature and pressure on water permeability are studied using MD simulations.

Using Density Functional Theory to examine the specific adsorption of ions to electrode surfaces

Ionic species in electrolyte solutions can specifically adsorb to catalytic electrode surfaces, impacting electro-catalytic reaction kinetics and mechanisms. For example, bromide can specifically adsorb to platinum at low potentials and reduce the rate of the hydrogen oxidation reaction. (Bi)sulfate and organic sulfonate adsorbs to platinum at high potentials and can affect the rate of the oxygen reduction reaction. In addition to spectator species, the adsorption strength and coverage of hydroxide, a hydrogen oxidation reaction intermediate, also affects the rate of hydrogen oxidation in alkaline solutions. Hydrogen oxidation and oxygen reduction are fuel-cell relevant reactions; having a clear understanding of the effects of ionic species is necessary to design better performing and lower cost fuel cells. Density Functional Theory (DFT) is an *ab-initio* method which can be used to probe ion specific adsorption. Modeling the reactant and product states can be used to determine under what conditions (concentrations, electrode potential) specific adsorption is favorable and to elucidate the effects of the adsorbed species on catalytic reactions. Difficulties include calculating the chemical potential of the ionic species in solution, determining the effects of solvation of the specifically adsorbed state, and calculating or controlling the electrochemical potential within the simulation environment. We will present and discuss an approach using experimentally measured solvation energies or dissolution potentials to determine the ionic species chemical potential and an explicit water molecule method to estimate the effects of solvation on the free energy of the specifically adsorbed state. Methods to include the electrochemical nature of the interface include estimations of surface dipole/field interactions and full electrification of a fully solvated interface by varying the system charge. Using these methods, the electrochemical potential at which a given ionic species will adsorb to a surface can be calculated. The application of these methods to examine the adsorption of alkali cations and their effect on hydrogen oxidation and the adsorption of halides and their effect on carbon dioxide electroreduction will be discussed.

Optimum Excipient Selection and Design for Protein Drugs

Haider Tarar, Brock C. Roughton, Kyle V. Camarda

There is a high demand for protein drugs but any positive consequences of protein drugs are inaccessible if these drugs cannot be stored and delivered in a stable form. Protein drugs have a tendency to aggregate which adversely affects their shelf life and delivery as a stable drug formulation. Therapeutic effects of the protein drugs are dependent on the specific primary structure of the proteins. During Aggregation the protein irreversibly loses its unique shape and thus its function. Lyophilization is the primary method used in the industry to improve the stability of protein drug formulations but even after this process proteins may lose their native structure due to aggregation.

The aim of this project is to find an effective, cheap and efficient method to develop stable formulations for protein drugs which are not vulnerable to aggregation and in turn have stable shelf life and effective delivery. A method using a combination of computer-aided molecular design and molecular docking simulations is proposed to find a specific optimum formulation for a specific protein minimizing aggregation probability.

The choice of excipient in the protein drug formulation can lower the probability of protein aggregation through interactions between the excipient and hot spots on the protein. A pool of candidate excipients is chosen based on certain desired properties. For example an excipient which exhibits a higher glass transition temperature in lyophilized products is preferred as a glass can immobilize the protein. Furthermore the lower the protein loss after lyophilization, the better the excipient-protein combination will function. Another factor to consider is the water content in a lyophilized excipient-protein solid. Lower water content is preferred because removal of water slows down reactions and reduces protein mobility. The next step is to perform molecular docking simulations with these excipients. The excipients looked at in this work are designed to provide stabilizing interactions for the protein myoglobin. In order to apply molecular docking, two commonly used excipients, the sugars sucrose and mannitol, were investigated. Octyl glucoside, Tween 40 and Tween 80 are the surfactants which will also be used in the molecular docking simulations in the future. Molecular docking simulations were performed using Autodock to predict which residues were most likely to interact with the excipient in question. The computational results from the above procedure were compared to experimental hydrogen-deuterium exchange (HDX) mass spectrometry experiments (done at Purdue University) using lyophilized formulations of myoglobin with either sucrose or mannitol. The regions on the protein which exhibit free exchange between hydrogen and deuterium are uncovered and not protected by the protein-excipient interactions. Therefore, protein regions which show low deuterium uptake are protected. These regions of low deuterium uptake were then correlated with protein-excipient interactions predicted by molecular docking, to validate those simulation results.

For the hotspot regions of the protein predicted by Aggrescan, the docking simulations are to be used to optimally select excipients which have the highest number of interactions with such regions. The selection process for the excipient with the most interactions is executed for the entire hotspot sequence. Using Computer aided molecular design the excipient pair which provided the most number of interactions is to be selected as the optimal drug formulation. The aim is to expand this method to consider many protein-excipient interactions and develop into an excipient selection and design tool.

Towards the development of a model for particle nucleation

Jeffrey S. Lowe[1], Paolo Elvati[2], Angela Violi[1,2]
[1]Department of Chemical Engineering, [2]Department of Mechanical Engineering
University of Michigan, Ann Arbor, MI USA

Background

Above: Graphical representation of soot formation in a vehicle engine. Details are exaggerated to emphasize the formation process.

- Carbon nanoparticles emitted from engines and other combustion processes are a hazard to human health and the environment
- Soot particle formation is not well understood at a molecular level

Introduction

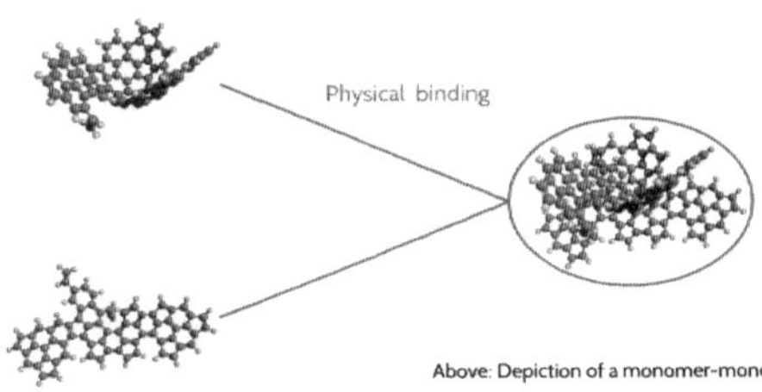

Above: Depiction of a monomer-monomer coagulation event

- The current study is aimed at a simple physical growth model - dimerization
- Our starting point is to model monomer-monomer coagulation from structural factors of molecules

Methods

- Molecular dynamics simulations coupled with a free energy algorithm to assess the coagulation propensity between two monomers
- The current study quantifies the contribution of aliphatic chains

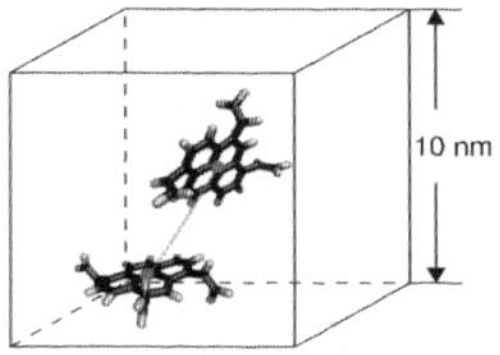

Simulation Parameters
LAMMPS Molecular Dynamics
PLUMED Well-Tempered Metadynamics
Herdman/Miller (2008) + OPLS

Left: Simulation setup showing the center of mass distance between monomers as the independent variable for calculating the free energy surface

Molecules

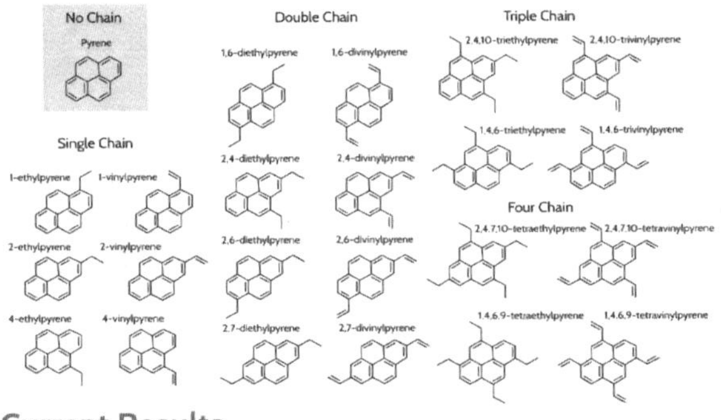

Current Results

Dimer Stability Summary

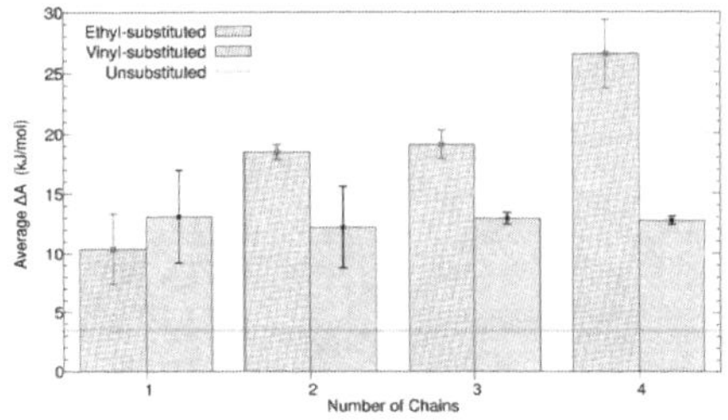

Above: Dimer stability dependence on chain type and the number of chains at 500 K. Statistical error bars are shown.

- Ethyl chains: Dimer stability increases with an increase in the number of chains
- Vinyl chains: Dimer stability is roughly constant with respect to the number of chains

The major difference between ethyl and vinyl chains is that vinyl chains are more polar, resulting in a larger separation of charges as compared with ethyl chains

Out-of-plane Torsion Between Monomers

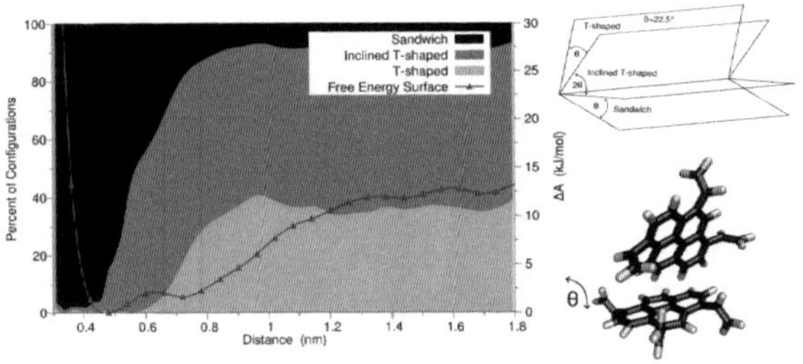

Above: Out-of-plane torsion for 1,4,6-trivinylpyrene as a function of the center of mass distance between monomers and the corresponding free energy surface. Torsional angles are grouped into three categories as shown above: sandwich, T-shaped, and inclined T-shaped.

- The test molecule forms dimers with inclined T and T-shaped geometries that are energetically-competitive with sandwich configurations

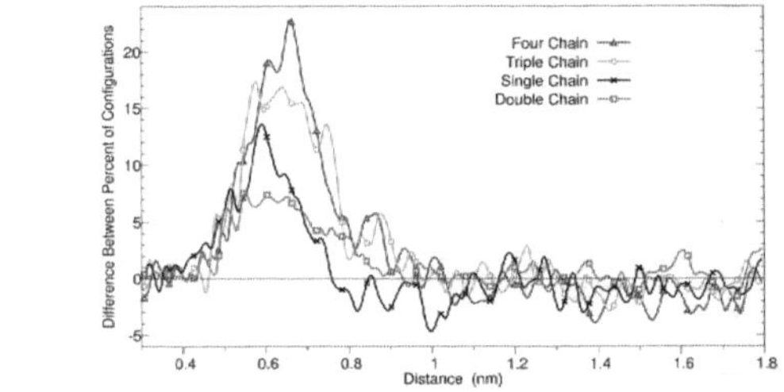

Above: Difference in the percent of sandwich configurations in a molecular dynamics study between ethyl- and vinyl- substituted compounds. The ethyl-substituted compounds were more frequently observed in the sandwich configuration.

- There is a general increase in the favorability for non-sandwich orientations with an increase in the number of chains for vinyl-substituted compounds with respect to ethyl-substituted compounds

Conclusions and Future Work

- The increased importance of Coulombic repulsion for compounds with a larger separation of charges may lead to reduced dimerization propensities in a physical model for particle coagulation
- Further work will quantify the effect of Coulombic and van der Waals interactions

MICHIGAN ENGINEERING
UNIVERSITY OF MICHIGAN

Acknowledgements
This material is based upon work supported by the U.S. Department of Energy, Office of Science, Office of Chemical Sciences, Geosciences, and Biosciences under Award Number DE-FG02-09ER16112

Nanophase-Segregated Structures and Transport Properties in Polysuflonated (190-PSU-S_n) based Ionomers for Proton Exchange Membrane Fuel Cell

Young Hun Lee, Khaldoon Abu-Hakmeh, Kim Ki-_ChHul Kim and Seung Soon Jang

Computational NanoBio Technology Laboratory, School of Materials Science and Engineering, Georgia Institute of Technology, 771 Ferst Drive NW, Atlanta, GA 30332-0245

Nafion is commonly used in polymer electrolyte membrane fuel cells (PEMFC) due to its high proton conductivity and thermal, chemical, and mechanical stability. However due to several limitations it has, such as high methanol crossover and poor performance under low humidity and at high temperature condition, polysulfonated monomers have been considered as alternatives with the specifications that is essential for the fuel cell membranes. This specification includes high proton conductivity, reduction-oxidation stability, and excellent mechanical, thermal, and chemical stability with its 'aryl-SO_2-aryl' backbone. In this study, mechanical properties of polysulfonated ionomers are investigated using full-atomistic MD simulation methods. In order to determine how the structure of sulfonate groups affects the mechanical properties, we considered three different functional groups (-$SCF_2CF_2SO_3H$, -OCF_2 SO_3H and –CF_3 $2(C_4H_4CF_2$ $SO_3H)$ attached to polysulfone. Various water contents were implemented to achieve effective results on the relationship between structure and transport of hydrated polysulfonated membranes. Each of PSU-S_4, PSU-S_5 and PSU-S_6 membranes are equilibrated with two different water contents (10 wt. % and 20 wt. %) through an annealing procedure. The equilibrated states were assumed to be reached with the generally standardized density and potential energy distribution curves. The pair correlation function was examined in order to analyze the local structure of various atoms in the membrane, which determines the solvation. The diffusion coefficient of water and hydronium were determined to calculate the position probability diffusing over time with the mean squared displacement equation.

Abstract

Nafion is commonly used in polymer electrolyte membrane fuel cells (PEMFC) due to its high proton conductivity and thermal, chemical, and mechanical stability. However due to several limitations it has, polysulfonated monomers have been considered as alternatives, which satisfy such limitations of Nafion. In this study, mechanical properties of poly (arylene ether sulfone)-based ionomers are investigated using full-atomistic MD simulation methods. In order to determine how the nature of the structures of sulfonates affect the mechanical properties, we considered three different types of perfluoroalkyl sulfonates (PSU-S_4, PSU-S_5 and PSU-S_6). Also we implemented with various water contents to achieve its effect on the structure and transport of hydrated polysulfonated membranes. Each of PSU-S_4, PSU-S_5 and PSU-S_6 membranes are equilibrated with two different water contents (10wt% and 20wt %) through an annealing procedure. The equilibrated states were assumed to be reached with the generally standardized density and potential energy distribution curves. The analysis on the simulated characteristic correlation length, the pair correlation function and the diffusion coefficient of water and hydronium are done for the comparison with each different membranes and also with Nafion.

HYDROCARBON AND CARBON DIOXIDE MIXTURES IN SILICA PORES: RELATION BETWEEN STRUCTURE AND DYNAMICS

Thu Le*, Stephanie Ogbe and Alberto Striolo

Department of Chemical Engineering, University College London, London, United Kingdom

David R. Cole

School of Earth Sciences, Ohio State University, Columbus, Ohio 43210, United States

MOTIVATIONS AND OBJECTIVES

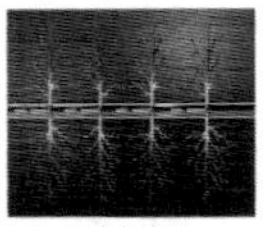

http://fracfocus.org/hydraulic-fracturing-process

Understanding the molecular features related to hydrocarbon and carbon dioxide adsorption and diffusion in narrow pores might be useful for (1) quantifying the molecular phenomena that occur during natural gas production following hydraulic fracturing; (2) developing long-term CO_2 capture strategies; and (3) optimizing natural gas production.

The objectives are to use equilibrium MD simulations to:

- Study **sub**, **near** and **super-critical propane adsorption isotherms, molecular structure** and **mobility in confinement**;
- Investigate **the effect of temperature** and **pressure** on the **adsorption, preferential adsorption** and **mobility** of confined CO_2/n-hydrocarbon mixtures at a wide range of concentrations and compositions.

Cross-sectioned Woodford shale specimens

Curtis, M. E., Ambrose, R. J., Sondergeld, C. H., Rai C. S. 2010. CUSG/SPE 137693

SIMULATION METHODOLOGY

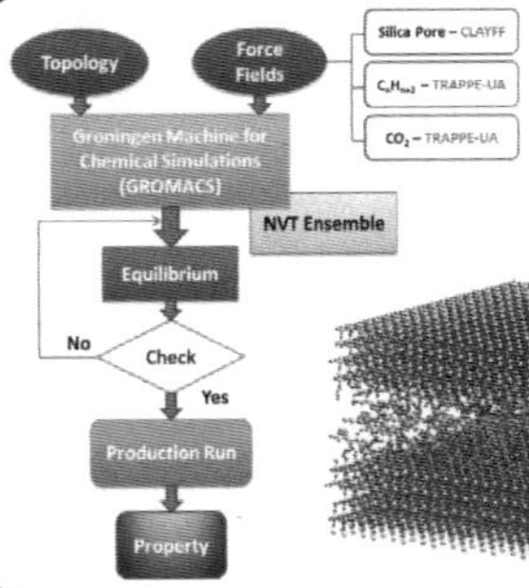

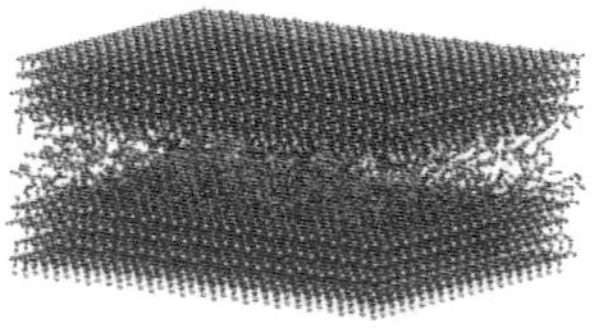

CO_2 and n-octane Confined in 2 nm Silica Pore | **CO_2 and n-butane Confined in 2 nm Silica Pore**

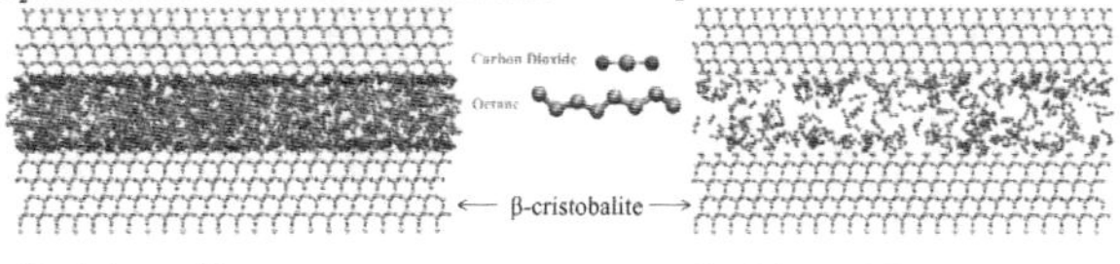

Simulation conditions:

- T = 300K
- Constant density of n-octane (282 molecules)
- Increasing density of CO_2

Simulation conditions:

- T = 290K, 343K and 430K
- Constant mixture density (100, 300 or 500 molecules) but different composition

STRUCTURAL AND DYNAMIC RESULTS

Density Profiles

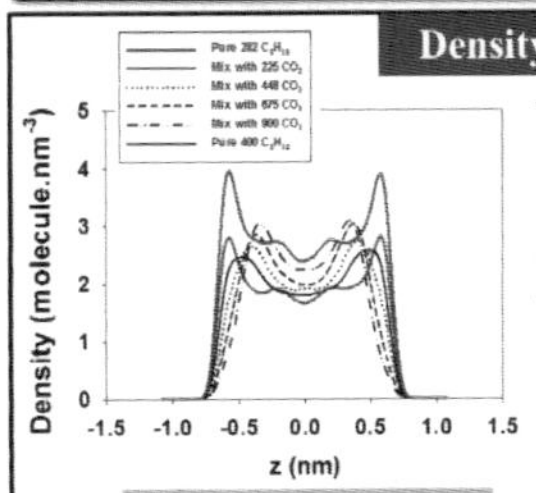

Density profiles of C_8H_{18}

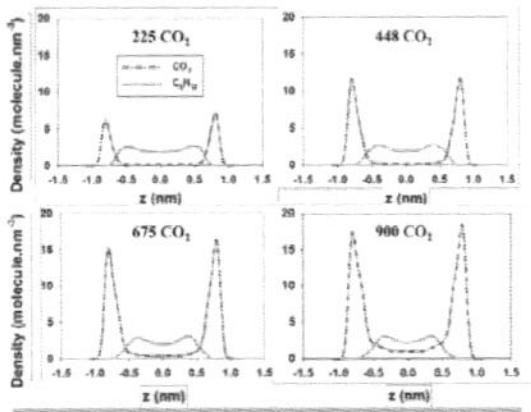

Density profiles of C_8H_{18} versus CO_2

Planar Density Profile

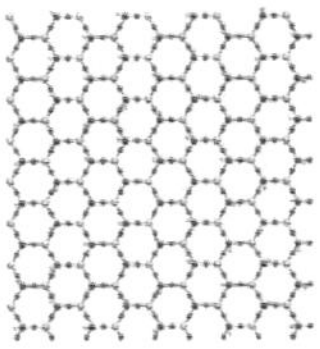

Silica topology

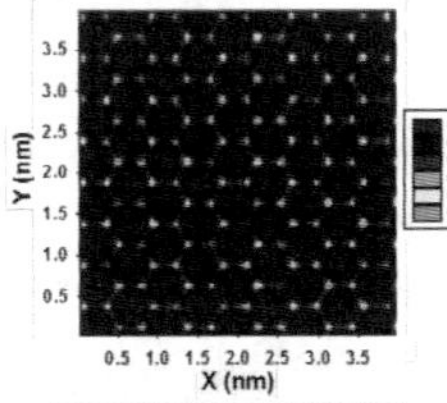

CO_2 planer density profile

Orientation

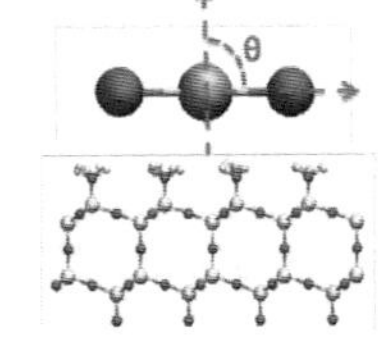

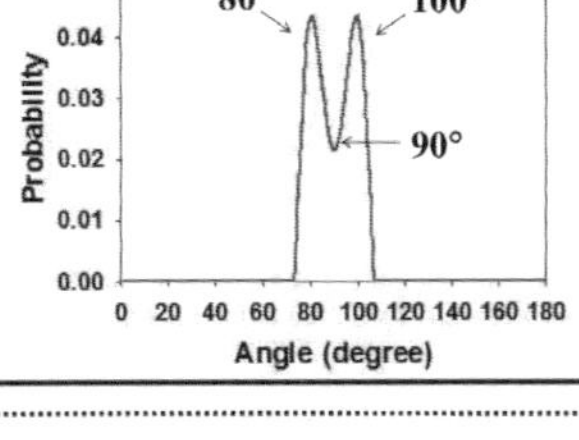

Adsorption Sites

Activation Energy

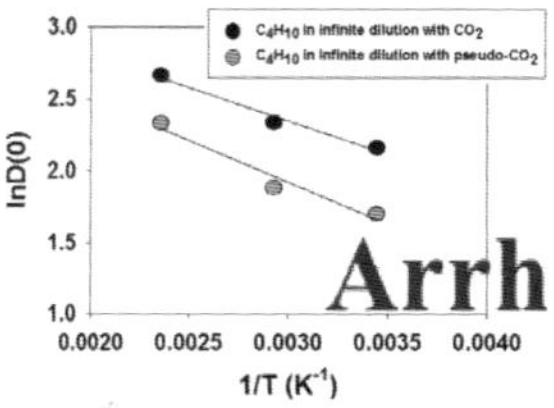

- $E_{a\text{-mix}}$ = 3.79 kJ/mol
- $E_{a\text{-pure}}$ = 4.73 kJ/mol

Mean Square Displacement

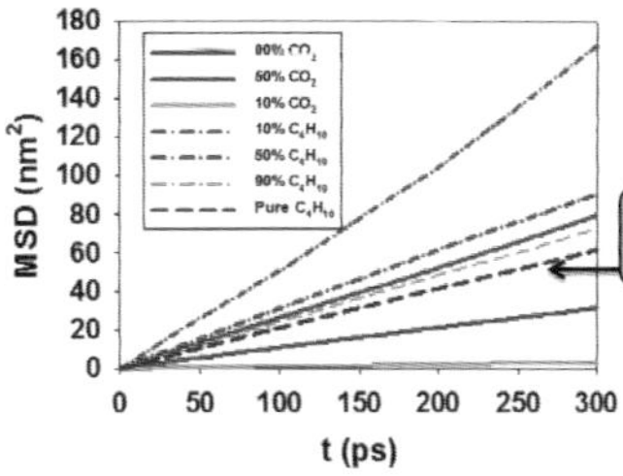

The presence of carbon dioxide induces an increase in diffusivity of n-butane

CO_2-C_4H_{10} system

Calculated Diffusion Coefficient (10^{-9} m²/s)

N-Octane Molecules	Carbon Dioxide Molecules	$D_{s\text{-n-octane}}$	$D_{s\text{-Carbon Dioxide}}$
282	–	5.7 ± 0.2	–
	225	10.1 ± 0.7	1.4 ± 0.2
	448	11.5 ± 0.2	2.7 ± 0.25
	675	11.2 ± 0.8	3.75 ± 0.05
	900	9.9 ± 0.3	5.2 ± 0.6
400	–	4.5 ± 0.8	–

Maximum in n-octane diffusivity

CO_2-C_8H_{18} system

Le, Thu; Ogbe, Stephanie; Cole, D; Striolo, A. Short Communication: N-Octane Diffusivity Enhancement via Carbon Dioxide in Silica Slit-Shaped NanoPores – A Molecular Dynamics Simulation. [In preparation]

Acknowledgements

This work was supported, in part, by the U.S. Department of Energy and by the Sloan Foundation (Deep Carbon Observatory). Generous allocations of computing time were provided by the Oklahoma Supercomputer Center for Education and Research (OSCER) and by the US National Energy Resources Supercomputer Center (NERSC).

376368 Studying Conformational Changes of Mhp1 Using Unbiased All-Atom Molecular Simulation

Abstract:

Studying Conformational Changes of Mhp1 using Unbiased All-

atom Molecular Simulations

Pouyan Khakbaz and Jeffery Klauda

Department of Chemical and Biomolecular Engineering,

University of Maryland College Park, MD 20742

Abstract

Mhp1 is a secondary active transporter from nucleobase/cation symporter-1 (NCS1) family. We have developed the implicit-explicit membrane (IM-EX) hybrid simulation method to study conformational changes in secondary active transporters (1). This novel approach starts with IM simulations and certain conformations from the implicit simulations are placed in an explicit environment to obtain the final conformation. Studying conformational changes of Mhp1 is an important test to our IM-EX method. Three different conformations of Mhp1 are known, i.e., inward-open, outward-occluded and outward-open (2). The implicit simulation was applied by using the SGLD method (3). The main advantage of the SGLD method is to overcome the energy barrier in a reasonable amount of simulation time. Crystal structures of different conformation were used as initial conformations for the SGLD simulations. Regular molecular dynamics and SGLD method was used for explicit membrane simulations to determine the efficiency of SGLD in explicit membrane environment. Separate sets of MD simulations starting with crystal structures were carried out as "control simulations" to test the effectiveness of our hybrid simulation method.

References:

1. Pendse & Klauda. JMB, 2010, 404:506

2. Shimamura & Weyand, Science, 2010, 328, 470-473

3. Wu & Brooks, J. Chem. Phys. , 2011, 135, 204101

380468 Novel, Efficient, and Accurate Methods for Calculating Pressure in Polymer Lattice Simulations

Abstract:

Pressure calculation in polymer lattice simulations is an important but nontrivial subject. The three existing methods—thermodynamic integration, repulsive-wall, and hydrostatic equilibrium methods—all have their limitations and cannot be used to *accurately* calculate the pressure at *all* polymer volume fractions φ. Here we propose two novel methods. In the first method, we combine the Wang-Landau--Optimized Ensemble (WL-OE) sampling with the chain insertion/deletion to accurately calculate the pressure over a range of polymer volume fraction in a single simulation run, which is very efficient at low to intermediate φ and exhibits negligible finite-size effects. In the second method, we introduce a repulsive plane with bridging bonds, which is similar to the repulsive-wall method but eliminates its confinement effects, and estimate the density of states in terms of the number of bridging bonds and the contact number, respectively, from two independent WL-OE simulations, and thus the pressure. This works well at all φ, especially at *high* φ where all the methods involving chain insertion/deletion fail. Finally, combining both our methods further gives the complete thermodynamics (e.g., chain chemical potential, isothermal compressibility, etc.) over the entire range of φ (from 0 to nearly 1) with negligible finite-size effects.

385000 Computation Ready Experimental (CoRE) Metal-Organic Frameworks: A Critical Tool to Enable High-Throughput Screening of Nanoporous Crystals

Abstract:

Computational screening of the over 6,000 experimentally reported MOF structures can streamline efforts to identify materials suitable for applications ranging from gas storage to chemical sensing. To date, the availability of simulation-ready crystal structures is a major impediment to applying high-throughput computations to MOFs. The central result of this work is a nearly comprehensive, open-access database of simulation-ready MOF structures that are derived directly from experimental data. As an example of using our CoRE MOF database, we have correlated the simulated methane storage capacity of over 4,000 porous MOFs with structural characteristics of each material.

PURDUE UNIVERSITY

National Institutes of Health

Accelerated Stochastic Simulation

Presented by: Vu Tran

Dr. Doraiswami Ramkrishna

School of Chemical Engineering

Purdue University, West Lafayette

Motivation

- Weather prediction
- Stock market
- Chemical reactions
- Biological systems

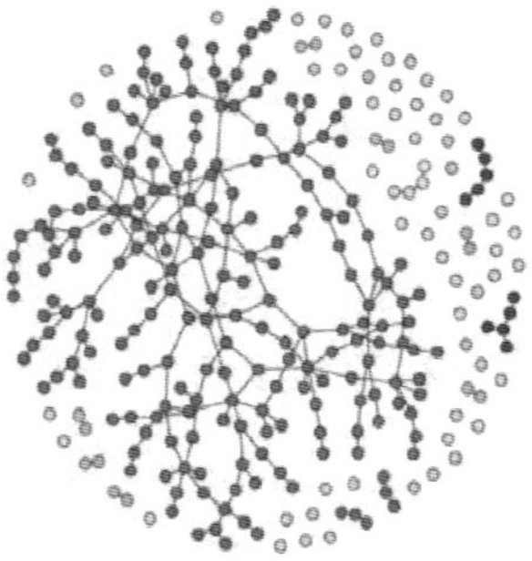

http://preview.tinyurl.com/nucey73

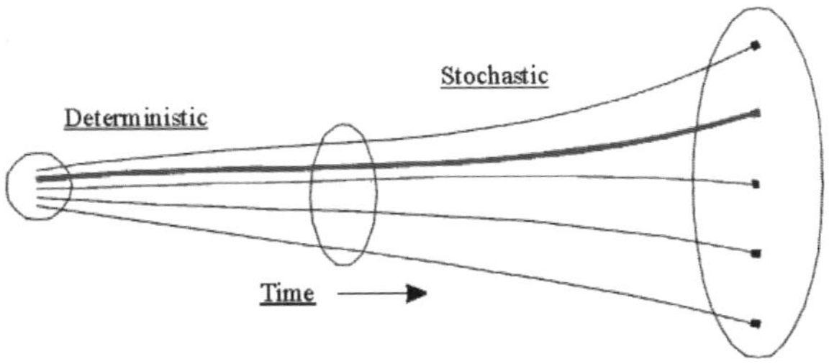

Different modeling approaches

DETERMINISTIC

- Assumes that its outcome is certain, for a fixed set of inputs
- Involves a large number of species/units
- Can be solved numerically or analytically

STOCHASTIC

- The outcome can be described as some distribution
- Involves one or more stochastic elements
- Can be described using RNG with a great number of trials (Monte Carlo)

Deterministic models can be deduced from stochastic models

Description of the problem

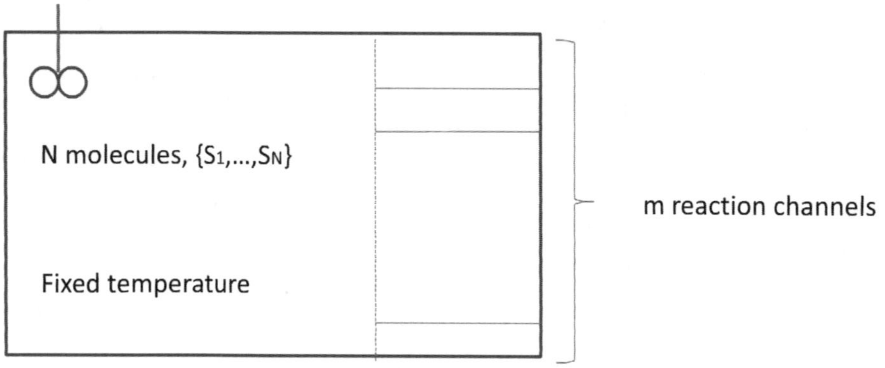

Dynamic states $X(t) \equiv (X_1(t),\ldots,X_N(t))$

Given: Initial State $X(t_0)=x_0$
Goal: Track the evolution of $X(t)$

Master Equation and SSA

$$\frac{\delta}{\delta t}P(x,t\,|\,x_0,t_0)=\sum_{j=1}^{M}[a_j(x-v_j)P(x-v_j,t\,|\,x_0,t_0)-a_j(x)P(x,t\,|\,x_0,t_0)]$$

$$P(\tau,j\,|\,x,t)=a_j(x)\exp(-a_0(x)\tau)(\tau\geq 0,j=1,\ldots,M)$$ Poisson distribution

$$a_0(x)=\sum_{j=1}^{M}a_j(x)$$

- SSA or uses rigorous Monte Carlo techniques to generate random pair (τ,j) then augments t and x by τ and v_j, respectively
- Shah and Dr. Ramkrishna (1977) developed time interval quiescence method
- This process is repeated for a suitably large number of sample paths

1. Gillespie, D. *J. Chem. Phys*, 1977
2. Shah et al., *AIChe*, 1977

Is SSA all we need?

- SSA steps along the history and tracks down every reaction event when occurring
- SSA construction gives us a complete and detailed history of X(t) but is very a time-consuming process

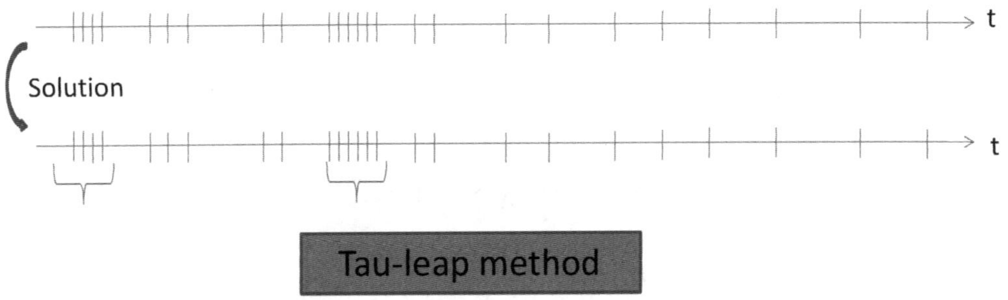

Tau-leap method

- <u>Leap condition:</u>

  τ is small enough that the change during [t,t+τ] will be so little that no propensity function will suffer an appreciable change in its value

- <u>Question</u>: How small is small enough?

"Leap condition" (Gillespie and Petzold, 2006)

$$a_j(x+\Lambda(\tau;x))-a_j(x)\le \varepsilon a_o(x)$$

where

$$a_o(x)=\sum_{j=1}^{N}a_j(x)$$

......or in other words, this change needs to be bounded, within each step size and each propensity takes a small jump

Using Taylor series expansion, we have

$$a_j(x+\Lambda(\tau;x)) - a_j(x) \equiv \Delta a_j(\tau;x) = \sum_{i=1}^{N} \frac{\delta a_j(x)}{\delta x_i} \Lambda(\tau;x) + H.O.T \approx \sum_{j'=1}^{M} f_{jj'} P(a_{j'}(x);\tau)$$

- *The random variable $\Delta a_j(x)$ has expectation and variance:*

$$E\Delta a_j(\tau;x) \equiv \mu_j(x)\tau \qquad V\Delta a_j(\tau;x) \equiv \sigma_j^2(x)\tau$$

- *The interval τ is then chosen as below*

$$\tau = \min_{j=1,\ldots,M} \{\tau_j\} \qquad \tau_j = \min\left(\frac{a_o \varepsilon}{|\mu_j|}, \frac{a_o^2 \varepsilon^2}{\sigma_j^2}\right)$$

Cao et al., D. *J. Chem. Phys*, 2006

Complications

- Uncertainty with which the tau-leap condition is satisfied

$$a_j(x+\Lambda(\tau;x)) - a_j(x) \leq \varepsilon a_o(x)$$

Our proof...

- Triangular inequality:

$$|\Delta a_j(\tau;x)| \le |\Delta a_j(\tau;x) - \mu_j(x)\tau| + |\mu_j(\tau;x)\tau|$$

- Imposed to:

$$|\Delta a_j(\tau;x) - \mu_j(x)\tau| + |\mu_j(\tau;x)\tau| \le \varepsilon a_0(x)$$

- So

$$|\mu_j(\tau;x)\tau| \le \varepsilon a_0(x)$$

- or $\tau \le \dfrac{\varepsilon a_o(x)}{|\mu_j(x)|}$

Chebyshev Inequality

- For any random variable X with finite expectation EX and variance VX, this inequality holds:

$$\Pr\{|X - EX| \ge \varepsilon\} \le \frac{VX}{\varepsilon^2}$$

- It is equivalent to

$$\Pr\{|X - EX| < \varepsilon\} > 1 - \frac{VX}{\varepsilon^2}$$

Our proof(cont.)

- Using Chebyshev inequality $\Pr\{|X - EX| < \varepsilon\} > 1 - \frac{VX}{\varepsilon^2}$
- And replacing ε by ($a_o(x)\ \varepsilon - |\mu_j(x)|\tau$), we have

$$\Pr\{|\Delta a_j(\tau;x) - \mu_j(x)\tau| < a_o(x)\varepsilon - |\mu_j(x)|\tau > 1 - \frac{\sigma_j^2(x)\tau}{[a_o(x)\varepsilon - |\mu_j(x)|\tau]^2}$$

- Since $$1 - \frac{\sigma_j^2(x)\tau}{[a_o(x)\varepsilon - |\mu_j(x)|\tau]^2} > 0$$
- We arrive at $$\tau \le \frac{a_o^2 \varepsilon^2}{\sigma_j^2}$$

Proof (cont.)

- Put everything together, we have:

$$\tau = \min_{j=1,\ldots,M} \{\tau_j\}$$

$$\tau < \tau_j = \min\left(\frac{a_o \varepsilon}{|\mu_j|}, \frac{a_o^2 \varepsilon^2}{\sigma_j^2}\right)$$

- We successfully prove the two candidates for the tau value from Gillespie's and Penzold's proposal

Introduction of δ

- Let make the right hand side of the equation below close to 1...

$$\delta = 1 - \frac{\sigma_j^2(x)\tau}{[a_o(x)\varepsilon - |\mu_j(x)|\tau]^2} \rightarrow 1$$

- We arrived an quadratic equation of τ

$$\tau^2 - 2[\frac{a_o\varepsilon}{|\mu_j|} + \frac{1}{2(1-\delta)}\{\frac{\sigma_j}{|\mu_j|}\}^2]\tau + [\frac{a_o\varepsilon}{|\mu_{j}|}]^2 = 0$$

Our choice of picking tau-leap

- Using triangular inequality and Chebyshev inequality, we proposed:

$$\tau = \min_{i=1,\dots,N}\{\frac{a_i}{|\mu_i|} + \frac{1}{2(1-\delta)}[\frac{\sigma_i}{|\mu_i|}]^2 - \sqrt{(\frac{a_i}{|\mu_i|} + \frac{1}{2(1-\delta)}[\frac{\sigma_i}{|\mu_i|}]^2)^2 - [\frac{a_i}{|\mu_i|}]^2}\}$$

$$a_i \equiv \max(\varepsilon x_i / g_i, 1), \mu_i \equiv \sum v_{ij} a_j(x), \sigma_i^2(x) \equiv \sum v_{ij}^2 a_j(x)$$

g_i: highest order of rxn with respect to species i

$$\tau < \tau_j = \min(\frac{a_o\varepsilon}{|\mu_j|}, \frac{a_o^2\varepsilon^2}{\sigma_j^2}) \qquad \text{For all j}$$

Our choice of picking tau-leap (cont.)

$$\tau = \min_{i=1,\ldots,N}\left\{\frac{a_i}{|\mu_i|} + \frac{1}{2(1-\delta)}\left[\frac{\sigma_i}{|\mu_i|}\right]^2 - \sqrt{\left(\frac{a_i}{|\mu_i|} + \frac{1}{2(1-\delta)}\left[\frac{\sigma_i}{|\mu_i|}\right]^2\right)^2 - \left[\frac{a_i}{|\mu_i|}\right]^2}\right\}$$

δ: probability at which the leap condition is satisfied

ε	δ	τ	Accuracy	Simulation Time
Fixed	Close to 1	Smaller	Increase	Longer
	Close to 0	Larger	Decrease	Shorter

Interpretation of Older Methods

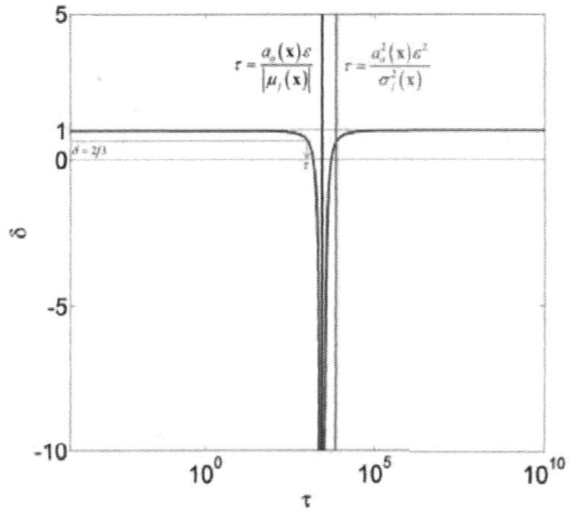

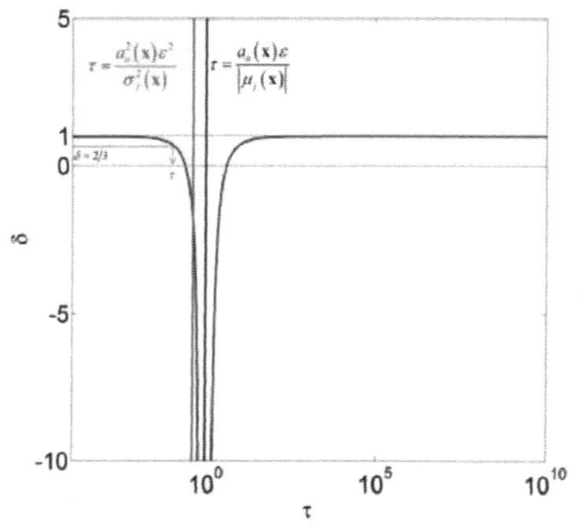

Example

$$X_1 \xrightarrow{c_1} X_2 \xrightarrow{c_2} X_3$$

$c_1=1$, $c_2=1$
$X_1(0)=10^4$, $X_2(0)=1$, $X_3(0)=0$
$t=0$, $t_f=0.1$

Binomial distribution

$a_1=c_1*X_1$, $a_2=c_2*X_2$
$X_3(0.1)=?$

Binomial methods

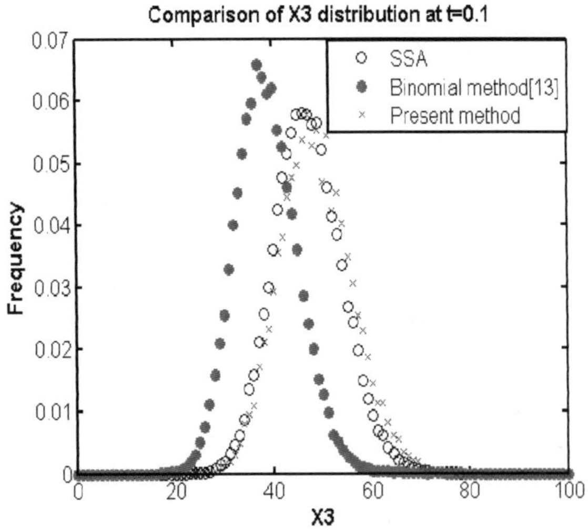

Peng at al., D. *J. Chem. Phys*, 2007

Result comparison

	Binomial Method	Present Method
Epsilon	0.02	0.02
Delta	----	0.667
Histogram Error	.0090	.0018
Sim. Time(s)	5.417e3	3.269e3 (40% faster)

Example 2 (Schlogi's model)

$$B_1 + 2X \underset{c_2}{\overset{c_1}{\rightleftarrows}} 3X$$

$$B_2 \underset{c_4}{\overset{c_3}{\rightleftarrows}} X$$

N_1=1.0*10^5, N_2=2*10^5

C_1=3*10-7, c_2=10-4, c_3=10-3, c_4=3.5

X(0)=250, X(4)=?

Cao et al.

Poisson distribution

Mid-point Gillespie

Our method

Comparison of different trajectories

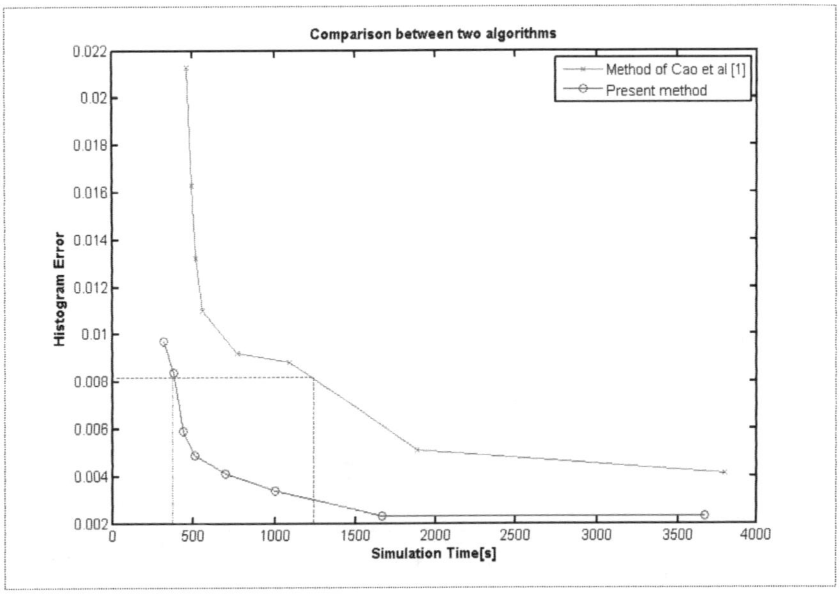

Comparison of different numbers of trajectories

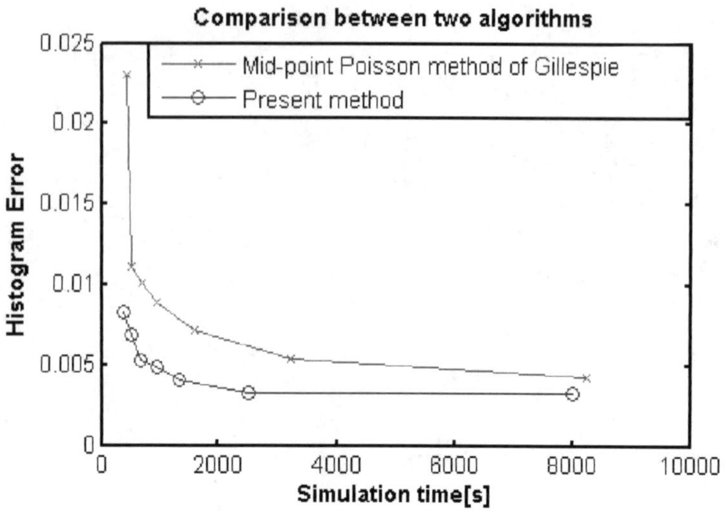

Comparison with the same epsilon

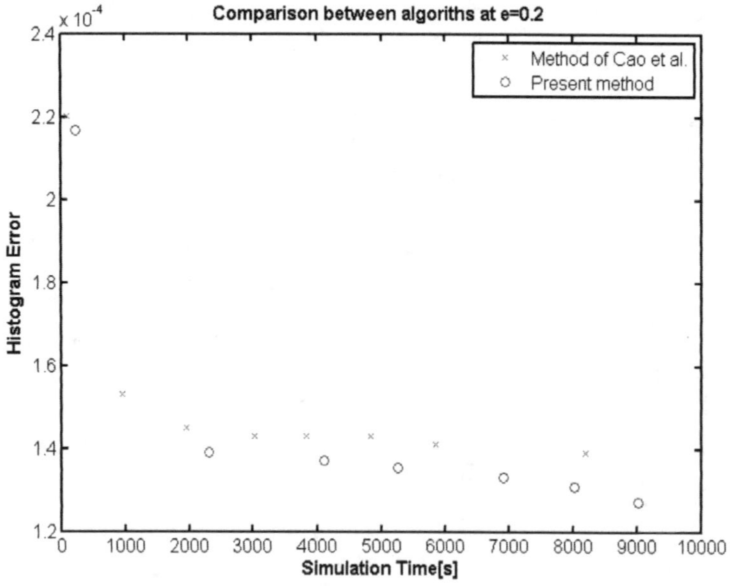

Conclusions

- δ gives another parameter to control the accuracy
- The new algorithm takes use of triangular inequality and Chebyshev to choose ε in a more systematic way
- The new method improves both speed and accuracy for stochastic simulation

Acknowledgement

- National Institute of Health (NIH)
- Purdue University
- Advisor: Dr. Doraiswami Ramkrishna
- Cooperation groups from University of Minnesota
- Group members
- Friends and family

386496 Development of a Dynamic Cutoff Method and Interface Detection Algorithm for Long-Range Dispersion Interactions

Abstract:

We discuss the development and implementation of a dynamic cutoff method (DCM) for molecular simulations which exhibits a linear-time complexity as well as good scalability properties. The DCM, which is intended for simulations of large interfacial systems to be performed on modern supercomputers, adjusts the cutoff for dispersion potentials for atoms as a function how far removed the atoms are from an interface. This design choice allows particles in a bulk phase to benefit from the reduced computation with limited loss of accuracy associated with traditional dispersion cutoffs, while incorporating more interactions for particles near an interface without the need for communications-intensive Ewald-like methods.

Moreover, the dynamic cutoff method requires rapid computation of the location of the interface throughout the course of a simulation. We have therefore also developed a real-time interface detection method which is perfectly parallelizable and scales linearly with the number of particles. This method is based on the work of Berkels et al., who developed an efficient image segmentation algorithm.

We have implemented the dynamic cutoff method as part of the open-source LAMMPS molecular dynamics package. Our results for a liquid/vapor system of Lennard-Jones particles, shows that the accuracy of the DCM is comparable to that of PPPM. A performance comparison indicates that the DC method is preferable for large systems because of the poor scaling of FFTs within the PPPM algorithm. We also demonstrate the good weak- and strong-scaling behavior of this method on the BlueGene/Q supercomputer at the Forschungszentrum Jülich.

389517 Extending a Linear Basis Function Approach to Alchemically Solvating Charged Particles and Changing Atomic Identity

Abstract:

Our linear basis function approach is an efficient tool for inserting or removing atomic sites in free energy calculations. We have previously shown how splitting the potential energy function into a sum of pairs of basis functions, which depend only on coordinates, and alchemical switches, which depend only on the coupling variable can minimize the variance and statistical error of alchemical solvation. Here we present the optimal alchemical switch and basis functions for solvating charged particles in dense fluid. The most optimal sequence of nonbonded forces to alchemically couple to the environment is also examined to further reduce statistical error.

Changes in chemical identity at a fixed atomic site are also examined with the linear basis function approach. The basis function approach allows predicting the free energy surface across a wide range of nonbonded properties requiring very minimal sampling across the entire surface. The basis function approach removes the need to re-run the inner force loop of the simulation package at every point of the free energy surface and can be run entirely in post-processing. This is done only with Lennard-Jones spheres as changing identity often requires geometry changes in bonded molecules, which we do not explore. This formalism can accelerate future nonbonded parameterizations.

384405 The Need for New Tools to Explore Hypothetical Molecular Systems

Abstract:

Dramatic increases in computer power combined with steady improvements in simulation algorithms have significantly reduced a key bottleneck in modeling large, complex molecular systems: hardware costs. However, the waning of the hardware cost bottleneck has revealed other bottlenecks that were not previously apparent. For small systems humans have been able to manually construct trial molecular configurations (i.e, choosing initial coordinates for a handful of atoms in a small molecule) and then relax their structures (or observe their time evolution) using various simulation methods. However, for systems involving thousands of atoms creating trial configurations is much more challenging. Whereas in the case of modeling *naturally* occurring or *already experimentally synthesized* systems one can obtain reasonable trial coordinates from x-ray crystallography (or analogous measurement techniques), this is not possible for large *hypothetical* systems. Since it is, in principle, a fundamental strength of molecular simulations to be able to explore hypotheses in advance of comparatively much more expensive experiments, the inability to efficiently generate trial coordinates for large hypothetical systems is a painful constraint.

Certain tools, such as Packmol [1], which generate trial coordinates for tightly packed molecules in user-defined volumes partially address this need, but in this presentation I will discuss other challenges that have yet to be addressed. Analogies to tools developed in mechanical engineering, architecture, and computer graphics will be discussed, and I will explore the idea that there is significant room for innovation in designing new tools for exploring hypothetical molecular systems.

[1] L. Martínez, R. Andrade, E. G. Birgin, J. M. Martínez. Packmol: A package for building initial configurations for molecular dynamics simulations. Journal of Computational Chemistry, 30(13):2157-2164, 2009.

386644 HOOMD-Blue - Development of a Highly Scalable General Purpose Molecular Dynamics Code

Abstract:

I will present on the development of the latest release of HOOMD-blue, an open-source, high performance particle simulation framework. Its capabilities are motivated in part by large-scale computational research on block copolymers, and have resulted in a highly scalable molecular dynamics code. I will discuss strong- and weak scaling performance on over 3000 GPUs of the Titan supercomputer, as well as scalability using latest GPUDirect RDMA technology on an academic cluster.

390467 Methods for Free Energy Calculations and Other Enhancements in the Gromacs Molecular Simulation Package

Abstract:

I discuss a number of molecular simulation techniques recently included in the GROMACS molecular dynamics package for calculating free energy differences between two user-defined end states. These include Hamiltonian replica exchange with all-to-all state swaps and the capability to perform expanded ensemble simulations.

We also discuss the validation efforts used to check the implementation of these free energy calculation methods in GROMACS, and show how these methods can be used to very finely test the sensitivity of calculated free energies to different methods for performing long range force cutoffs and or other types of simulation approximations.

I also address a number of other recent additions to GROMACS, such as GPU computing capabilities and PME-based long-range dispersion calculations.

387846 Catmap: A Software Package for Descriptor-Based Micro-Kinetic Mapping of Catalytic Trends

Abstract:

Descriptor based analysis is a powerful tool for understanding the trends across various catalysts. In general, the rate of a reaction over a given catalyst is a function of many parameters - reaction energies, activation barriers, thermodynamic conditions, etc. The high dimensionality of this problem makes it very difficult and expensive to solve completely and even a full solution would not give much insight into the rational design of new catalysts. The descriptor based approach seeks to determine a few "descriptors" upon which the other parameters are dependent. By doing this it is possible to reduce the dimensionality of the problem - preferably to 1 or 2 descriptors - thus greatly reducing computational efforts and simultaneously increasing the understanding of trends in catalysis.

The "CatMAP" Python module seeks to standardize and automate many of the mathematical routines necessary to move from "descriptor space" to reaction rates. The module is designed to be both flexible and powerful. A "reaction model" can be fully defined by a configuration file, thus no new programming is necessary to change the complexity or assumptions of a model. Furthermore, various steps in the process of moving from descriptors to reaction rates have been abstracted into separate Python classes, making it easy to change the methods used or add new functionality. This talk will explain the general structure of the code, provide an introduction to the implementation, and show examples of catalytic rate and selectivity maps generated with the code.

382881 Accelerating the Discrete Element Method for Faceted Particles Using HOOMD-Blue

Abstract:

Although studies of static, equilibrium properties of nanoscale anisotropic particles have been highly successful through the use of Metropolis Monte Carlo algorithms, efficient methods of studying kinetic processes such as glassy dynamics and nucleation and growth for anisotropic particles have not been as plentiful. In this talk, we discuss a GPU-accelerated adaptation of the discrete element method to incorporate polyhedral particles into HOOMD-Blue. This formulation allows us to efficiently simulate conservative and nonconservative dynamics of faceted shapes within a classical molecular dynamics framework.

390707 Automated Global and Local Optimization Methods for Atomistic Force Field Development

Abstract:

Generally molecular dynamics simulations require the adaptation of different atom or molecular models to desired physcial or chemical properties. In this procedure the appropriate choice of force fields and of corresponding parameters is a crucial issue.

In recent years, we developed modular program packages, which can be simply tuned to the different occurring optimization problems. Once started, the programs run mostly without user interaction until the force field ist generated. In this way, force field parameters can be determined for complicated atomistic models of any chemical matter. In this way, we address various inter-linked resolutions of molecular modeling to solve one of the primary goals of our research: To develop accurate and reliable molecular parameters and models in a reasonable time and as error-free as possible. For intramolecular interactions, we have created a scientific "Workflow for force-field optimization package" (**Wolf$_2$Pack**) that incorporates our approach for transfering knowledge gained from QM to Newtonian- based models. We define a scientific workflow as a series of independent steps that are linked together according to the data flow and the dependencies between them. For intermolecular interactions we developed a systematic optimization workflow, based on efficient gradient-based numerical algorithms called **GROW**. GROW is a modular tool kit of programs and scripts. It is a generic implementation and can be easily extended by other developers. Very recently, GROW was extended with a global optimization module called **COSMoS** („Calibration/Optimization by Simultaneous Modeling of Simulated data").

Both programs facilitate: a) the development and optimization of molecular parameters for a given simulation engine, b) the transfer of parameters from one software package to another, and c) testing of the parameters using a standard test suite and protocol via a semi-automated iterative parameterization process.
In the presentation, we will show how the programs are developed and set-up along the scientific aspects. Consequently, we will show some example molecules. The physical properties to be fitted are e.g. saturated liquid density, enthalpy of vaporization, and vapor pressure, simultaneously at different temperatures. The resulting force fields will be assessed with respect to their applicability to the prediction of other physical properties over a range of temperatures.

367838 Scalable Monte Carlo Simulations of Millions of Particles on Thousands of GPUs

Abstract:

We develop a scalable parallel algorithm for Metropolis Monte Carlo simulations of particles with short-range interactions. We use a checkerboard decomposition to allow massively parallel trial moves on GPUs. A second level of domain decomposition enables scaling to systems of millions of particles on thousands of GPUs. A CPU code path efficiently scales to thousands of CPU cores, with as few as one hundred particles per rank. We implement this algorithm for hard anisotropic particles as a plugin to our open-source particle simulation toolkit HOOMD-blue (http://codeblue.umich.edu/hoomd-blue). Our implementation is production-ready with all the features needed for general use in a large research group. It supports a variety of shapes (spheres, ellipsoids, convex (sphero)polygons, simple polygons, convex (sphero)polyhedra, and general polyhedra), NVT and NPT ensembles, pressure measurement, and free energy calculations along with all the file I/O and scripting capabilities already present in HOOMD-blue.

370953 Polymatic: A General Simulated Polymerization Algorithm

Abstract:

Generating representative structures of amorphous polymers for simulations is a non-trivial task due to their disordered nature. This problem is exacerbated for (a) glassy polymers with bulky structures and slow dynamics, and (b) network polymers with complex connectivity. To this end, we have released an open-source code called Polymatic [1,2] for structure generation of amorphous polymers with a wide range of structures and connectivity. The code implements a general simulated polymerization algorithm, where bonds are formed between repeat units during molecular dynamics simulations according to a set of bonding criteria to construct well-relaxed polymeric structures. It requires only the description of the monomer repeat unit and definition of the bonding criteria, and works in conjunction with the LAMMPS package to perform energy minimization and molecular dynamics simulations.

Here, we will present the basic algorithm of Polymatic and its application to a variety of different examples, including glassy linear polymers, hypercrosslinked polymers, and thermally post-crosslinked copolymers. Definition of the structural connectivity between pairs of "reactive" atoms allows for the construction of copolymers and complex ladder backbones. In addition, networked structures can be built in a single stage, where bonds and crosslinks are formed simultaneously, or in two stages, where linear polymers are first made and then post-crosslinked, which allows different synthetic routes for network polymers to be mimicked. As such, use of the Polymatic algorithm provides not only a realistic model of amorphous polymers, but also a unique view into the evolution of the structure throughout the virtual synthesis.

[1] Abbott, L. J.; Hart, K. E.; Colina, C. M. *Theor. Chem. Acc.* **2013**, 132, 1334.

[2] Abbott, L. J. Polymatic: A Simulated Polymerization Algorithm, 2013, https://nanohub.org/resources/17278.

CASSANDRA: A Novel, Open-Source Monte Carlo Molecular Modeling Platform for Materials Modeling

With the identification of computing as the third pillar of science, the use of *in silico* prediction of thermophysical, transport and phase-equilibria properties from molecular simulations is becoming increasingly important for novel materials discovery. A number of open-source molecular dynamics packages such as LAMMPS, NAMD, Gromacs and DL-POLY are available for the computation of thermophysical and transport properties of yet-to-be synthesized materials. Though Monte Carlo (MC) simulations are ideally suited for the calculation of phase-equilibria properties, only a small number of open-source simulation codes such as *ms2* and Towhee are available. This is due to the fact that MC moves are dependent on the types of molecules under consideration resulting in most groups developing specialized codes for a particular problem. Additional challenges arise as there are only a few Monte Carlo algorithms available that properly and efficiently sample systems with multiple coupled intramolecular degrees of freedom.

In an effort to fill this void, we have developed a general purpose Monte Carlo molecular modeling framework, CASSANDRA in Fortran90. In this presentation, we will discuss algorithms we have developed which enable molecules of arbitrary complexity to be simulated using biased MC procedures. We will also describe the main features of the code such as modular approach to design, easy of simulation set-up, computational efficiency and ease of extensibility, and ability to perform molecular simulations using the state-of-the-art multicore computer architecture. In addition, we will provide details of CASSANDRA dissemination efforts and roadmap.

We will demonstrate the capability of CASSANDRA for simulation of the phase behavior of polar and non-polar species such as water, alcohols and alkanes.

Development of a GPU Optimized Gibbs Ensemble Monte Carlo Simulation Engine

In this work, we discuss the development of a high performance, open source, Gibbs Ensemble Monte Carlo simulation engine that is capable of simulating complex molecular systems containing over 100,000 atoms at speeds up to two orders of magnitude faster than existing serial codes. GPU Optimized Monte Carlo (GOMC) supports simulations in the canonical, isobaric-isothermal, grand canonical and Gibbs ensembles for molecules of arbitrary geometry (linear, branched, rings). A variety of advanced configurational-bias Monte Carlo techniques are supported[1, 2]. GOMC utilizes CHARMM style input files (topology, parameter, PSF, PDB), allowing the use of a single set of input files for CHARMM, NAMD[3] or GOMC. Output from GOMC is compatible with VMD[4]. Parallelization of the pairwise energy calculation is performed on the GPU using energetic decomposition. Optimization strategies unique to simulations with changing numbers of interaction sites, e.g. Gibbs Ensemble and grand canonical Monte Carlo, and the implementation of configurational-bias methods on the GPU are discussed. Other optimization strategies, such as cell lists, and GPU specific features, such as warp shuffle, are presented.

1. Martin, M.G. and J.I. Siepmann, *Novel configurational-bias Monte Carlo method for branched molecules. Transferable potentials for phase equilibria. 2. United-atom description of branched alkanes.* Journal of Physical Chemistry B, 1999. **103**(21): p. 4508-4517.
2. Wick, C.D. and J.I. Siepmann, *Self-adapting fixed-end-point configurational-bias Monte Carlo method for the regrowth of interior segments of chain molecules with strong intramolecular interactions.* Macromolecules, 2000. **33**(19): p. 7207-7218.
3. Phillips, J.C., et al., *Scalable molecular dynamics with NAMD.* Journal of Computational Chemistry, 2005. **26**(16): p. 1781-1802.
4. Humphrey, W., A. Dalke, and K. Schulten, *VMD: Visual molecular dynamics.* Journal of Molecular Graphics & Modelling, 1996. **14**(1): p. 33-38.

388079 A Parallel Fortran Code to Perform Density Derived Electrostatic and Chemical (DDEC) Analysis

Abstract:

One of the primary challenges in the study of complex materials is to extract meaningful chemical descriptors that can be correlated to experimental properties. Net atomic charges (NACs), atomic spin moments (ASMs), and effective bond orders (EBOs) are examples of chemical descriptors that are commonly used to explain electrostatic, magnetic, and chemical properties of materials. NACs, ASMs, and EBOs computed with the Density Derived Electrostatic and Chemical (DDEC) method have good correlations to a wide range of spectroscopic experiments. In this talk, we explain the development of a parallel Fortran program to compute DDEC NACs, ASMs, and EBOs. This methodology can be applied to materials with 0, 1, 2, or 3 periodic boundary conditions with no magnetism, collinear magnetism, or non-collinear magnetism. The computational cost scales linearly with increasing number of atoms per unit cell. Shared memory parallelization achieved using OpenMP directives. When run on a single processor, the new Fortran program runs approximately 8 times faster than the previous version (which was coded in Matlab). When run on multiple processors, the new Fortran program has a parallelization efficiency of approximately 70-85% on 16 processors and approximately 60-70% on 32 processors. As applications, we studied a B-DNA decamer with 733 atoms in the unit cell, the Mn_{12} acetate single molecule magnet, and diisopropylammonium bromide (DIPAB) molecular crystals. We compute the electric polarization of the DIPAB ferroelectric phase and compare it to the recent experimental and computational results of Fu et al. (Science, Jan. 2013, Vol. 339, pp. 425-428).

364723 Molecular Dynamics Investigation of the Products of Alkoxysilane Condensation: Bulk Gels and Surface Coatings

Abstract:

We characterize silica gels and organo-silicon surface coatings using reactive molecular dynamics simulations. To model the chemical reactions, we use a reactive force field (ReaxFF) which we have optimized in a parallelized semi-automatic way to model hydrolysis and condensation reactions. The morphologies of silica gels obtained from tetra- and tri-alkoxysilanes are determined by allowing the system to condense while simultaneously removing water and replacing it with precursor solution. It is found that the gels obtained from trialkoxysilanes are more loosely bonded, and that the chemistry of the headgroup is important to the gel morphology. We furthermore simulated the chemisorption of alkoxysilanes with organic headgroups to hydroxylated silica surfaces. We observe a competition between alkoxysilanes condensing with themselves or with the silica surface.

379686 Hybrid Molecular-Continuum Simulations Using Smoothed Dissipative Particle Dynamics

Abstract:

A broad range of mesoscale hydrodynamic problems are characterized by multiple time and length scales, motivating "hybrid" simulation techniques that involve multiple spatial simulation domains of differing resolution. Such multiscale simulation approaches typically decouple length scales by employing a detailed description (e.g. molecular dynamics) for domains with processes occurring at the molecular level of detail, while coarse-graining bulk regions far away. For many problems, the coarse regions may be treated using the continuum approximation, with some intermediate overlap or buffer region between the atomistic and continuum domains that ensures quantities such as mass and momentum are appropriately conserved and transferred between the two domains.

We discuss a novel hybrid simulation approach based on a continuum technique called "smoothed dissipative particle dynamics" (SDPD), which is a stochastic particle method for solving the fluctuating Navier-Stokes equations. SDPD is ideally suited for coupling to molecular dynamics because it is particle-based and offers a characteristic length scale that can be tuned between molecular and continuum regimes. Our approach makes it possible to embed an atomically-resolved molecular dynamics (MD) domain within an SDPD fluid, or within a hierarchy of SDPD fluids, each at a different resolution. Because SDPD is a particle-based continuum solver, it is easily reconciled with MD since it is only necessary to mediate particle-particle interactions. The coupling is achieved by introducing a "buffer" region separating the atomistic and continuum domains, where forces are continuously interpolated using a switching function. This mixture of forces results in density deviations within the buffer which can be removed by introducing a so-called "thermodynamic force", which we reformulate into a pairwise interaction using the SDPD approach. We demonstrate that these techniques are successful in reproducing both equilibrium and non-equilibrium scenarios for simple fluids, such as the start up of shear flow. The results from these tests are in agreement with analytical theory and independent of the placement of the buffer region within the domain. We also discuss the generalization of this multiscale methodology to multicomponent systems and more complicated domain geometries and modeling setups.

383158 Enhanced Sampling Methods Provide Insights into Macromolecular Interactions As Drug Targets

Abstract:

Biophysical theory, modeling, and simulation techniques rooted in statistical mechanics are playing an increasingly important role in elucidating molecular signatures of cellular processes. Given the large number of degrees-of-freedom in biomolecules, conventional simulation methods remain limited in providing information on conformational statistics, particularly metastable intermediates, at longer time-scales. A challenging biophysical problem is to understand the mechanisms by which small molecule drugs can *surprisingly* gain access to deeply buried protein cavities thereby deactivating a specific target protein or its interactions with other biomolecules. Interestingly, rare biomolecular events such as slow conformational changes in proteins can potentially play a major in role in facilitating drug access. In this work, we study using atomistic simulations the mechanism of action of a potent inhibitor of the regulators of G-protein signaling (RGS) proteins. Particularly, we report how the dynamical motions of a specific RGS protein (named RGS4) appear to play a critical role in exposing a buried cysteine residue that is required for drug action. The dynamic nature of RGS4 involving large-scale conformational changes in a pair of alpha-helices only became apparent upon use of all-atom explicit-solvent temperature-accelerated molecular dynamics (TAMD) simulations to carry out enhanced conformational sampling of apo-RGS4 structures. Modification of exposed cysteine residue by the drug molecule then prevents restoration of the native-like RGS-fold as well as RGS4/G-alpha interface that is required for functional activity. We also provide NMR chemical-shift data that corroborates predictions from our enhanced sampling simulations. These results have potential implications for understanding how the conformational dynamics among RGS proteins, revealed via novel simulation approaches, may play a key role in inhibitor sensitivity.

385886 Molecular Simulation of Homogeneous Nucleation of Ionic Liquids from the Melt

Abstract:

Crystallization of ionic liquids (ILs) can be used in the synthesis of optically-active (fluorescent) and magnetic nanomaterials based on ILs. However, very little is understood about the mechanism of nucleation of crystal phases, especially for ILs as most of their applications involve liquid phases. In this work we studied the homogeneous nucleation of crystals of the ILs 1,3-dimethylimidazolium chloride, [dmim$^+$][Cl$^-$], and 1-butyl-3-methylimidazolium chloride, [bmim+][Cl-], from their supercooled liquid phase. We used the string method in collective variables [1] and the Voronoi tessellated milestoning method [2] to study the nucleation process of these ILs. Recently developed order parameters [3]for molecular crystals were used in conjunction with those methods, in order to detect local order in nm-sized regions and to distinguish between different crystal polymorphs and liquid/amorphous phases. Minimum free energy paths for the nucleation process, free energy barriers, mechanisms of nucleation, sizes of the critical nucleus and rates of nucleation for these ILs will be presented and discussed.

---

[1] Maragliano, L.; Fischer, A.; Vanden-Eijnden, E.; Ciccotti, E. *J. Chem. Phys.* **2006**, *125*, 024106

[2] Vanden-Eijnden, E.; Venturoli, M. *J. Chem. Phys.* **2009**, *130*, 194101; Maragliano, L.; Vanden-Eijnden, E.; Roux, B. *J. Chem. Theory. Comput.* **2009**, *5*, 2589-2594

[3] Santiso, E. E.; Trout, B. L. *J. Chem. Phys.* **2011**, *134*, 064109

University at Buffalo
The State University of New York
Chemical and Biological Engineering Department

Free Energy of Rigid-Molecule Crystals:
Application to clathrate hydrates

Sabry Moustafa, *Lin Weisong, Andrew J. Schultz, and David A. Kofke*

Motivation

University at Buffalo

- Accurate estimation of the free energy is needed to locate the phase boundaries between crystalline phases.
- It can also be used to suggest new crystal structures for experimentalists.
- Harmonic approximation is exact at $T \to 0$; thus, it can work as a starting point to compute the full free energy (*harmonic* + *anharmonic*) at finite temperatures using, let's say, thermodynamic integration.
- In this treatment, the quantum free-energy is computed at no extra cost.
- We apply this approximation for clathrate hydrate systems.

Introduction

- Clathrate hydrates are water-based crystalline solids in which guest molecules are trapped inside polyhedron cages.
- Guest molecules (CH_4, O_2, N_2, H_2, ... etc.) can be gas or liquid.
- *Empty* hydrates are thermodynamically unstable relative to ice.
- They are of growing importance for energy transportation and storage applications.

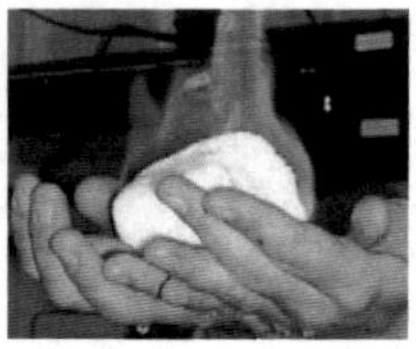

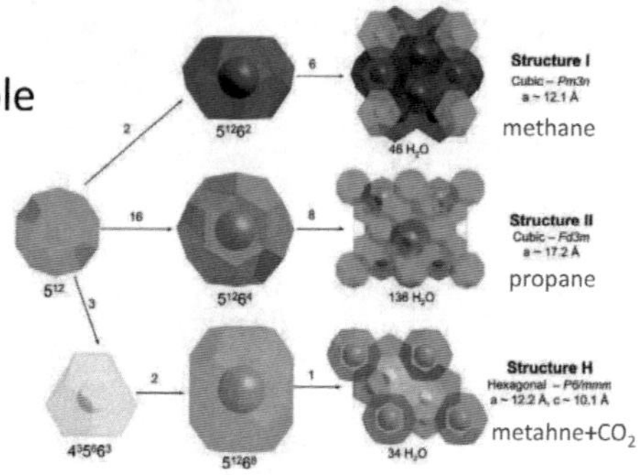

Strobel, T. A. and *et al.*, Chemical Physics Letters, vol. **478**, pp. 97-109, 2009.

Introduction

Challenges:

- Water molecules can be treated as rigid (TIP4P); this adds an extra complexity computing the harmonic free energy.
- Finite-size effects can be important.
- The proton-disorder nature of hydrates must be accounted for.

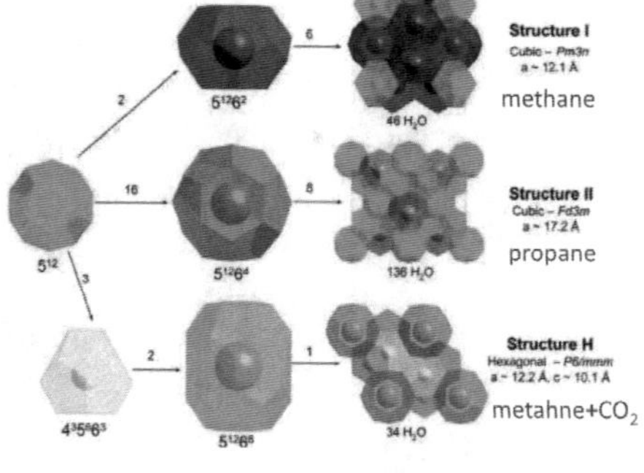

Strobel, T. A. and *et al.*, Chemical Physics Letters, vol. **478**, pp. 97-109, 2009.

Introduction

University at Buffalo

○ Proton disorder

- The hydrate crystals have residual entropy due to their proton-disordered structures.
- Two ice rules must be satisfied:
    1. One H atoms between each O-O link,
    2. Two H atoms associated with each O.

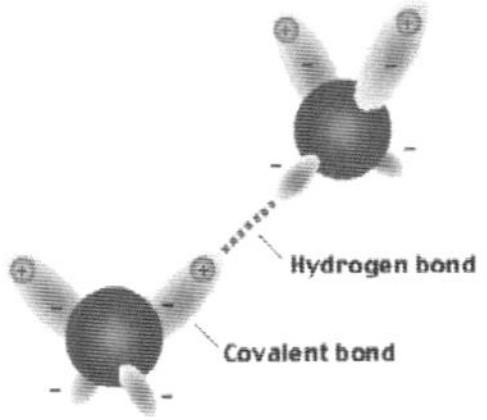

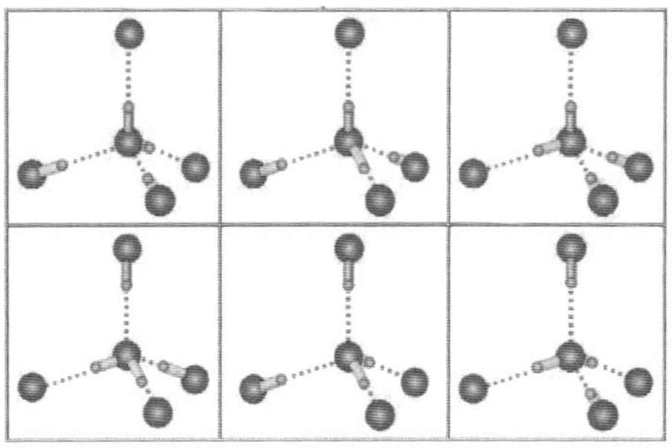

http://www.physicsofmatter.com/

Harmonic approximation of rigid molecules

University at Buffalo

○ For *rigid-molecule* model, the potential energy, U, can be Taylor expanded in terms of molecular displacements around the nominal ("0") positions/orientations:

$$U = U_0 + \sum_{k\alpha}\left(\frac{\partial U}{\partial u_\alpha^t(k)}\right)_0 u_\alpha^t(k) + \sum_{k\alpha}\left(\frac{\partial U}{\partial u_\alpha^r(k)}\right)_0 u_\alpha^r(k)$$

translation-translation; translation-rotation

$$+\sum_{kk',\alpha\beta}\left(\frac{\partial^2 U}{\partial u_\alpha^t(k)\partial u_\beta^t(k')}\right)_0 u_\alpha^t(k)u_\beta^t(k') + \sum_{kk',\alpha\beta}\left(\frac{\partial^2 U}{\partial u_\alpha^t(k)\partial u_\beta^r(k')}\right)_0 u_\alpha^t(k)u_\beta^r(k')$$

$$+\sum_{kk',\alpha\beta}\left(\frac{\partial^2 U}{\partial u_\alpha^r(k)\partial u_\beta^t(k')}\right)_0 u_\alpha^r(k)u_\beta^t(k') + \sum_{kk',\alpha\beta}\left(\frac{\partial^2 U}{\partial u_\alpha^r(k)\partial u_\beta^r(k')}\right)_0 u_\alpha^r(k)u_\beta^r(k') + \text{higher order terms}$$

rotation-translation; rotation-rotation

- k is the molecule number.
- u_α^t is the αth component of the translational displacement.
- u_α^r represents the extrinsic (fixed frame) rotation angle around α axis.

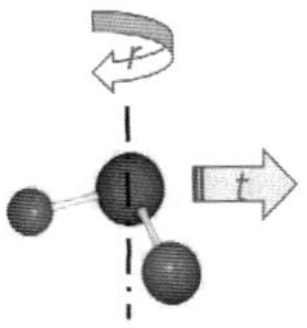

Harmonic approximation of rigid molecules

University at Buffalo

- For *rigid-molecule* model, the potential energy, U, can be Taylor expanded in terms of molecular displacements around the nominal ("0") positions/orientations:

$$U = U_0 + \sum_{k\alpha} \left(\frac{\partial U}{\partial u_\alpha^t(k)}\right)_0 u_\alpha^t(k) + \sum_{k\alpha} \left(\frac{\partial U}{\partial u_\alpha^r(k)}\right)_0 u_\alpha^r(k)$$

All first derivatives (*i.e.* forces and torques) vanish at the equilibrium configuration.

$$+ \sum_{kk',\alpha\beta} \left(\frac{\partial^2 U}{\partial u_\alpha^t(k)\partial u_\beta^t(k')}\right)_0 u_\alpha^t(k) u_\beta^t(k') + \sum_{kk',\alpha\beta} \left(\frac{\partial^2 U}{\partial u_\alpha^t(k)\partial u_\beta^r(k')}\right)_0 u_\alpha^t(k) u_\beta^r(k')$$

$$+ \sum_{kk',\alpha\beta} \left(\frac{\partial^2 U}{\partial u_\alpha^r(k)\partial u_\beta^t(k')}\right)_0 u_\alpha^r(k) u_\beta^t(k') + \sum_{kk',\alpha\beta} \left(\frac{\partial^2 U}{\partial u_\alpha^r(k)\partial u_\beta^r(k')}\right)_0 u_\alpha^r(k) u_\beta^r(k') + \text{higher order terms}$$

translation-translation; translation-rotation; rotation-translation; rotation-rotation

- k is the molecule number.
- u_α^t is the αth component of the translational displacement.
- u_α^r represents the extrinsic (fixed frame) rotation angle around α axis.

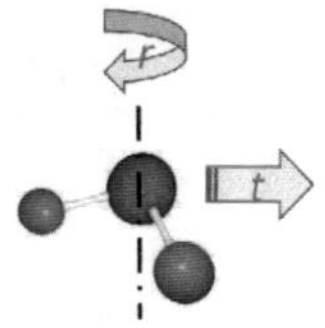

Harmonic approximation of rigid molecules

University at Buffalo

- Ignoring the higher-order terms, we are left with only the *second-order* contributions – known as harmonic energy:

$$U_{\text{harm}} = \frac{1}{2} \sum_{kk',\alpha\beta,ii'} \phi_{\alpha\beta}^{ii'}(k,k')\ u_\alpha^i(k) u_\beta^{i'}(k')$$

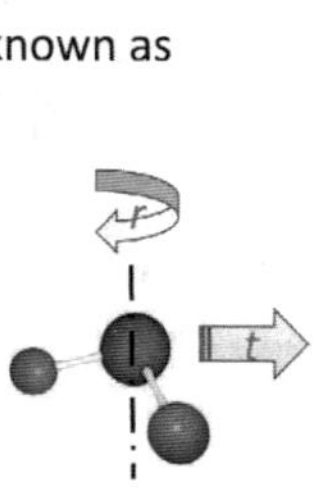

where, $\phi_{\alpha\beta}^{ii'}(k,k')$ is the force constant between mode i of molecule k and mode i' of molecule k':

$$\phi_{\alpha\beta}^{ii'}(k,k') = \left(\frac{\partial^2 U}{\partial u_\alpha^i(k)\partial u_\beta^{i'}(k')}\right)_0$$

- The Helmholtz free energy of harmonic system (in the center-of-mass) is given by:

Classical

$$\beta A_{\text{harm}} = \sum_j \ln\left(\frac{\hbar\omega_j}{k_B T}\right)$$

Quantum

$$\beta A_{\text{harm}} = \sum_j \ln\left[2\sinh\left(\frac{\hbar\omega_j}{2\,k_B T}\right)\right]$$

where j runs over all 6N-3 degrees-of-freedom (dof) and $\{\omega_j\}$ are the system frequencies given by:

$$\mathbf{\Phi}_{6N\times 6N}\mathbf{v}_{6N\times 1} = \omega^2\, \mathbf{M}_{6N\times 6N}\, \mathbf{v}_{6N\times 1}$$

Atomic-to-molecular force constants transformation

University at Buffalo

- We use the 4-sites TIP4P rigid-molecule model of water.
- For such site-site interactions, the molecular force-constants, $\boldsymbol{\phi}^{ii'}(k,k')$, can be expressed in terms of their respective atomic force-constants, $\boldsymbol{\varphi}(ka,k'a')$. (*)

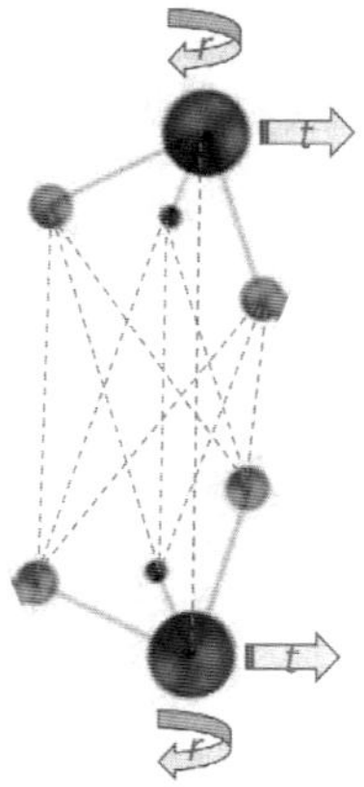

$$\boldsymbol{\phi}^{tt}(k,k') = \sum_{aa'} \boldsymbol{\varphi}(ka,k'a')$$

$$\boldsymbol{\phi}^{tr}(k,k') = \sum_{aa'} \boldsymbol{\varphi}(ka,k'a')\,\mathcal{R}(k'a')$$

$$\boldsymbol{\phi}^{rt}(k,k') = \sum_{aa'} \mathcal{R}^{T}(ka)\boldsymbol{\varphi}(ka,k'a')$$

$$\boldsymbol{\phi}^{rr}(k,k') = \sum_{aa'} \mathcal{R}^{T}(ka)\boldsymbol{\varphi}(ka,k'a')\mathcal{R}(k'a') + \delta_{kk'}\sum_{a}[\mathbf{X}(ka)\cdot\mathbf{F}(ka)\mathbf{I} - \mathbf{X}(ka)\otimes\mathbf{F}(ka)]$$

where $\mathcal{R}(ka)$ is 3×3 anti-symmetric tensor in the atomic positions (w.r.t. COM of molecule k), $\mathbf{X}(ka)$.

- Ewald-sum is used for the Coulomb interactions while lattice-sum is used for the LJ.
- The atomic force constants are given in an exact analytical form; thus, doing finite-difference approximations to get the molecular second derivatives is avoided.

* G. Venkataraman and V. C. Sahni, *Rev. Mod. Phys.* **42**, 409 (1970).

Phonon density of states results

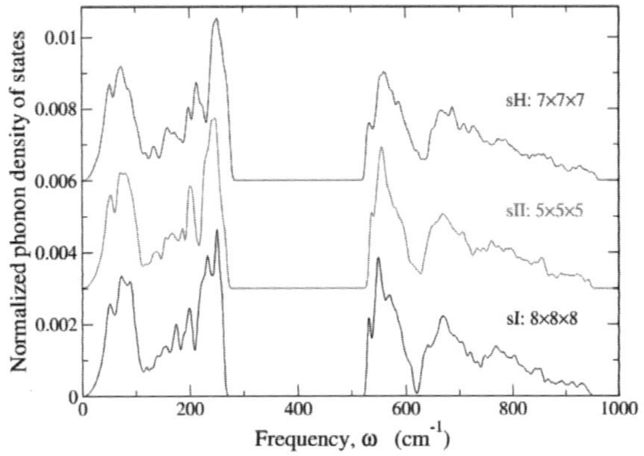

- Based on eigenvector arguments, the lower and upper bands are mostly for translation and rotation modes, respectively. (*)

* T. Inerbaev and *et al.*, Comput. Mater. Sci. **36**, 229 (2006).

Phonon density of states results

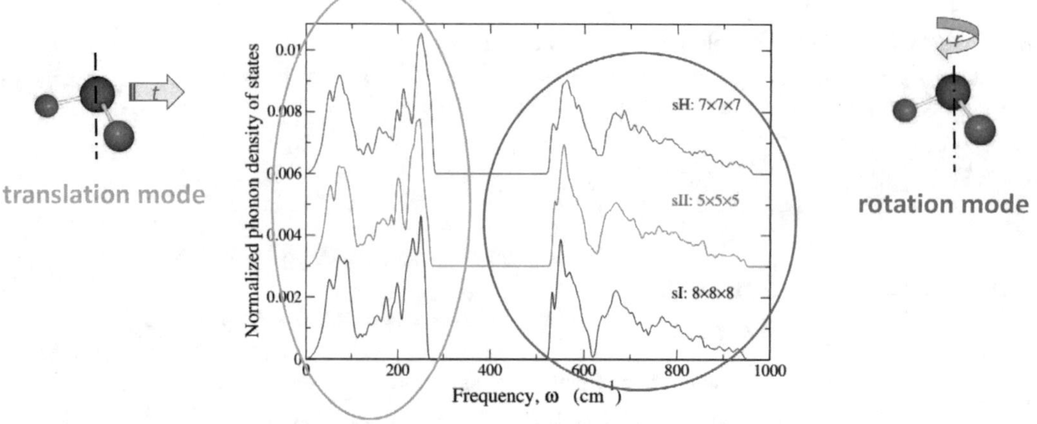

- Based on eigenvector arguments, the lower and upper bands are mostly for translation and rotation modes, respectively. (*)

* T. Inerbaev and *et al.*, Comput. Mater. Sci. **36**, 229 (2006).

Generating proton-disordered isomers

University at Buffalo

- We used Buch and *et al.* (*) algorithm to generate thousands of proton-disordered isomers that satisfy Bernal-Fowler ice rules.
- The bulk hydrate has zero dipole moment.
- Thus, only 100 structures with the lowest dipole moment are chosen for each hydrate.
- Relaxation of the molecular coordinates is then performed to get the minimum structure needed for the harmonic analysis.

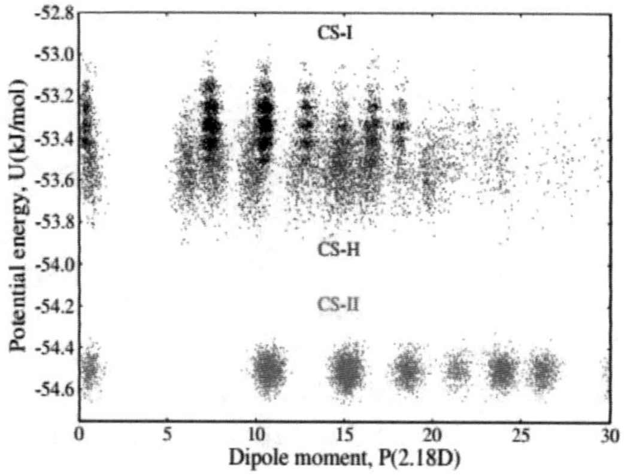

* V. Buch, P. Sandler, and J. Sadlej, The Journal of Physical Chemistry B **102**, 8641 (1998).

Proton-disorder averaging

University at Buffalo

- The exact way to get the FE a system of N molecules having proton-disorder is given as a sum over all possible isomers, $\sim(3/2)^N$ according to Pauling.

$$\beta A = -\ln \sum_i \exp[-\beta A(i)]$$

where, $A(i) \equiv U_0(i) + A_{\text{harm}}(i)$. However, in practice, only M ($\ll (3/2)^N$) isomers can be generated.

- **Averaging Approximations:**

(1) $< A_{harm} + U_0 >_{exp}$: $\beta A = -\ln\left(\frac{1}{M}\sum_i^M \exp[-\beta A(i)]\right) - N\ln 3/2$

(2) $< U_0 >_{exp}$: $\beta A \approx \beta A_{\text{m}} = -\ln\left(\frac{1}{M}\sum_i^M \exp[-\beta U_0(i)]\right) + \beta A_{\text{harm}}(m) - N\ln 3/2$

(3) $< U_0 >$: $\beta A \approx \beta A_{\text{m}} = \beta\frac{1}{M}\sum_i^M U_0(i) + \beta A_{\text{harm}}(m) - N\ln 3/2$

(4) No averaging: $\beta A \approx \beta A_{\text{m}} = \beta U_0(m) + \beta A_{\text{harm}}(m) - N\ln 3/2$

Proton-disorder averaging

University at Buffalo

- The error due to not using full exponential-average is temperature-dependent.

- Using only one isomer introduced the largest error, while using exponential-average of U_0 resulted in the smallest error.

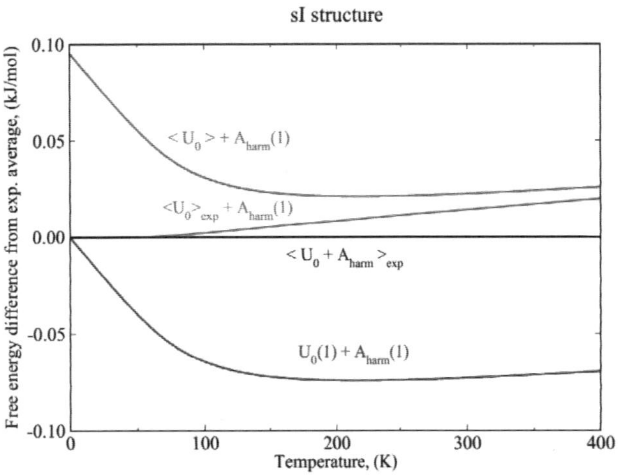

Finite-size effects

- Doing full (*i.e.* $U_0 + A_{\text{harm}}$) exponential average is very expensive for large systems; thus, an approximation is needed.
- **Exponential Averaging Correction (EAC):** is the free energy difference between an isomer and the full exponential average.

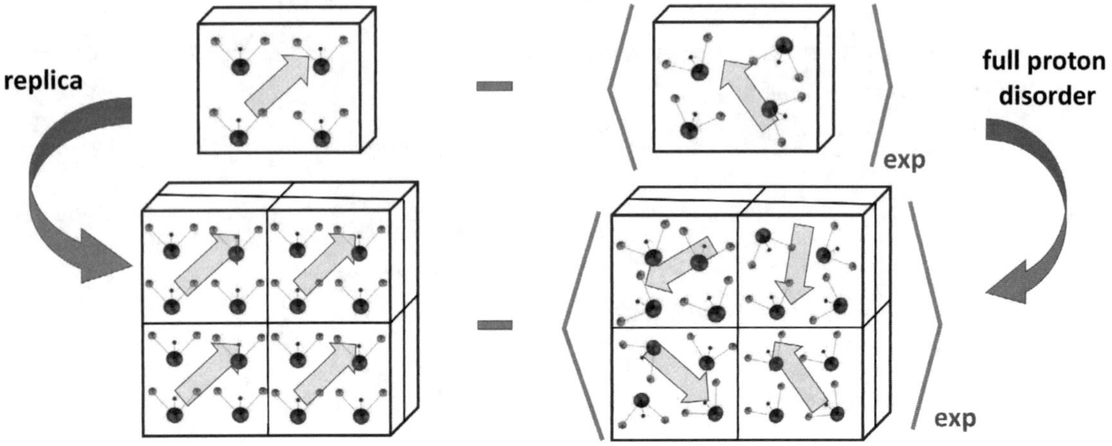

Finite-size effects

- The EAC for $1 \times 1 \times 1$ unit-cell is close to that of $2 \times 2 \times 2$ supercell for $T >\approx 100K$.

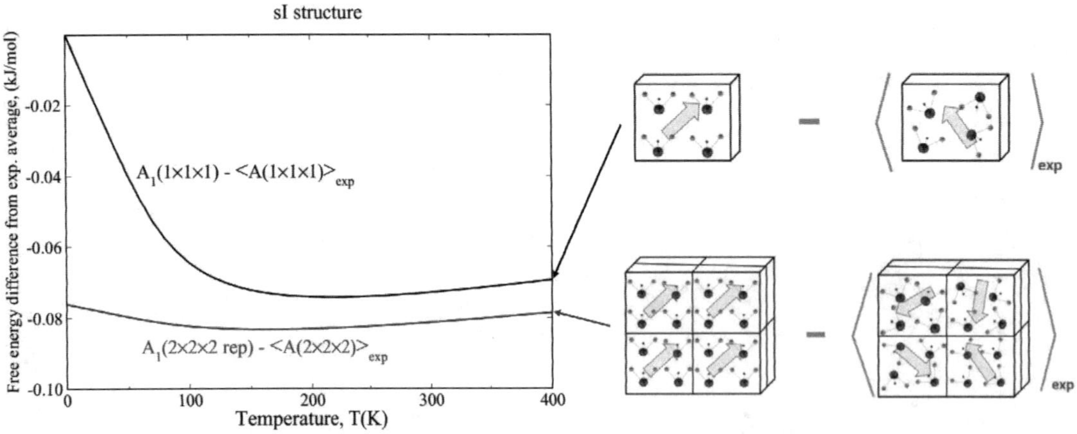

- Thus, only one isomer can be used for all supercell sizes. Then, we can correct for the proton disorder using the EAC measured in the $1 \times 1 \times 1$ supercell.

Finite-size effects

University at Buffalo

- The minimum energy isomer is used.
- Subtracting the $\ln N/N$ finite-size correction helps get more linear trend.
- Same behavior exists in atomic systems (*); thus, $\ln N/N$ is independent of the number of dof per molecule.
- Using only one unit-cell introduces error due to different hydrate structures sizes (46, 136, 68 for sI, sII, sH, respectively).
- In the quantum treatment the extrapolation is temperature dependent; so fitting has to be done for each T.

$$\beta A_{\text{harm}}^{\text{quan}} = \sum_j \ln\left[2\sinh\left(\frac{\hbar\omega_j}{2\,k_B T}\right)\right]$$

$$\beta A_{\text{harm}}^{\text{class}} = \sum_j \ln\left(\frac{\hbar\omega_j}{k_B T}\right)$$

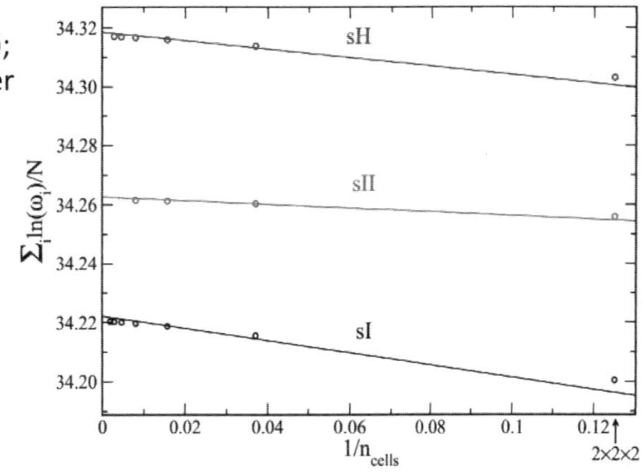

* W. G. Hoover, J. Chem. Phys. **49**, 1981 (1968).

Finite-size effects

University at Buffalo

- The minimum energy isomer is used.
- Subtracting the $\ln N/N$ finite-size correction helps get more linear trend.
- Same behavior exists in atomic systems (*); thus, $\ln N/N$ is independent of the number of dof per molecule.
- Using only one unit-cell introduces error due to different hydrate structures sizes (46, 136, 68 for sI, sII, sH, respectively).
- In the quantum treatment the extrapolation is temperature dependent; so fitting has to be done for each T.

$$\beta A_{\text{harm}}^{\text{quan}} = \sum_j \ln\left[2\sinh\left(\frac{\hbar\omega_j}{2\,k_B T}\right)\right]$$

$$\beta A_{\text{harm}}^{\text{class}} = \sum_j \ln\left(\frac{\hbar\omega_j}{k_B T}\right)$$

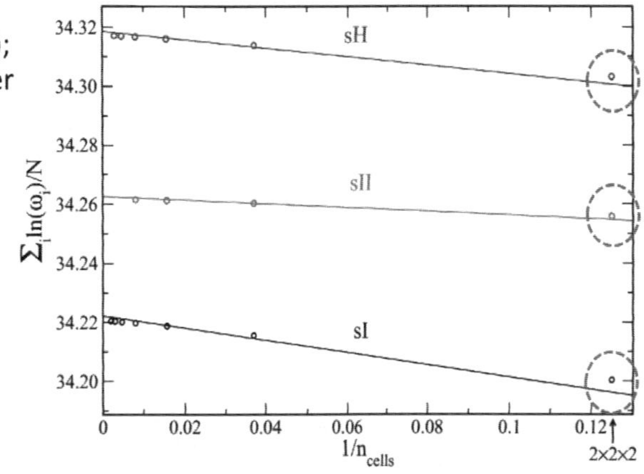

* W. G. Hoover, J. Chem. Phys. **49**, 1981 (1968).

Finite-size effects

University at Buffalo

- The minimum energy isomer is used.
- Subtracting the $\ln N/N$ finite-size correction helps get more linear trend.
- Same behavior exists in atomic systems (*); thus, $\ln N/N$ is independent of the number of dof per molecule.
- Using only one unit-cell introduces error due to different hydrate structures sizes (46, 136, 68 for sI, sII, sH, respectively).
- In the quantum treatment the extrapolation is temperature dependent; so fitting has to be done for each T.

$$\beta A_{\text{harm}}^{\text{quan}} = \sum_j \ln\left[2\sinh\left(\frac{\hbar\omega_j}{2\,k_B T}\right)\right]$$

$$\beta A_{\text{harm}}^{\text{class}} = \sum_j \ln\left(\frac{\hbar\omega_j}{k_B T}\right)$$

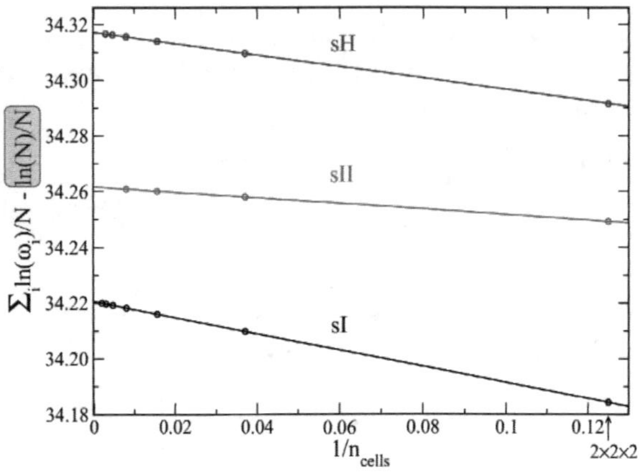

* W. G. Hoover, J. Chem. Phys. **49**, 1981 (1968).

Harmonic free energy

University at Buffalo

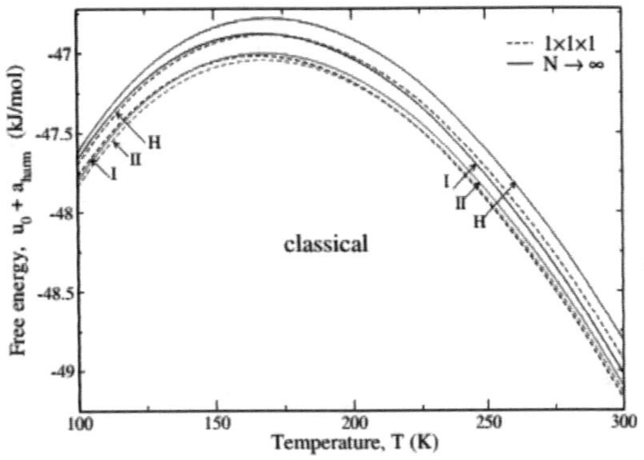

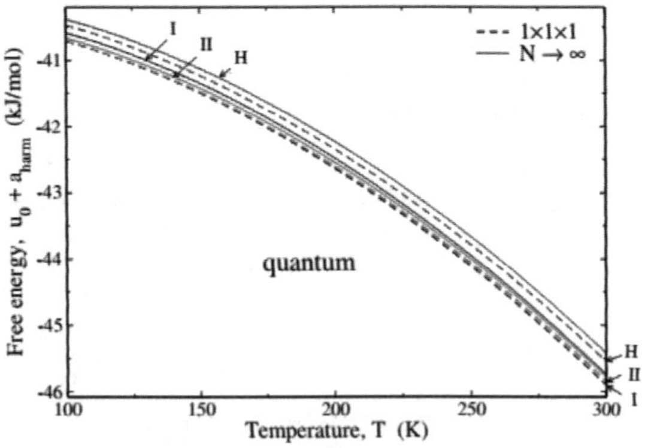

- sII hydrate has the lowest free energy in both classical and quantum treatments.
- Using only one unit cell changes the relative free energy differences.

Harmonic free energy

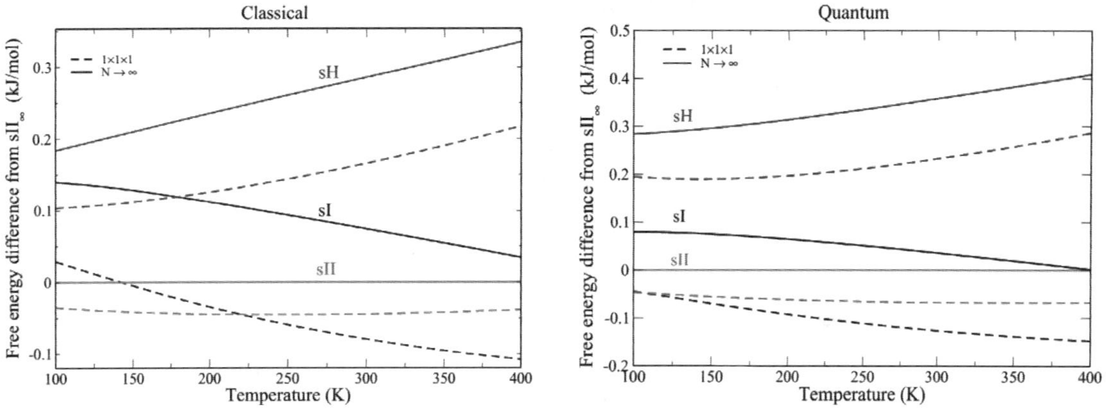

- sII hydrate has the lowest free energy in both classical and quantum treatments.
- Using only one unit cell changes the relative free energy differences.

Accuracy of the local-harmonic approximation (LHA)

- In the LHA(*), only the self interactions of the force constants matrix are non vanishing.
- It helps reduce the computational cost needed to diagonalize that matrix.
- However, the absolute free energy is inaccurate by at least 0.8 kJ/mol compared to the full analysis.
- Although the relative ordering of the structures is preserved, the free energy differences changed.

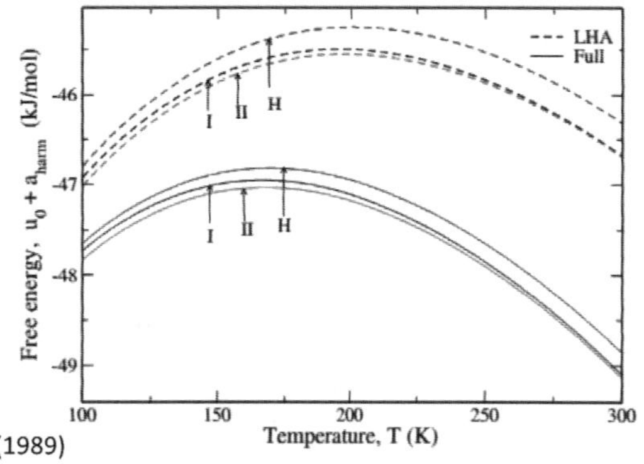

* R. LeSar, R. Najafabadi, and D. J. Srolovitz, Phys. Rev. Lett. **63**, 624 (1989)

Conclusions

- Classical and quantum free energies of sI, sII, and sH hydrates were computed.
- The molecular force constants are given in analytical forms as a function of their atomic contributions; so, finite-difference approximations are avoided.
- Although using only one proton disorder isomer introduced temperature-dependent error, the nature of that error appears to be system-size independent.
- The free energy was computed in the thermodynamic limit by extrapolating it to $1/N \to 0$ for only one isomer and then add the proton-disorder correction of the smallest system.
- Quantum effects are of the order of 5 kJ/mol, and differ between structures by about 0.1 kJ/mol.
- The local-harmonic-approximation introduced inaccuracy by about 0.8kJ/mol.

Acknowledgments

Advisor: David A. Kofke, *Distinguished Professor*

Andrew J. Schultz, *Research Assistant*

ACS PRF

University at Buffalo IMPACT Award

Center for Computational Research University at Buffalo *The State University of New York*

Thank you for listening...

Questions ?

Heterogeneity and reduction for complex network dynamics

Ioannis G. Kevrekidis, Carlo Laing, Alexander Holiday, Tom Bertalan
Chemical and Biological Engineering & PACM, Princeton University, USA

Summary. In the study of dynamic, large-scale, complex networks, coarse-grained (reduced, low-dimensional) descriptions of the network can be useful not only in identifying significant network features, but also in enabling reduced, systems-level analysis of the evolution dynamics. Obtaining such a coarse-grained description for networks is a challenging problem. Additionally, the implementation of coarse-grained computational algorithms requires a systematic approach for translating low-dimensional network features to full-scale network realizations. We discuss recent advances, and experience with applications, in both these directions for complex network evolution.

Complex, large scale networks are often dynamically evolving in time. One can discriminate several different forms for such an evolution: (a) it may be that the connectivity of the network is fixed, but properties of the nodes evolve in time (e.g. concentrations in a complex biochemical reaction network); this we term dynamics *on* networks; (b) it may be that the only thing evolving is the connectivity of the network - edges either form or disappear (alternatively, their weights evolve) in time; this is termed dynamics *of* networks; finally (c) it is possible that *both* properties of the nodes and existence/weights of edges evolve, giving us dynamics *of and on* networks or, as it is also sometimes termed, *adaptive* network dynamics. In all these cases we have complex, large scale evolving dynamical systems, and this work presents several efforts towards their algorithmic model reduction.

We begin by describing and illustrating an approach to coarse-graining the evolution of networks whose connectivity changes dynamically. The approach is formulated within our *equation-free* framework given a full, detailed model of the network evolution dynamics. We assume that we know what the crucial macroscopic network features are, the "dependent variables" in terms of which we can write effective closed evolution equations (e.g. the network degree distribution). We then perform short bursts of detailed network evolution simulations from which we estimate the right-hand-sides (and the actions of the Jacobians) of these unavailable closed evolution equations on the fly. The approach involves repeated use of lifting and restriction operators that translate between actual network realizations and their (appropriately chosen) coarse observables - the more "difficult" of the two is the lifting: creation of networks that have prescribed statistical features. The equation-free framework will be used here to accelerate temporal simulations (through coarse projective integration), and to implement coarse-grained fixed point algorithms (through matrix-free Newton-Krylov GMRES). The approach will be illustrated through a simple network evolution example, for which analytical approximations to the coarse-grained dynamics can be independently obtained, so as to validate the multiscale computational results. The scope and applicability of the approach will be discussed.

So far we have assumed that the appropriate coarse-grained observables (the "right" variables in terms of which coarse network evolution equations can be expressed) are known *a priori*. We now discuss the detection of such variables based on data-mining tools, in particular manifold-learning techniques like Diffusion Maps. We extend these data mining approaches to cases in which every data point in a time series is a "snapshot" of an evolving graph. We are thus able to detect intrinsic low-dimensionality in ensembles of graphs, which is the first step in coarse-graining the detailed graph information; the second step is the detection of the appropriate coarse variables that parameterize the network evolution. One of the main challenges in mining graph evolution data is the definition of a suitable pairwise similarity metric in the space of graphs. We explore two practical solutions to solving this problem: one based on finding subgraph densities, and one using graph spectral information. We demonstrate the data-mining process by detecting and parametrizing low-dimensional families of graphs arising from graph construction algorithms as well as from dynamic graph evolution laws.

The last component of the work addresses heterogeneous networks, and the construction of good coarse-grained observables for such networks. This involves a "technology transfer" from modern uncertainty quantification methods, namely from generalized polynomial chaos techniques and from noninvasive, collocation-type implementations of them. There, the response of a system to a parameter whose value comes from a known probability distribution is expanded in terms of orthogonal polynomials weighted by this distribution. Here, the behavior of a system of heterogeneous units, whose heterogeneity parameter comes from a known probability distribution, is again expanded in terms of orthogonal polynomials weighted by this distribution. This allows the collective behavior of large numbers of coupled heterogeneous oscillators to be described in terms of "just a few" evolving expansion coefficients, in the spirit of spectral methods. What makes the approach particularly appealing, however, is the ability to use it both in cases where the heterogeneity is intrinsic to the nodes (e.g. neurons whose kinetic property values are sampled from a known distribution), as well as in cases where the heterogeneity is characteristic of the network structure (e.g. neurons whose degree in a network is sampled from a known degree distribution). Remarkably, the approach works equally well, at least in principle, when joint intrinsic as well as structural heterogeneities are considered. What now becomes interesting is the case where these heterogeneity distributions are not independent, but rather correlated; then the simple and powerful noninvasive collocation-type methods are not easily applicable any more.

Our illustrative examples range from "textbook" Kuramoto-on-a-network type dynamics to more detailed heterogeneous neuronal network models, where both intrinsic and structural heterogeneity arises. "Lifting" (the construction of networks consistent with one or more statistical properties) in the context of Mixed Integer Linear Programming is discussed and contrasted to stochastic optimization techniques (like simulated annealing); this component of the work has been done in collaboration with C. Floudas (Princeton) and C. Gounaris (CMU).

References

[1] A. C. Tsoumanis, Rajendran, K., Siettos, C. I., and Kevrekidis I. G., Coarse-graining the dynamics of network evolution: the rise and fall of a networked society, N. J. Phys. 14 (083037) (2012)

[2] K. A. Bold, K. Rajendran, B. Rath and I. G. Kevrekidis, An equation-free approach to coarse-graining the dynamics of networks, J. Comp. Dyn. **1**(1) (2014), also http://arxiv.org/abs/1202.5618

[3] K. Rajendran, A. C. Tsoumanis, C. I. Siettos and I. G. Kevrekidis, On the Selection of Collective Variables in the Coarse-graining of Dynamics on Networks, submitted to Phys. Rev. E (2013)

[4] K. Rajendran and I. G. Kevrekidis, Mining data in the form of graphs using pair-wise graph similarity measures, submitted to Phys Rev E (2013)

386919 Multiscale Modeling of Rod-like Structures Using Molecular Dynamics and Continuum Mechanics

Abstract:

Molecular structures can be adequately studied using molecular dynamics (MD), although the utility of MD is limited by the large high computational costs of atomistic simulations. One possible way to reduce these costs is to replace the atomistic description of the system under consideration with a continuum mechanical one.

In this work, we focus in particular on rod-like structures where the extension in one direction is clearly larger than in the remaining two dimensions. Well-known examples of such materials on the nanoscale include carbon nanotubes and macromolecules such as DNA and cellulose bundles. For these materials, the concept of geometrically exact beams introduced by Simo (also known as Reissner-Simo beam theory) is a promising candidate for a continuum formulation.

Using atomistic molecular simulations to parameterize a continuum model, we can obtain numerical results for a variety of geometric configurations and system conditions much more efficiently than using exclusively atomistic MD simulations. To this end, we outline a systematic procedure that estimates the necessary parameters (the initial geometry and strain-stress relationships) at finite, constant temperatures. By using benchmark problems that bridge the molecular and continuum scales, we can demonstrate the validity of the estimated parameters in finite element models of the beam model. We show how this model can be used to derive continuum representations of carbon nanotubes under various conditions.

388982 Continuum/Mesoscale/Molecular Level Hybridization in Complex Fluid Flows

Abstract:

Continuum/Mesoscale/Molecular Level Hybridization in Complex Fluid Flows

Sesha Hari Vemuri, Yu Liu, Myung S. Jhon, and Lorenz T. Biegler

Department of Chemical Engineering, Carnegie Mellon University, Pittsburgh, PA 15213

Considerable effort is being dedicated to understand the range of time and length scale phenomena that take place in a broad range of systems including colloidal suspensions, polymer solutions, micro- and nano hydrodynamic flows, especially in biological systems. These systems are characterized by multiscale interaction between "large and slow" polymer particles, and "small and fast" solvent particles [1]. A detailed modeling of such systems can be provided by a molecular dynamics (MD) simulation where the motion of all the particles is tracked via Newton's equations of motion. However, this approach has severe drawbacks due to the limitations on typical length scale covered in the MD simulations and the computational time scaling as $O(N^2)$, where N is the number of particles. On the other hand, the continuum based theory breaks down for these systems as the confinement size is comparable to or smaller than the molecular/pseudo-particle mean free path.

Mesoscopic models, and notably those arising from kinetic theory, are natural candidates to fill this gap because they operate precisely at intermediate scales between the atomistic and continuum levels. During the past two decades, lattice Boltzmann method (LBM), which originates from Boltzmann transport equation has proven successful to model characteristic features of nano/microscopic flows, such as the occurrence of slip boundary conditions near solid walls [2]. Furthermore, being a pseudo particle simulation methodology, it has the advantages of easily implementing easily the rotational degrees of freedom of the particles [3], and ability of simulating multi-physical phenomena in complex geometries. In this work, we present a hybrid MD/LBM method for the simulation of hydrodynamic phenomena in polymer solutions and colloidalsystems.

This will be achieved by generating a buffer zone at the component interface (e.g., polymer-solvent interface, fluid-boundary interface etc.) where both MD and LBM solutions are valid. Here, the particle velocities and forces obtained are exchanged between particles modeled by LBM and those modeled by MD [3]. The constrained dynamics algorithm forces the instantaneous mean particle velocity to equate the continuum solution at the other boundary. The equations of motion in the MD are also modified to match the external force obtained from the LBM via proper scaling.

To validate our hybrid model, we examine our results of the benchmark complex fluid flow systems including including colloidal solution & flow in a confined geometry and will extend our analysis towards the complex multi-physical phenomena various applications.

[1] A. Chatterji and J. Horbach, *J. Chem. Phys.*, **122**, 184903 (2005)

[2] J. Horbach and S. Succi, *Phys. Rev. Lett.*, **96**, 224503 (2006)

[3] P. Ahlrichs and B. Dünweg, *Int. J. Mod. Phys. C*, **09**, 1429 (1998)

Quantifying the Free Energy Landscape for Curvature Induction and Sensing in Lipid Bilayer Membranes Through a Mesoscale Model

Background: Several proteins recruited to the cell membrane have been found to induce and stabilize curvature in the lipid bilayer. Curvature generation in the cell membrane is an essential part of several key biophysical processes including cell motility, endocytosis, and invadopodia formation. In addition to the colocalization of curvature-inducing proteins to sites of curvature *in vivo*, these membrane proteins have been shown to both migrate along mean curvature gradients and tubulate liposomes *in vitro* [1, 2]. Curvature-inducing membrane proteins are thought to generate curvature through either the insertion of an amphipathic helix, or through the electrostatic scaffolding of the bilayer [3,4,5]. Both these mechanisms create a local deformation in the bilayer. It is not clear how these proteins sense mean curvature gradients with different affinities, and what major determinants impact their segregation, curvature induction, or the morphological phase transitions they drive.

Methods: In this work, the lipid/cell membrane is modeled with a continuum Monte Carlo model evolved with the Helfrich Hamiltonian [6, 7]. In the model, the membrane is described by an infinitesimally thin triangulated sheet, which can access length scales on the order of 100 nm and above. Membrane proteins interact with the membrane through a prescribed intrinsic curvature field and can move along the triangulated mesh. The intrinsic curvature field is approximated as a Gaussian function, with parameters derived from relating the curvature energy to isothermal titration experiments measuring Epsin association energy with the bilayer. Some membrane proteins such as those comprising BAR domains are thought to induce an anisotropic curvature field, and are modeled by an anisotropic two-dimensional Gaussian function. In addition to single proteins, a larger step function is also used to approximate an entire assembly of proteins such as those present in a clathrin coat during clathrin-mediated endocytosis. The excess area present in the membrane can also be controlled in our model. This excess area gives rise to a renormalized tension, which can be quantified through a non-linear fit of the undulation spectrum.

Several free energy methods have been coupled to the membrane model in order to determine a membrane proteins chemical potential. In addition to the Widom particle insertion method [8], two free energy perturbation methods are utilized to track the free energy for the addition of a curvature inducing protein to the bilayer. The Widom method computes the excess chemical potential upon insertion of a membrane protein, but this method is known to fail at high densities. For this reason, both thermodynamic integration and Bennett acceptance methods are used to validate Widom calculations at low densities and access the free energy landscape at higher protein densities in simulation [9,10].

Results: Using our mesoscale membrane model, the chemical potential of curvature inducing proteins is quantified for planar bilayers. The planar geometry represents a 400 by 400 nm patch of the cell membrane. For low protein concentrations, the chemical potential required to bend the membrane to a certain curvature field is shown to depend on both the extent as well as the amplitude of the curvature field. A larger curvature field senses mean curvature gradients to a greater degree. The chemical potential also scales linearly with bending rigidity. Anisotropic curvature inducing proteins are shown to have preferential alignment with background mean curvature fields.

At higher densities, these curvature inducing membrane proteins cluster to form tubules. Our free energy methods show a pitchfork bifurcation in the excess chemical potential at the point of tubulation, which is a signature of a morphological phase transition. This morphological phase transition is further quantified in its dependence on tension, with higher tensions requiring a higher critical concentration of proteins. This model connects the strength of an individual protein curvature field to the mesoscale morphological phase transitions seen *in experiments*.

The free energy of vesicle nucleation in clathrin-mediated endocytosis is also quantified with our model. A free energy barrier is shown to be present during the formation of the vesicle neck. Our results suggest that tension hinders curvature induction, and the formation of the vesicle neck. These results suggest a novel mechanistic interpretation to experimental studies reporting actin localization to some sites of endocytosis, which is necessary to overcome membrane tension.

References

[1] Ford, M., et al. Curvature of clathrin-coated pits driven by epsin. Nature. 2002.

[2] Capraro, B., Yoon, Y., Cho, W., Baumgart, T. Curvature Sensing by the Epsin N-TerminalHomology Domain Measured on Cylindrical Lipid Membrane Tethers. J. Am. Chem. Soc. 2010.

[3] Campelo, F., McMahon, H., and Kozlov, M.The Hydrophobic Insertion Mechanism of Membrane Curvature Generation by Proteins. Biophysical Journal. 2008.

[4] McMahon, H., and Gallop, J. Membrane curvature and mechanisms of dynamic cell membrane remodeling Nature. 2005.

[5] Zimmerberg, J., and McLaughlin, S. Membrane Curvature: How BAR Domains Bend Bilayers Current Biology. 2004.

[6] Helfrich, W. Elastic Properties of Lipid Bilayers - Theory and Possible Experiments. Z.Naturforsch. 1973.

[7] Ramakrishnan, N., Kumar, P. B. and Ipsen, J. H. Monte Carlo simulations of fluid vesicles with in-plane orientational ordering. Phys. Rev. E. 2010.

[8] Widom, B. Some Topics in the Theory of Fluids. J. Chem. Phys. 1963.

[9] Frenkel, D., and Smit, B. *Understanding Molecular Simulation*, 2nd ed. 2001.

[10] Bennett, C. H. Efficient estimation of free energy differences from Monte Carlo data. J. Comp. Phys. 1976.

374538 Rippling Instabilities on the Surface of Stressed Crystalline Solids

Abstract:

Morphological instabilities on surfaces of stressed solids and the resulting pattern formation are of major fundamental and technological significance toward improving materials function and reliability and developing innovative directed assembly processes for nanotechnology. Such a well-known phenomenon is the Asaro–Tiller/Grinfeld (ATG) instability, which leads to the formation of a regular pattern of crack-like grooves emanating from the surface of a stressed elastic solid and deepening by surface diffusion.

We have carried out a systematic analysis of morphological evolution of stressed solid surfaces under the action of an electric field, or a temperature gradient, or the simultaneous action of both and reported that sufficiently strong external fields can inhibit the ATG instability and prevent surface cracking; a linear stability theory predicted fairly accurately the current-induced stabilization of surface morphology in stressed elastic solids. However, complex aspects of surface morphological evolution and pattern formation are not accounted for by the linear theory.

Using self-consistent dynamical simulations according to a well-tested fully nonlinear model of surface morphological evolution, we have studied the surface morphological response of stressed solids both in the absence of and under the simultaneous action of external fields focusing on low-amplitude long-wavelength perturbations from the planar surface morphology. We find that in addition to the ATG instability, a long-wavelength tip-splitting instability may be triggered forming a pattern of secondary ripples on the surface; rippling occurs in the absence of external field application, as well as for weaker-than-critical applied external fields. We have developed a nonlinear wavelength selection theory to analyze the formation of such secondary surface ripples. We have also carried out a comprehensive simulation study to characterize the ripple formation, identify the critical wave-length for rippling, and determine the dependence of the critical wave number for the onset of rippling on the applied external field strength.. Finally, we have examined the effect of the Arrhenius temperature dependence of the surface diffusivity, the strongest temperature dependence of all the relevant material properties, on surface rippling under the action of an externally applied temperature gradient.

382383 Robust Control of Multiscale Process Systems Under Uncertainty: An Application to Epitaxial Thin Film Growth

Abstract:

Multiscale processes are characterized by inherently coupled chemical and physical phenomena occurring over different temporal and spatial scales [1]. In such systems, control of the events evolving at the microscopic scale is essential to product quality while efficient operations require manipulated variables at the macroscopic scale for real-time feedback control [2]. Accordingly, to describe the behavior of these processes, it is common to employ a multiscale model encompassing macro-scale continuum equations (e.g., Partial Differential Equations) embedded with micro-scale simulations (e.g., Kinetic Monte Carlo). Although such a model is capable to provide a fair representation of a system, the microscopic processes are subject to model parameter uncertainty that can significantly affect the control and optimization objectives. As such, to achieve the desired objectives, it is crucial to develop control strategies that are robust to these uncertainties.

This study explores a systematic framework to analyze model parameter uncertainty for the purpose of robust control and optimization in multiscale models. The difficulties in analysing uncertainty in such systems arise due to the computational intensity of the fine-scale simulations (typically KMC simulations). Moreover, the lack of a closed formulation between the process optimization objectives and the model parameters, and absence of sensors to measure microscopic features at fine scales make this analysis even more challenging. Motivated by these observations in this work, Power Series Expansion (PSE), which obtains a simpler representation of the actual system, is employed to analyze model uncertainty propagation [3].

The thin film growth process has been used in this study to effectively exemplify a multiscale process system. Strong dependence of the electrical and mechanical properties of thin films on their microstructure requires accurate control of the product's quality (e.g., film's roughness and thickness) at the fine scale. For uncertainty analysis of the thin film process, the probability distributions of microscopic events are determined using the PSE-based models that rely on the prior distribution of the uncertain parameters. Subsequently, the upper and lower bounds on desired outputs (e.g., film roughness) are determined at a given confidence level of the distributions obtained for the states of the system [4]. Moreover, multiple reduced-order lattices are employed in KMC simulations to reduce the computational costs. The potential application of the proposed method can be illustrated in open-loop optimization or closed-loop purposes. In open-loop optimization, the optimal temperature profile is obtained to optimize the end-point properties of the thin film. For closed-loop purposes, on the other hand, measurements are not available as frequent as required for the development of an effective feedback control strategy. The proposed algorithm can therefore be applied to develop a robust estimator that is able to provide the estimate of the controlled output using online measurements. Furthermore, the feedback control of the film microstructure can be achieved via offline identification of a low-order model from lower and upper bounds on the process outputs in the presence of uncertainty.

[1] Vlachos, D. G., (2005) A review of multiscale analysis: Examples from systems biology, materials engineering, and other fluid-surface interacting systems. Advances in Chemical Engineering, 30, 1-61.

[2] Christofides, P. D., Armaou, A., Lou, Y. & Varshney, A., (2008) Control and optimization of multiscale process systems. Boston: Birkhäuser.

[3] Nagy, Z. K. & Braatz, R. D., (2007) Distributional uncertainty analysis using power series and polynomial chaos expansions. Journal of Process Control, 17, 229-240

[4] Rasoulian, S., Ricardez-Sandoval, L.A. "Uncertainty analysis and robust optimization of multiscale process systems with application to epitaxial thin film growth" Chemical Engineering Science, In Press.

388856 Lattice Boltzmann Method Based Multiscale Modeling of Two-Phase Flows through Microarchitectures with Microfeatures

Abstract:

Lattice Boltzmann Method based multiscale modeling of two-phase flows through microarchitectures with microfeatures

Agnieszka Truszkowska[1] and Goran Jovanovic[1]

[1]Oregon State University, Department of Chemical, Biological, and Environmental Engineering, Corvallis, OR 97330, USA

Emails: truszkoa@onid.orst.edu, goran.jovanovic@oregonstate.edu

Segmented two-phase flows in microscale-based structures display attractive characteristics such as enhanced mixing and intensification of mass and heat transfer, reduced Taylor dispersion, and opportunity to control and manipulate dispersed phase; bubbles or droplets. Two-phase flow in microscale-based structures is commonly used in emulsification and encapsulation operations, microreactors, material synthesis, bioassays and many other applications ([1],[2]). Apart for its purposeful presence, dispersed phase could occur in microarchitectures as a side effect emerging from phase transition, inadvertent inflow with process feed, or through equipment gaps ([3],[4]). Whether part of an intended operation or unwanted operating byproduct, second/discrete phase has to be properly controlled to achieve desired system performance.

Optimization of two-phase processes on microscale level results in numerous opportunities related to both operating conditions and design architecture. However, efficient and comprehensive optimization often requires recurring use of numerical models in addition to experimental data. Additionally, these computational models require high resolution computing due to dominant interfacial transport phenomena. Due to the need for excessive computational resources many researchers reduce modeling effort to a representative small segment of the system, or make overly simplifying assumptions and approximations.

In this paper we present a novel, multiscale-modeling approach for numerical representation of the two-phase flow in microscale-based structures containing micro features such as posts and pillars. Microscale-based architectures containing various micro features are relatively unexplored design options ([5], [6]) with vast potential for optimization due to their extreme design flexibility. We explore their characteristics and provide a novel multiscale modeling approach within the framework of Shan and Chen Lattice Boltzmann modeling method.

Example used in the numerical experiments is an array of cylindrical micropillars with staggered, equilateral arrangement. A multiscale operator in the form of field of discrete vector quantities, forcing terms, is developed. Discrete vector forcing terms are applied on the interface of the dispersed phase, which results in an equivalent average behavior of the dispersed phase in the geometrically simplified scale, just as in the geometrically fully resolved scale. We present several variations of the operator as well as its characteristics and robustness. Finally, we discuss applicability and limitations of the approach based on model system simulations and classical benchmark cases.

[1] Shui, Lingling, Jan CT Eijkel, and Albert van den Berg, "Multiphase flow in microfluidic systems--Control and applications of droplets and interfaces", Advances in Colloid and Interface Science, 133,1 (2007): 35-49

[2] Gunther, Axel, and Klavs F. Jensen, "Multiphase microfluidics: from flow characteristics to chemical and materials synthesis", Lab on a Chip, 6,12 (2006): 1487-1503

[3] Jensen, Mads Jakob, Goran Goranovic, and Henrik Bruus, "The clogging pressure of bubbles in hydrophilic microchannel contractions", Journal of Micromechanics and Microengineering, 14,7 (2004): 876

[4] Litterst, C., et al., "Improved gas bubble mobility in CHIC-type flow channels", Actuator, Messe Bremen, Germany (2004): 541-544

[5] Krishnamurthy, Santosh, and Yoav Peles, "Gas-liquid two-phase flow across a bank of micropillars", Physics of Fluids 19 (2007): 043302

[6] De Loos, S. R. A., et al. "Gas--liquid dynamics at low Reynolds numbers in pillared rectangular micro channels", Microfluidics and nanofluidics, 9,1 (2010): 131-144

371300 Characterizing Tip-Surface Contact Between Monolayer Functionalized Silica Surfaces through Molecular Dynamics Simulation

Abstract:

Chemisorbed alkylsilane monolayer coatings have been shown to reduce friction, adhesion, and wear of surfaces under shearing contact. Exploration of monolayer properties at the nanoscale is commonly performed through atomic force microscopy (AFM). However, inconsistencies in values and trends of friction properties may arise between separate studies as a result of variance in probe tip size. Flater et al. proposed that for tip contact of a monolayer-functionalized surface, plowing of monolayer chains provides an additional contribution to friction force.[1] Molecular dynamics simulations of Harrison and coworkers reveal an increased coefficient of friction during shearing of a buckyball tip compared to shearing of an infinitely flat amorphous surface, supporting the idea of molecular plowing presence and the dependence of shear behavior on tip size.[2] However, it remains unclear under which contact dimensions tip-surface contact behavior begins to deviate from that of infinitely flat (surface-surface) contact. Furthermore, although simulations of surface-surface contact may provide a good approximation to the contact behavior of many AFM tips, system size effects have yet to be fully explored. Using molecular dynamics simulations, we examine system size dependence during surface-surface contact for alkylsilane monolayers and study the influence of substrate size for both flat and curved tip-like contacting agents to help better define the tip-surface contact regime.

For systems of surface-surface contact, we find contact behavior to have a minimal dependence on system size; however, clear trends arise for systems of tip-surface contact. We observe a distinct transition in friction force as a function of substrate size for flat tip contacting agents. We attribute this to size influence on monolayer disorder arising from chain tilting at tip edges. A similar transition is found for contacting agents displaying curvature, where shear behavior is shown to transition from sliding to a plowing-like mechanism. This behavior is investigated through examination of penetration depth and local disorder of the countersurface monolayer, which reveals that less highly curved tips are unable to exhibit sufficient monolayer penetration and may be effectively modeled through surface-surface simulations.

[1] Flater EE, Ashurst WR, Carpick RW, Nanotribology of octadecyltrichlorosilane monolayers and silicon: self-mated versus unmated interfaces and local packing density effects, *Langmuir*, 23, (2008).

[2] Knippenberg MT, Mikulski PT, Harrison JA, Effects of tip geometry on interfacial contact forces, *Modelling and Simulation in Materials Science and Engineering*, 18, (2010).

386425 Understanding the Self-Assembly of Skin Lipids

Abstract:

The outermost layer of the skin (the stratum corneum) consists of skin cells embedded in a rich lipid matrix, whose primary role is to provide a barrier to foreign agents entering the body and to water leaving the body. This lipid system is unique in biological membranes in that it is composed of ceramides, cholesterol, and free fatty acids, with phospholipids, which are the major components of most biological membranes, being completely absent. This unique composition enables the organization of the stratum corneum lipids into lamella, which in turn is believed to control barrier function. While much is known about the nature of the skin lipids from extensive experimental studies, a clear understanding of how and why these molecules assemble into the structures observed through microscopy and biophysical measurements does not yet exist. In order to probe the molecular level arrangement, we are performing atomistic and coarse-grained model simulations of key stratum corneum lipids and water. The development and validation of the coarse-grained models will be discussed alongside the results of self-assembly studies in simple mixed lipid systems that enable us to validate the models developed and work towards the study of a realistic stratum corneum system.

386483 Entropic and Active Matter: Insights from Simulation

Abstract:

Our understanding of the role of steric interactions, crowding, dissipation, and entropy in colloidal soft matter systems has advanced greatly in recent years from computer simulations by our community. The emergence of directional entropic forces and entropic valence arising from particle shape drive ordering transitions to metastable, equilibrium, and persistent complex structures. In this talk, we present new results on phase transitions in hard particle systems both in, and away from, equilibrium. We compare our findings with enthalpically driven transitions, and describe new simulation tools under development for the study of these types of systems.

386363 How the Walls Come Crumbling Down: Elucidating Mechanisms of Cellulose-Active Enzymes Using Molecular Simulation

Abstract:

The microbial degradation of plant matter is one of the most important components of the global carbon cycle. Beyond the natural importance of biomass turnover, plant cell wall polysaccharides also offer significant promise as a renewable resource for the sustainable production of fuels and chemicals to partially offset mankind's fossil fuel usage, but the cost for using natural enzyme systems in industrial contexts remains high. Understanding the enzymatic mechanisms of cellulose deconstruction is key need to engineer enhanced biocatalysts, and computer simulation offers a powerful, complementary tool to experimental enzyme engineering approaches. In this talk, I will describe research efforts from our group aimed at understanding rate-limiting steps in cellulase enzyme action using a wide variety of computational methods including absolute ligand binding free energy calculations, transition path sampling, and QM/MM simulations. In concert with experimental measurements, computation has enabled the identification of the rate-limiting step in one of the most abundant and widely used cellulases in biofuels applications, and thus highlights a target for experimental protein engineering. I will also briefly highlight computational work from our group on understanding the molecular mechanisms of cellulose biosynthesis.

386505 Isolated Catalyst Sites on Amorphous Supports: A Systematic Ab Initio Route to Structure-Property Relationships

Abstract:

Methods for modeling catalytic sites on amorphous supports lag far behind methods for modeling catalytic sites on metal surfaces, zeolites, and other crystalline materials. One typical strategy for amorphous supports uses cluster models with arbitrarily chosen constraints to model the rigid amorphous support environment, but arbitrarily chosen constraints arbitrarily influence catalyst site activity. A common alternative strategy uses no constraints, but this results in catalytic sites with unrealistic flexibility. We present the first algorithm that systematically generates appropriately constrained structural models of active sites. The algorithm generates a family of structural models that span some key determinant of reactivity, e.g. an overall activation energy. The algorithm is illustrated for an empirical valence bond model energy landscape and for an Mo/SiO2 olefin metathesis catalyst. We also present a statistical framework for interpreting kinetics in systems with quenched disorder.

390113 Understanding and Applying Multistate Reweighting in a Broad Molecular Simulation Context

Abstract:

A recurring challenge in simulations of molecular systems of interest to chemical engineers is estimating physical quantities from sampled ensembles. Collecting sufficient statistics for reliable estimates often requires collecting multiple data sets under varying conditions, each from a different ensemble consistent with the different macrostate.

A large number of methods have been developed to analyze these data sets. The multistate Bennett Acceptance ratio method, based on recent advances in statistical sampling theory, unifies ideas from a number of different analysis methods for multiple ensembles. MBAR is provably the lowest variance unbiased estimator of both free energies and ensemble averages, and has a number of other features, such as no requirement for histograms, scaling with data set size rather than dimension of the problem, and an explicit error estimate.

In this talk, we analyze a number of insights about the nature of multistate reweighting that have emerged since the approach was introduced. We discuss how these methods can be applied to a wide variety of problems such as computing potentials of mean force, force field parameterization, and calculating free energy differences, with an emphasis on multidimensional parameter spaces. We resolve common confusions that may occur when setting up a problem to perform multistate analysis and discuss MBAR implementations that can drastically reduce the amount of time required and significantly increase robustness.

Multi-scale theory in the molecular simulation of electrolyte solutions

W. Zhang, X. You, and L. R. Pratt
Department of Chemical and Biomolecular Engineering,
Tulane University, New Orleans, LA 70118

To define a role for AIMD simulation on the limited time and space scales accessible to those demanding methods, this study organizes McMillan-Mayer theory[1], the potential distribution approach[2], and quasi-chemical theory[3] to provide theory for the thermodynamic effects associated with long-length scales. The theory treats composition fluctuations which would be accessed by larger-scale calculations, and also longer-ranged interactions that are of special interest for electrolyte solutions. The quasi-chemical organization breaks-up governing free energies into physically distinct contributions: packing, outer-shell, and chemical contributions. Here we study specifically the outer-shell contributions that express electrolyte screening. For that purpose we adopt a primitive model suggested by observation of ion-pairing in tetra-ethylammonium tetra-fluoroborate dissolved in propylene carbonate. Gaussian statistical models are shown to be effective physical models for outer-shell contributions, and they are conclusive for the free energies within the quasi-chemical formulation. For comparison, hypernetted-chain (HNC) theory and Bennett's method are implemented to evaluate these free energies. With the present data-set the Gaussian physical approximation obtains more accurate mean activity coefficients than does the Bennett's more ambitious direct evaluation of those free energies. Gaussian models have clear physical assumptions, and with accurate evaluation of distribution mean and variance, show better results than does HNC theory. HNC theory is more ambitious but does not fully evaluate the distribution mean and variance. System size effects are established to be small for these cases.

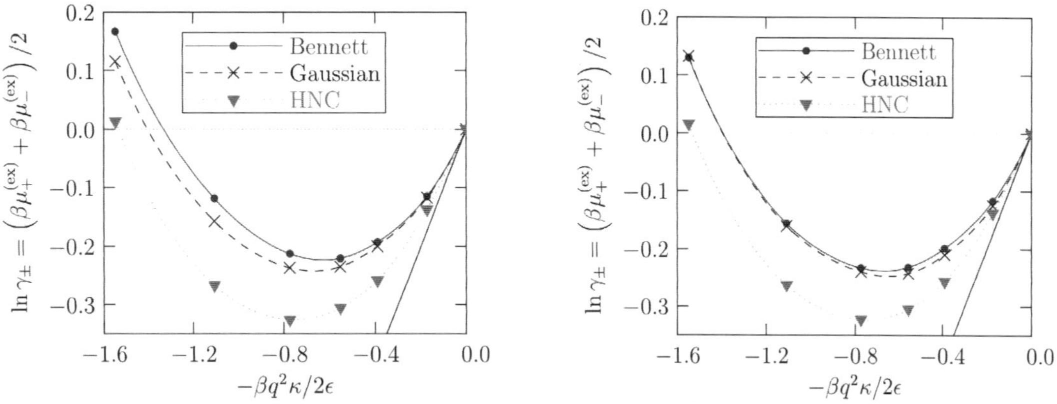

We estimate a non-electrostatic contribution directly by trial insertions, then electrostatic contribution on the basis of the observed distributions. The mean activity coefficients (left figure) obtained with the Gaussian approximation, the Bennett evaluation and HNC theory are qualitatively similar but quantitatively different from each other. With QCT conditioning (right figure), the mean activity coefficients evaluated by Bennett's method and the Gaussian approximation accurately agree, while HNC still disagrees with these. This suggests that Gaussian and also Bennett are physically reliable with this conditioning, but the more ambitious HNC errs by not fully evaluating the distribution mean and variance, and deviates from the simulation results.

---

[1] W. G. McMillan Jr and J. E. Mayer, J. Chem. Phys. 13, 276.
[2] T. L. Beck, M. E. Paulaitis, and L. R. Pratt, THE POTENTIAL DISTRIBUTION THEOREM AND MODELS OF MOLECULAR SOLUTIONS (Cambridge University Press, 2006).
[3] D. Asthagiri, P. D. Dixit, S. Merchant, M. E. Paulaitis, L. R. Pratt, S. B. Rempe, and S. Varma, Chem. Phys. Letts. 485, 1 (2010).

376857 Incorporation of Proton Hopping Mechanism into Dissipative Particle Dynamics

Abstract:

We suggest an approach that incorporates proton dissociation and diffusion in aqueous solutions into dissipative particle dynamics (DPD). Since standard DPD potentials are too simplistic for detailed description of proton transfer, we introduce proton as a separate particle that interacts with bases via the Morse potential, which effectively allows for dissociation of the proton-base pair. Bases are modelled as charged or neutral beads that describe groups which are able to protonate, such as acid anions, water, ammonia etc. The Morse potential is cut and shifted at cut-off distance and thus has a limited range. Therefore, the proton does not entirely "belong" to a particular base, but rather interacts via modified Morse potential with overlapping bases, which makes its stand-alone existence apart from a "base bead" extremely improbable. If a proton interacts with more than one base at the same time, it forms an intermediate complex that can dissociate via different pathways. As a result, the proton bead is transferred from one base to another in the course of simulation. We demonstrate the capabilities of our approach on several phenomena of different complexity.

1. Proton diffusion in solutions of strong acids. Our model of proton transfer artificially mimics the Grotthius proton hopping mechanism. By varying the Morse parameters, we are able to reproduce the experimental self-diffusion coefficient of proton (with water self-diffusion coefficient serving as a reference) for different bead sizes that include one to four water molecules. If one DPD bead represents one water molecule, the model is able to mimic the local dynamics, such as hopping frequency and characteristic time.
2. Acid dissociation in aqueous solution. By adjusting the Morse parameters, we model dissociation of benzenesulfonic acid, which is a fragment of sulfonated polystyrene proton-exchange ionomers.
3. Phase behavior of ionic surfactant in aqueous solution. As a characteristic example, we consider an aminooxide type of surfactant that can be protonated at low pH. By reducing the concentration of protons we cause a decrease in critical micelle concentration, which results in a partial precipitation of the surfactant, qualitatively reproducing the experimental behavior.

The approach is applicable in mesoscale modeling of various systems where proton transfer plays an important role, such as fuel cell membranes. We implemented our model in DL_MESO, which is an efficient and widely available software for DPD simulations. This work is funded by NSF grant number 1207239.

Surface tension of pure components and mixtures by molecular simulation

Stephan Werth[1], Katrin Stöbener[2], Martin Horsch[1], Hans Hasse[1]

[1]Laboratory of Engineering Thermodynamics, University of Kaiserslautern, Erwin-Schrödinger-Str 44, 67663 Kaiserslautern, Germany
[2]Department for Optimization, Fraunhofer Institute for Industrial Mathematics, Fraunhofer-Platz 1, 67663 Kaiserslautern, Germany

Keywords: Surface tension, Molecular Simulation, Multi-criteria Optimization
Corresponding author: Stephan Werth <stephan.werth@mv.uni-kl.de>

Abstract

The surface tension of fluids is a thermodynamic property which is important for many applications in chemical and process engineering. Despite the importance of mixtures, work on their surface tension is comparatively scarce in the literature.

The properties of real pure fluids and their mixtures are modeled here by force fields of the multi-center Lennard-Jones + electrostatic sites type. The models were taken from the literature [1] and were not modified. These models were adjusted only to vapor-liquid equilibrium properties of the bulk fluid, i.e. to vapor pressures, saturated liquid densities and enthalpy of vaporization. Hence, in the present work, their predictive power regarding the description of interfacial properties to which they were not adjusted is tested. The studied pure fluids include among others: CO_2, N_2, O_2, C_2H_6, CO. Furthermore several binary mixtures of these components were investigated.

Molecular dynamics simulations were carried out with the program *ls1* [2]. The density gradient at the interface makes it necessary to apply asymmetric long-range corrections for the pair potentials in these simulations to obtain meaningful results. Therefore, a slab-based long range correction [3] is combined with an angle averaging method for elongated molecules [4] in the present study.

It is shown that in many cases excellent predictions of the surface tension of the pure fluids and mixtures is achieved, even though the molecular models were not adjusted to such data. Where necessary, the quantitative agreement can be improved by including experimental data on the surface tension of the pure fluids in the parameterization of the force fields. It is shown that multi-criteria optimization based on the concept of Pareto fronts provides an excellent framework for this [5].

The simulation yields not only the surface tension but also detailed information on the properties of the interface including the adsorption at the interface and the resulting enrichment of components at the interface. It is shown that these effects, which can presently not be directly measured, can be readily obtained from the molecular simulations.

References

[1] J. Vrabec, J. Stoll, H. Hasse, J. Phys. Chem. B, 2001, 105, 12126-12133
[2] M. Buchholz, H.-J. Bungartz, J. Vrabec, J. Comput. Sci., 2011, 2, 124-129
[3] J. Janeček, J. Phys. Chem. B, 2006, 110, 6264-6269
[4] R. Lustig, Mol. Phys., 1988, 65, 175-179
[5] K. Stöbener, R. Klein, S. Reiser, M. Horsch, K.-H. Küfer, H. Hasse, Fluid Phase Equilib., 2014, 373, 100-108

Coarse-grained simulation of dsDNA-dendrimer complex formation

Abstract

To describe DNA compaction by dendrimers, which is a vehicle for gene delivery, we develop a new mesoscale model which nevertheless resolves the DNA double helix structure using the 3SPN.1 DNA model of the de Pablo group [1], and a model of the PAMAM dendrimer adapted from the work of Muthukumar and coworkers [2]. The predictions of our new model on effects of dendrimer generation, dendrimer surface charge density, and salt concentration on dendrimer – DNA complexes formation are consistent with both experiments and potential of mean force results from all-atom MD simulations, but give much more detail regarding the structure of the complex. The model predicts that DNA wraps a fully charged G5 (generation 5) or G6 dendrimer at low salt concentration (10 mM) similarly to a histone octamer, and for the G5 dendrimer, DNA super helices with both handednesses occur. At salt concentrations above 50 mM, or when a high fraction of dendrimer surface charges are neutralized by acetylation, DNA adheres, but does not compactly wrap the dendrimer, in agreement with experimental findings. We are also able to simulate pairs of dendrimers binding to the same DNA strand. Thus, this mesoscale simulation could be a good starting point of understanding chromatin formation. Moreover, this model can be extended to other cationic macroion – DNA systems which are also of great interest, such as, polylysine, micelles, and colloidal particles.

Reference

[1] E J Sambriski, D C Schwartz, J J de Pablo (2009) Biophysical journal 96 (5) p. 1675-1690

[2] P Welch, M Muthukumar (2000) Macromolecules 2 (1) p. 6159-6167

From Feedstock to Final Product

Toward Molecular Dynamics Simulations Encompassing All Stages of Single-Walled Carbon Nanotube Synthesis

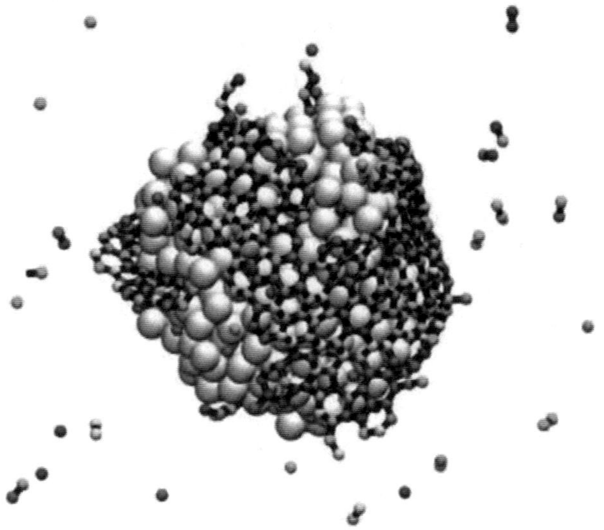

Jonathan E. Mueller| 19 November 2014
2014 AIChE Annual Meeting, Atlanta

ulm university universität uulm

1

Introductory Comments

1. Motivation:
   - Understand CNT growth in order to better control it.
2. Objective:
   - Describe CNT synthesis from start to finish using a single unconstrained model.
3. Caveat:
   - There is not one universal mechanism. Too many different systems work.

2

Outline

1. Method:
   - ReaxFF reactive force field
2. Previous work:
   - Feedstock decomposition
3. Recent progress:
   - Nanotube nucleation

3

ReaxFF Reactive Force Field

$$E_{system} = \boxed{E_{bond}} + \boxed{E_{val}} + \boxed{E_{tors}} + \boxed{E_{inv}} + E_{Hbond} + E_{vdWaals} + E_{Coulomb}$$

Bond-order dependent terms

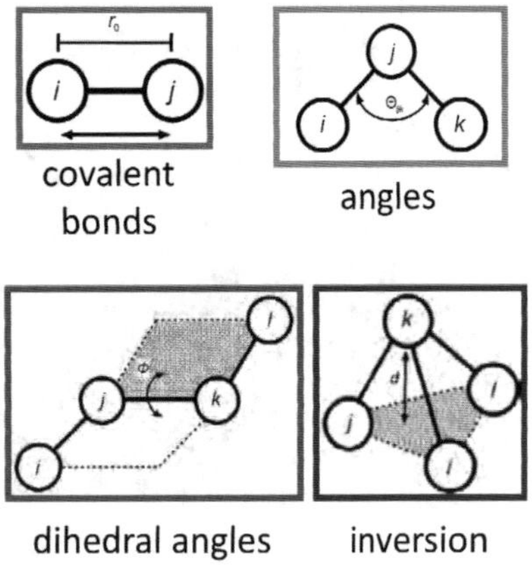

Bond-distance/bond-order relationship

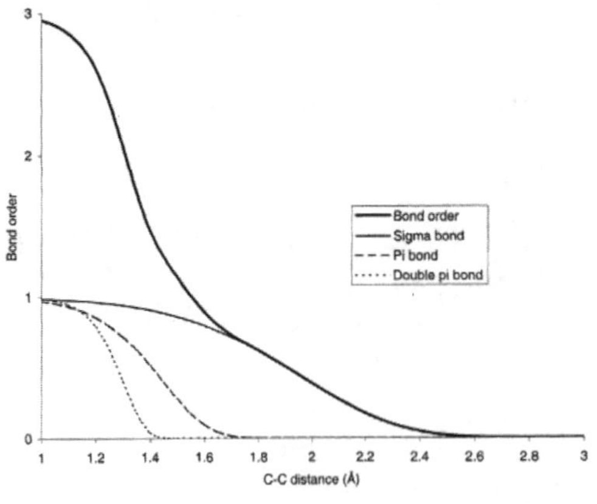

van Duin, A. C. T.; Dasgupta, S.; Lorant, F.; Goddard, W. A. *J. Phys. Chem. A*, **2001**, *105*, 9396-9409.

4

ReaxFF Reactive Force Field

$$E_{system} = E_{bond} + E_{val} + E_{tors} + E_{inv} + \boxed{E_{Hbond}} + \boxed{E_{vdWaals}} + \boxed{E_{Coulomb}}$$

Non-covalent bond terms

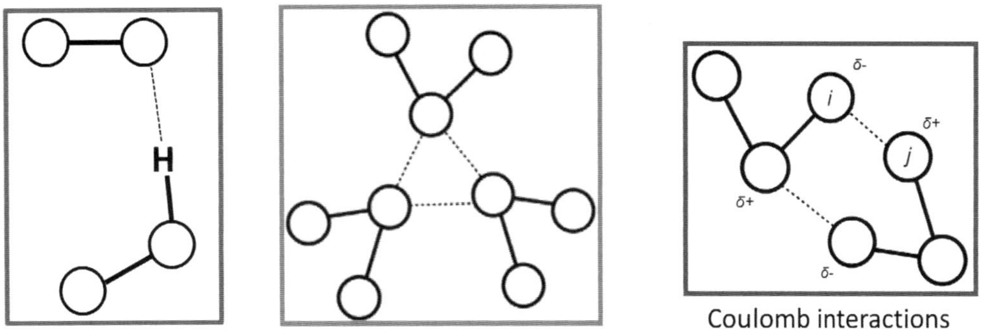

hydrogen bonding

van der Waals interactions

Coulomb interactions (time-dependent charges)

van Duin, A. C. T.; Dasgupta, S.; Lorant, F.; Goddard, W. A. *J. Phys. Chem. A,* **2001**, *105*, 9396-9409.

5

ReaxFF for Ni/C/H

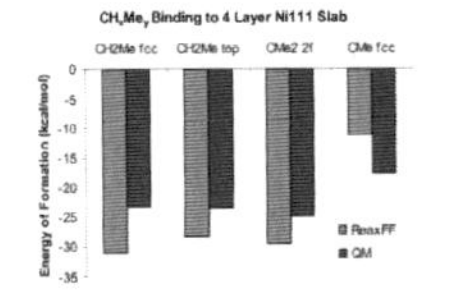

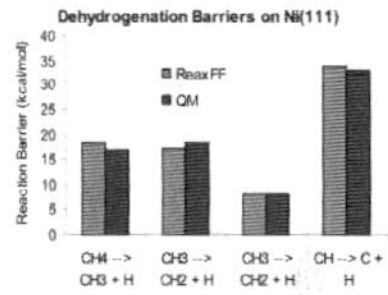

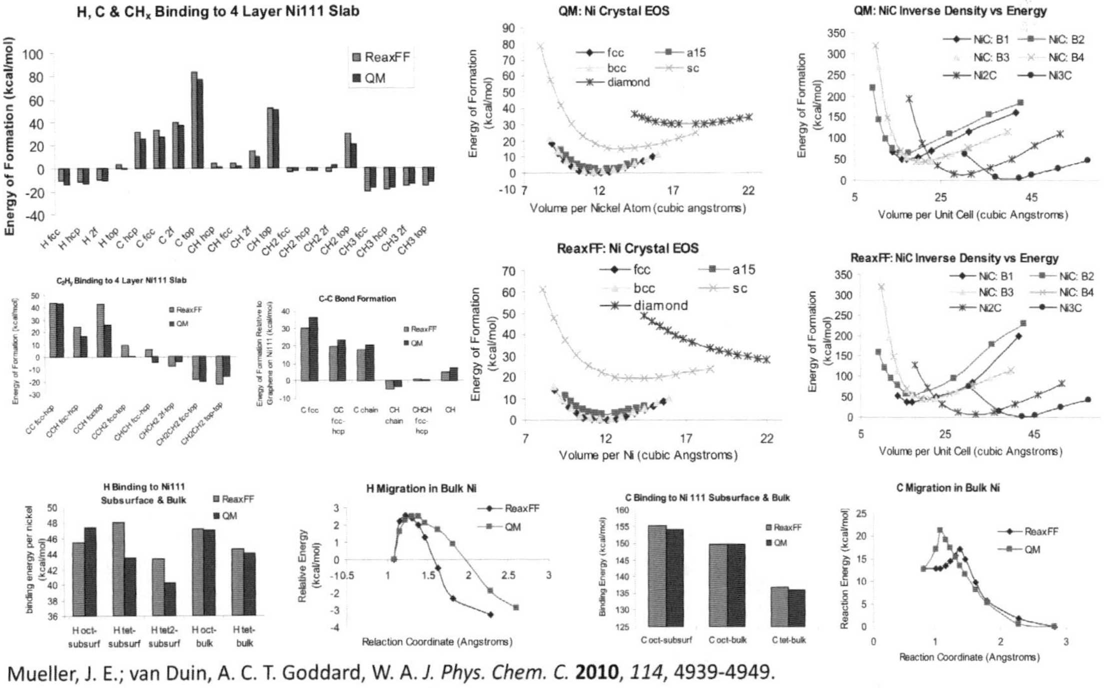

Mueller, J. E.; van Duin, A. C. T. Goddard, W. A. *J. Phys. Chem. C.* **2010**, *114*, 4939-4949.

Feedstock Decomposition

7

Temperature programmed reaction dynamics: MD studies of hydrocarbon decomposition

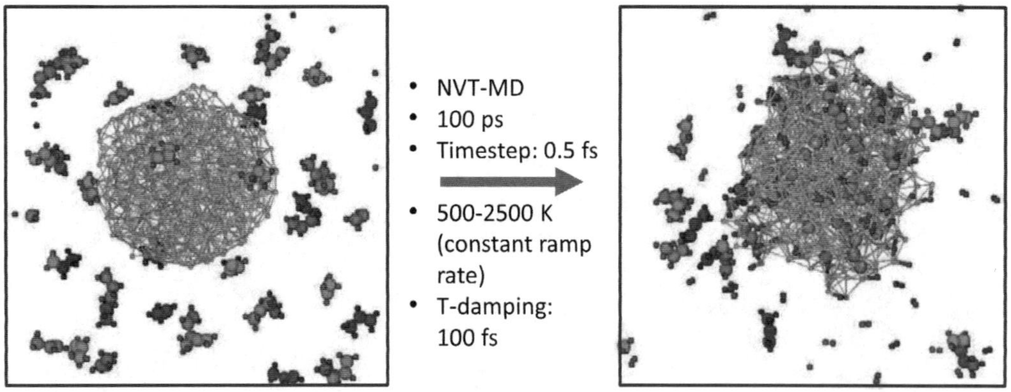

Reactive molecular dynamics were performed to investigate the adsorption and decomposition of various hydrocarbon species on a Ni_{468} nanoparticle.

8

Mueller, J. E.; van Duin, A. C. T. Goddard, W. A. *J. Phys. Chem. C.* **2010**, *114*, 5675-5685.

ReaxFF: acetylene on Ni nanocatalyst

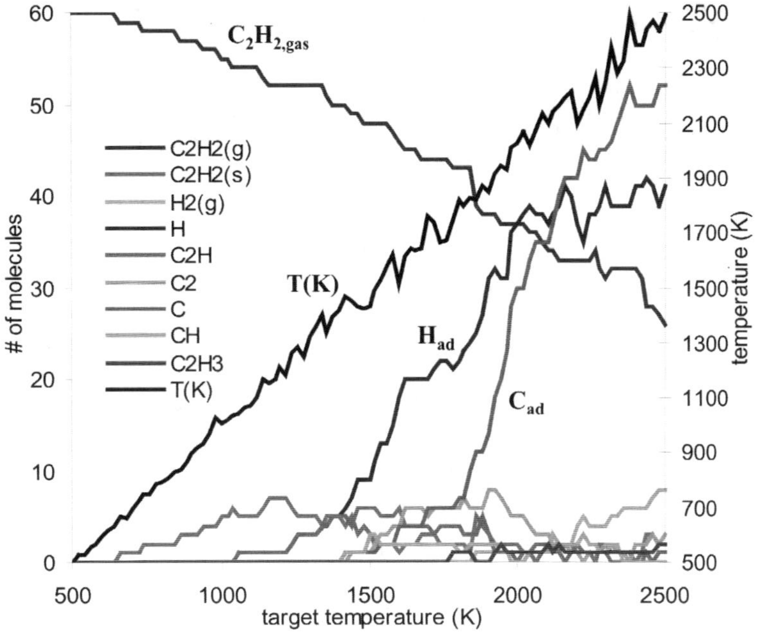

Mueller, J. E.; van Duin, A. C. T. Goddard, W. A. *J. Phys. Chem. C.* **2010**, *114*, 5675-5685.

Acetylene decomposition

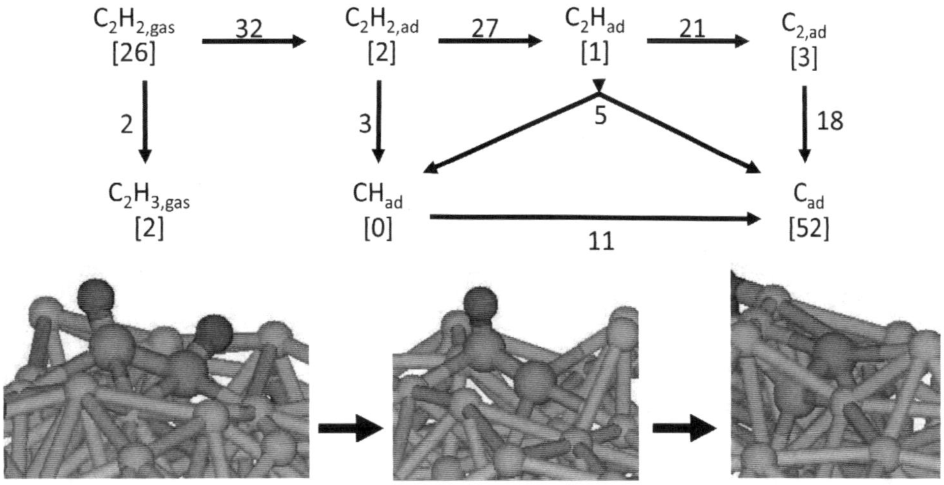

- Both C-H bonds typically break before the C-C bond.
- Formation of subsurface C helps break C-C bonds.

Mueller, J. E.; van Duin, A. C. T. Goddard, W. A. *J. Phys. Chem. C.* **2010**, *114*, 5675-5685.

Benzene decomposition

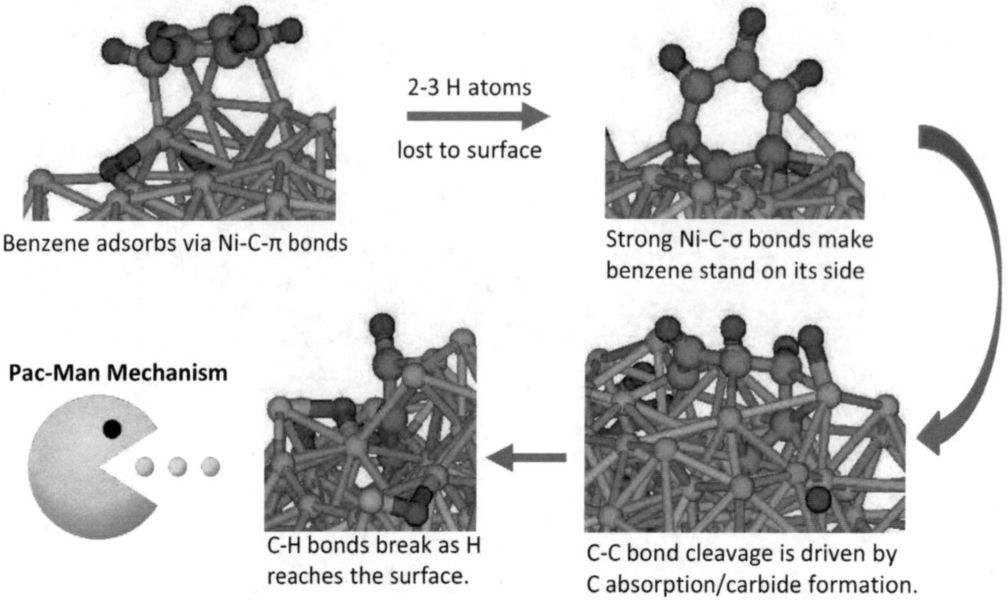

Decomposition path: $C_6H_6 \rightarrow C_6H_3 \rightarrow C_5H_3 \rightarrow C_5H_2 \rightarrow C_4H_2 \rightarrow C_4H \rightarrow C_3H \rightarrow C_3$

Mueller, J. E.; van Duin, A. C. T. Goddard, W. A. *J. Phys. Chem. C.* **2010**, *114*, 5675-5685.

11

Studying Nucleation

12

CNT nucleation from C_2H_2 on Ni_{468}

- Starting structure
  - 40Å x 40Å x 40Å periodic box.
  - Ni_{468} cluster (21Å)
  - 100 C_2H_2 gas phase molecules
- 2 or 5 ns of NVT-MD
- 0.5 fs time- step
- 50 gas phase C_2H_2 molecules added at each of the following: 200, 400, 600, 800 & 1000 ps.
- 4 temperatures:
  - 1000K (3 runs)
  - 1500K (3 runs)
  - 2000K (3 runs)
  - 2500K (3 runs)
- 100 fs T-damping

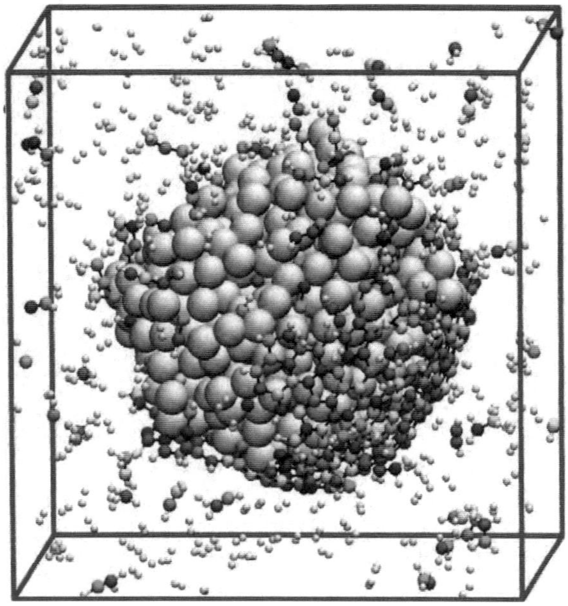

13

Structures after 2 ns

1000 K

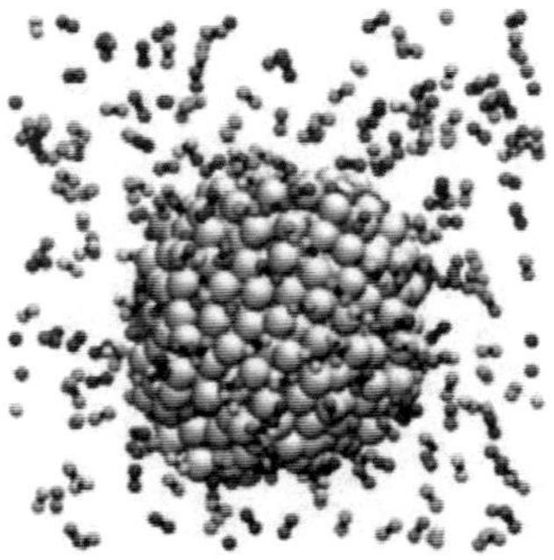

1500K

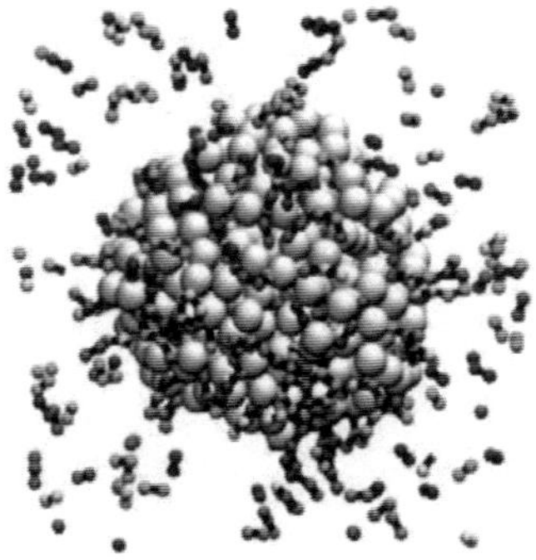

2000 K

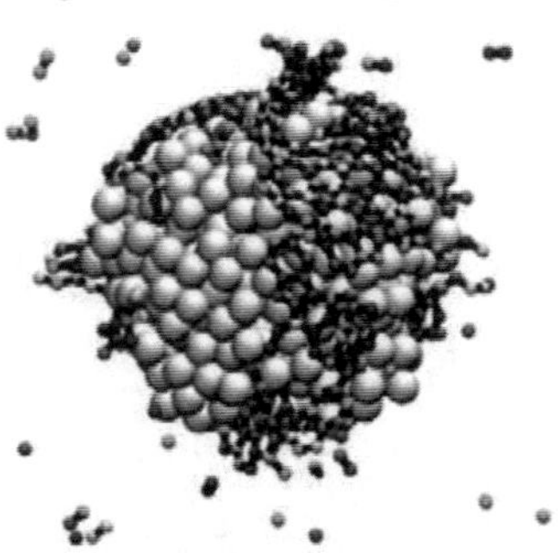

2500 K

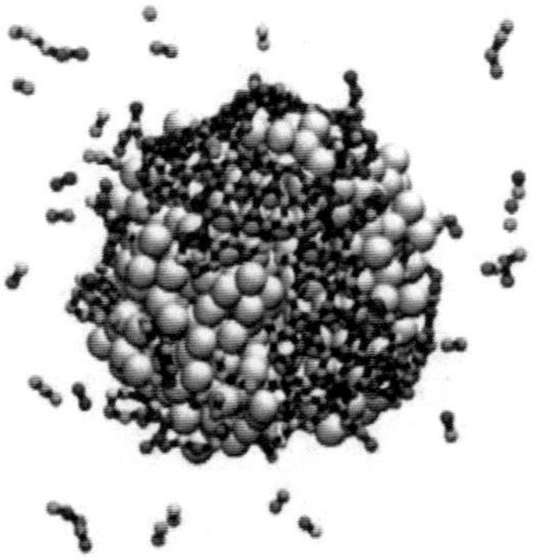

Temperature dependence

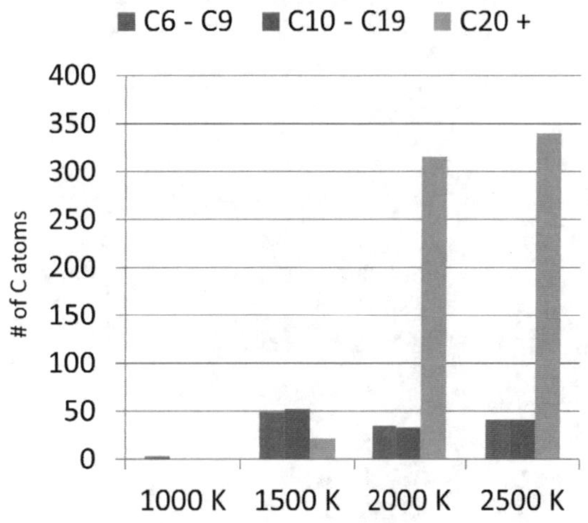

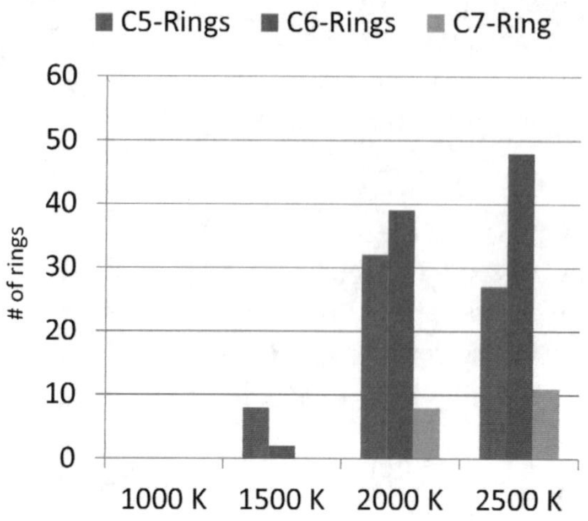

High temperatures are necessary for the formation of large carbon ring structures within a 2ns time frame.

15

C-Type Analysis: Acetylene (c_1)

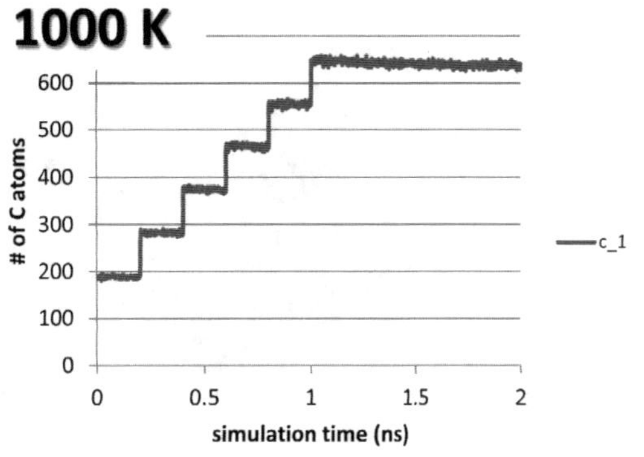

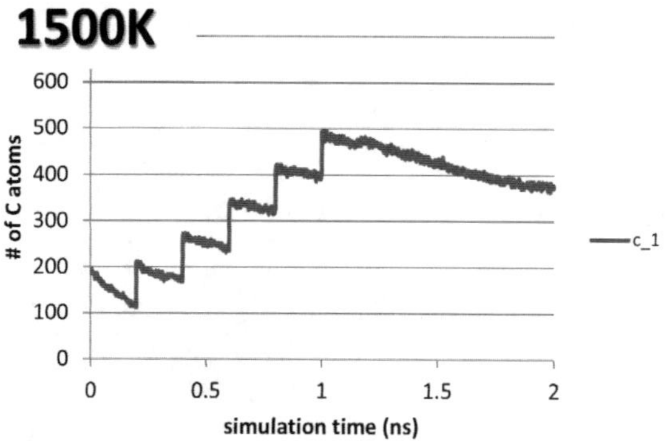

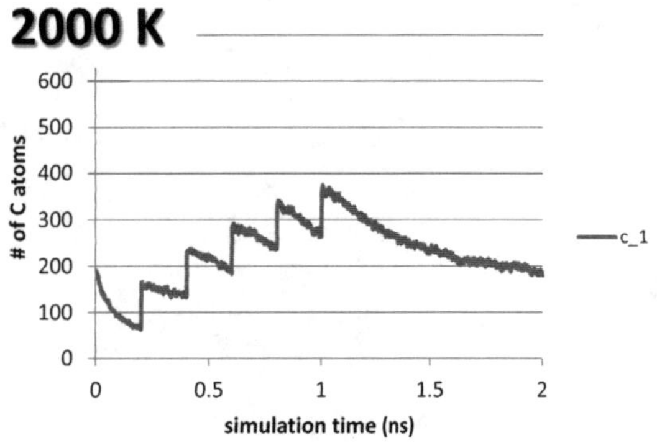

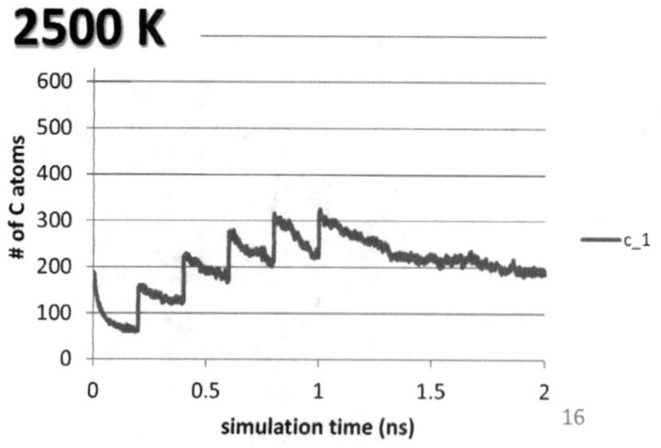

16

C-Type Analysis: Carbide (c_0)

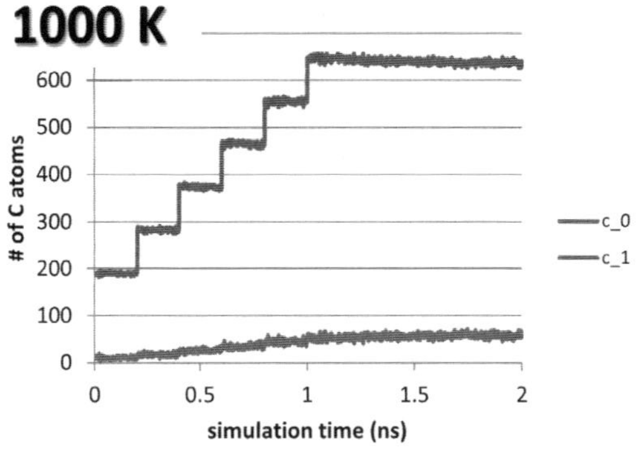

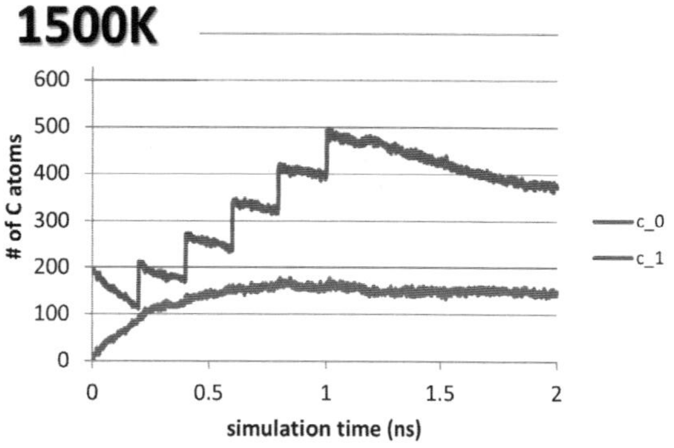

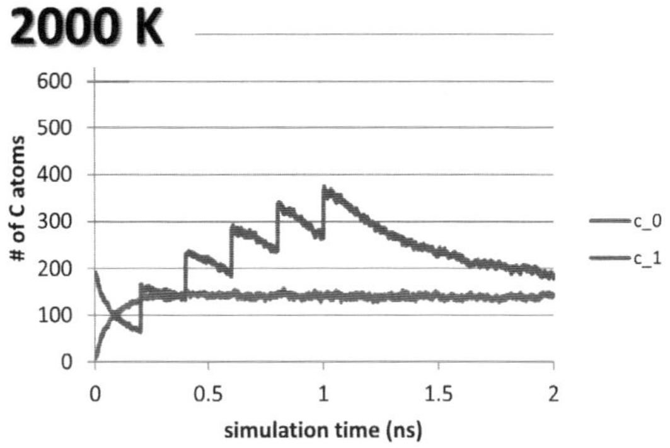

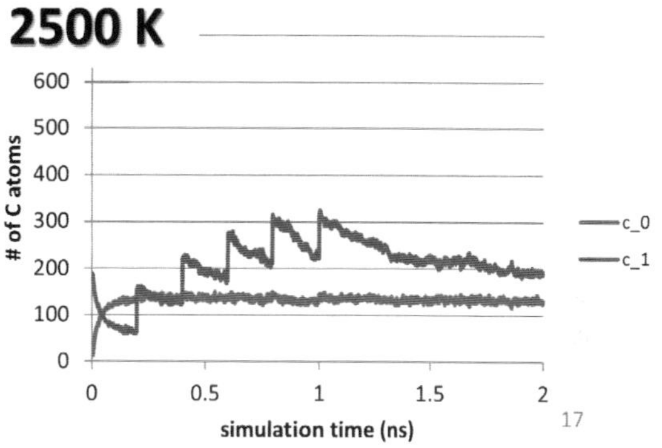

17

Cross sections showing carbide formation

Temperature dependence (after 1200 ps)

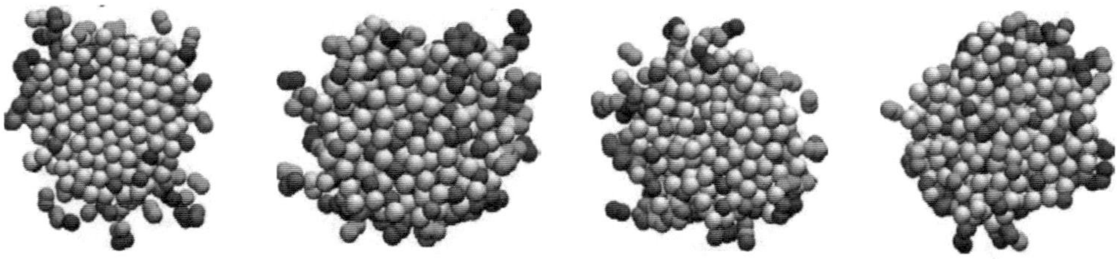

1000 K **1500 K** **2000 K** **2500 K**

Time dependence at 2000 K

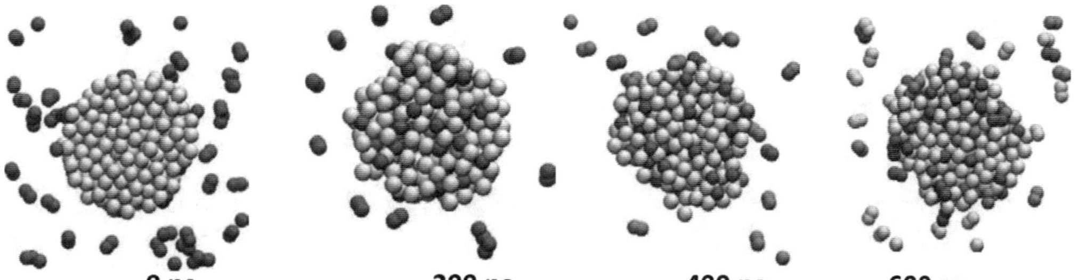

0 ps **200 ps** **400 ps** **600 ps**

18

C-Type Analysis: Chains (c_2)

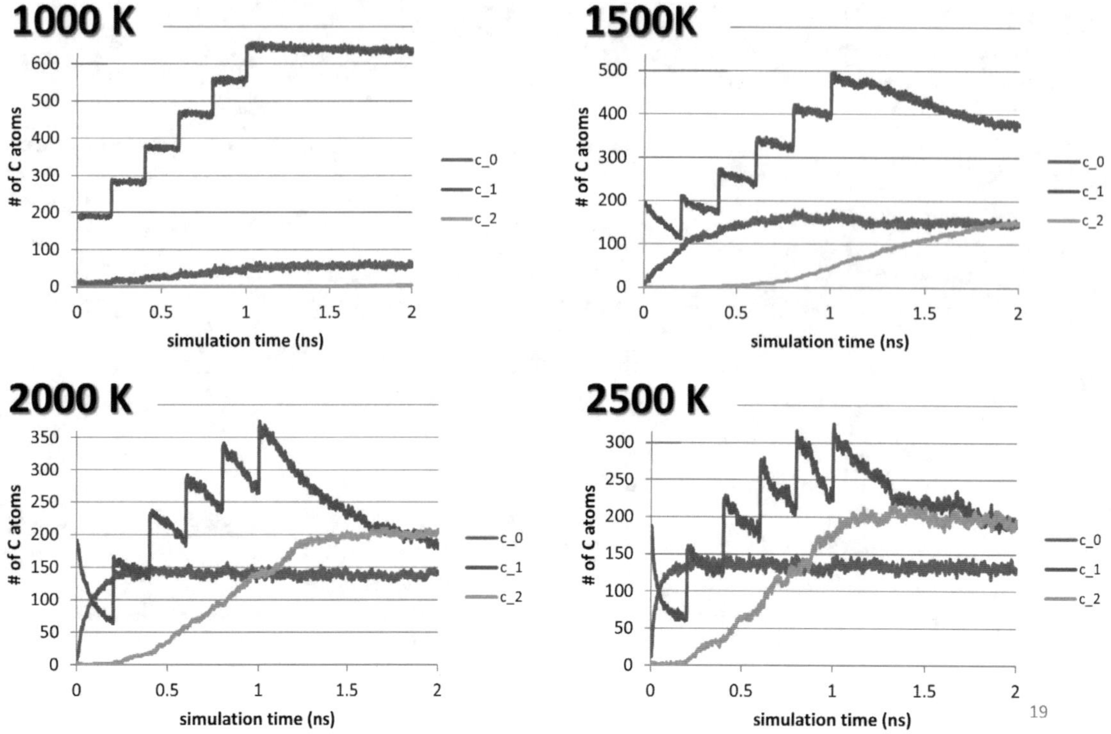

19

C-Type Analysis: 2-D (c_3)

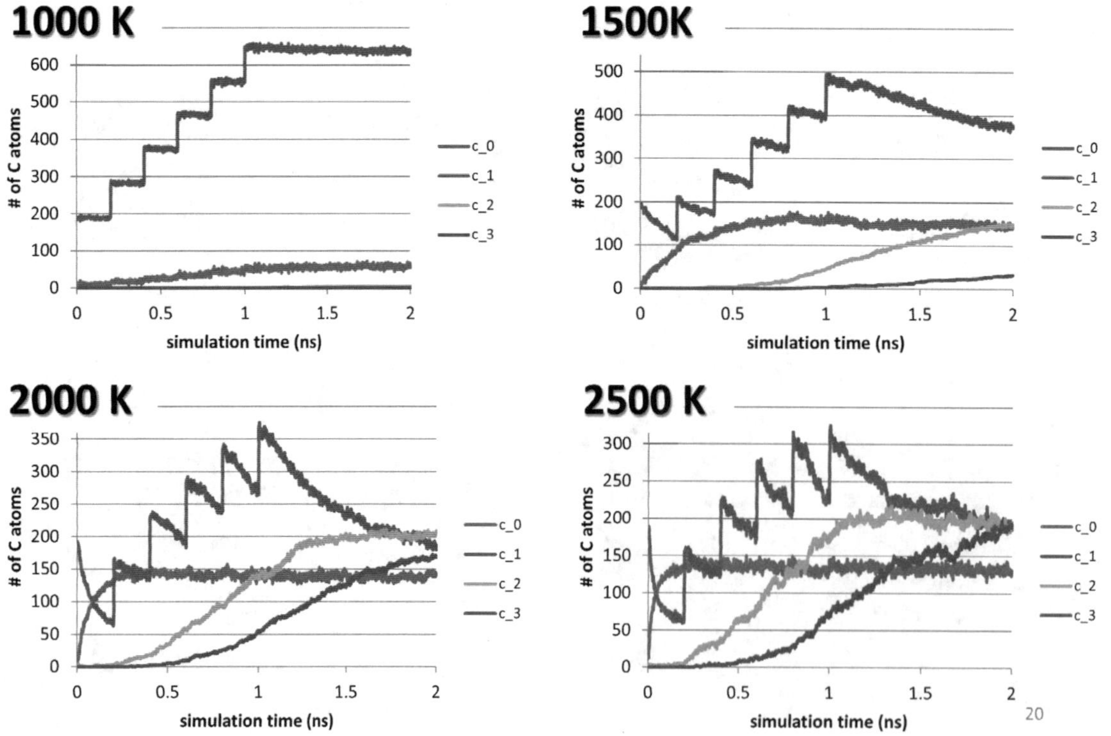

20

5 ns at 2000 & 2500 K

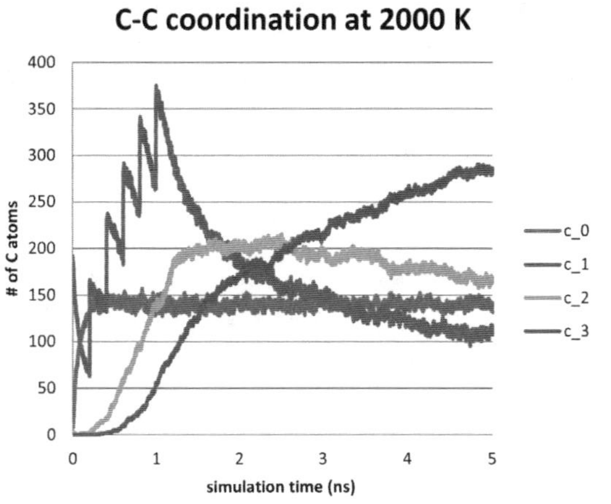

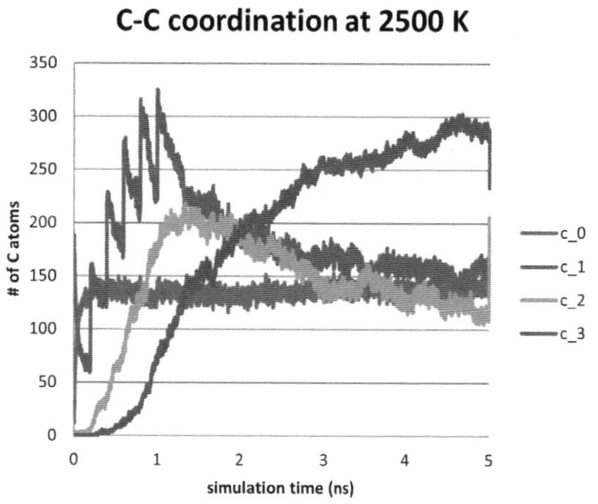

21

5 ns of ring formation

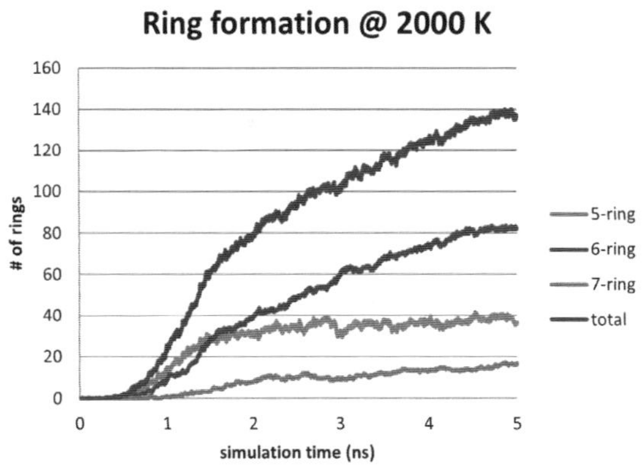

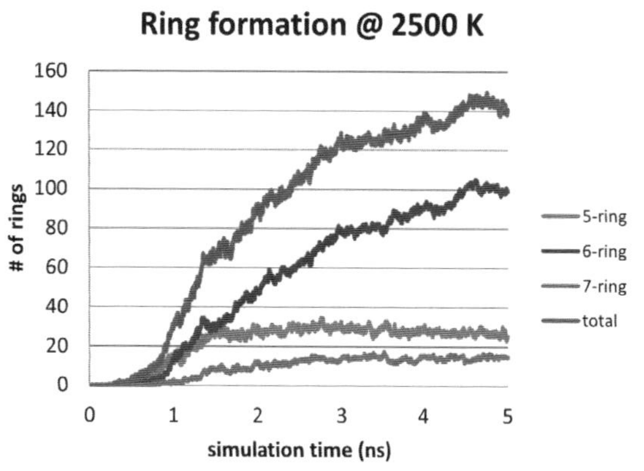

22

5 ns of C-type analysis

Processes observed

1. Formation of short chains, which wrap around to form rings.
2. Formation of multiple nuclei.
3. Nuclei grow individually and then combine together.
4. Healing of defects and further expansion of nucleus .

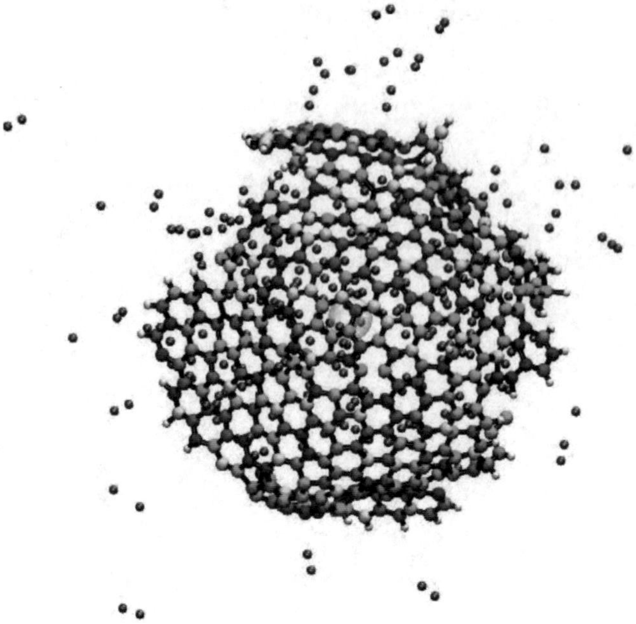

23

Structures after 5 ns

2000 K

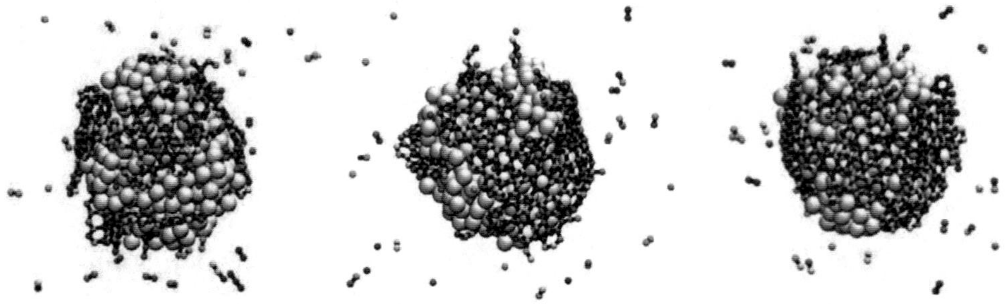

2500 K

24

Concluding Remarks

Steps from feedstock to CNT nucleus

1. Feedstock adsorption
2. C-H bond cleavage
3. C-C bond cleavage/ formation of carbide.
4. Formation of C-chains, which wrap around to form ring structures/nuclei.
5. Nuclei grow together to form a single CNT nucleus.

Challenges

1. Continuing the simulations through the „lift-off" phase and into the growth phase.
2. CNT growth simulations at lower temperatures (1500 K or lower).

25

Acknowledgements

Prof. William A. Goddard, Caltech

Prof. Adri van Duin, Penn State

Thank you for your attention!

26

Questions?

27

381805 Improving Intermolecular Force-Fields for Crystal Structure Prediction

Abstract:

A given molecule may crystallise in several crystal forms called *polymorphs*. Different polymorphs exhibit differences in physical properties and as a result *polymorphism* is a phenomenon of great industrial importance, especially for pharmaceutical and agrochemical industries where most solid products are crystalline. Thus the development of techniques for the prediction of the possible crystal structures of a given compound has great practical benefits, helping to avoid possible manufacturing problems and allowing the improvement of the properties of the stable structures.

There have been significant advances in recent years in computational methods for the prediction of the crystal structures of small organic molecules, based on lattice energy minimization at 0 K [1]. Reliable predictions of possible crystal structures have been obtained for many rigid molecules [2] and for flexible molecules of molecular weight up to 500 g/mol [3]. However, while known polymorphs are usually found as low-energy minima, there is evidence that in many cases the relative energy of the minima is not in agreement with experimental data [4]. The development of more accurate lattice energy models is thus necessary to enhance prediction capabilities, especially when relative stabilities of different polymorphs differ by only a few kJ/mol.

In our computational approach, based on the Crystal Predictor [5] and Crystal Optimizer [6] algorithms, the lattice energy is calculated following a classical approach in which the energy is partitioned into intramolecular and intermolecular pairwise contributions; the latter can be further partitioned into repulsive/dispersive and electrostatic terms. The intramolecular energy is derived using isolated molecule quantum mechanical calculations, the electrostatics are modelled through atomic charges or distributed multipole moments derived from the ab initio charge density [7] and the repulsive/dispersive energy using an empirical exp-6 potential with parameters taken from the literature [8, 9, 10, 11].

In this study, the impact of the modelling choices made for each component is investigated: the level of theory for the quantum mechanical calculations and the values of the repulsion/dispersion potential parameters used to model crystals of the ROY molecule are varied and the resulting changes in lattice energy, crystal structure and relative stability of two polymorphs are reported. The sensitivity analysis reveals the need to re-estimate the parameters of the exp-6 potential, which were originally derived based on levels of theory and electrostatic models that are now rarely used in crystal structure prediction. A re-parameterization is thus expected to produce an optimized set of parameters consistent with any changes in the lattice energy model and to capture remaining errors and approximations elsewhere in the model [12].

A method for deriving parameters for atom-atom exp-6 potential for crystals has been developed and is presented. The method is based on the minimization of deviations between experimental and minimized experimental crystal structures and deviations of their lattice energies from the experimental heats of sublimation. Experimental structures are compared with the minimized experimental structures using as measure of similarity the root mean square deviation of the 15 molecule coordination sphere of the two structures. The proposed method is applied to a set of hydrocarbons and the effect of the new set of parameters on the quality of the predictions is discussed.

References

[1] G.M. Day. Current approaches to predicting molecular organic crystal structures. *Crystallography Reviews*, 17(1):3-52, 2011.

[2] G.M. Day, T.G. Cooper, A.J. Cruz-Cabeza, K.E. Hejczyk, H.L. Ammon, S.X. M. Boerrigter, J.S. Tan, R.G. Della Valle, E. Venuti, J. Jose, S.R. Gadre, G.R. Desiraju, T. S. Thakur, B.P. van Eijck, J.C. Facelli, V.E. Bazterra, M.B. Ferraro, D.W. M. Hofmann, M.A. Neumann, F.J. J. Leusen, J. Kendrick, S.L. Price, A.J. Misquitta, P.G. Karamertzanis, G.W. A. Welch, H.A. Scheraga, Y.A. Arnautova, M.U. Schmidt, J. van de Streek, A.K. Wolf, and B. Schweizer. Significant progress in predicting the crystal structures of small organic molecules - a report on the fourth blind test. *Acta Crystallographica Section B*, 65(2):107-125, Apr 2009.

[3] A.V. Kazantsev, P. G. Karamertzanis, C. S. Adjiman, C.C. Pantelides, S.L. Price, P.T.A. Galek, G.M. Day,and A.J. Cruz-Cabeza. Successful prediction of a model pharmaceutical in the fifth blind test of crystal structure prediction. *International Journal of Pharmaceutics*, 418(2):168-178, 2011. A priori Performance Predictions.

[4] M. Vasileiadis, A.V. Kazantsev, P.G. Karamertzanis, C.S. Adjiman, and C.C. Pantelides. The polymorphs of ROY: application of a systematic crystal structure prediction technique. *Acta Crystallographica Section B*, 68(6):677{685, Dec 2012.

[5] P. G. Karamertzanis and C. C. Pantelides. Ab initio crystal structure prediction. II. Flexible molecules. *Molecular Physics*, 105(2-3):273-291, 2007.

[6] A.V. Kazantsev, P.G. Karamertzanis, C.C. Pantelides, and C.S. Adjiman. CrystalOptimizer: An Efficient Algorithm for Lattice Energy Minimization of Organic Crystals Using Isolated-Molecule Quantum Mechanical Calculations, pages 1-42. Wiley-VCH Verlag GmbH & Co. KGaA, 2011.

[7] A.J. Stone. *The Theory of Intermolecular Forces*. Clarendon Press, Oxford,1996.

[8] D. E. Williams and S. R. Cox. Nonbonded potentials for azahydrocarbons:the importance of the Coulombic interaction. *Acta Crystallographica Section B*, 40(4):404-417, Aug 1984.

[9] S. R. Cox, L.-Y. Hsu, and D. E. Williams. Nonbonded potential function models for crystalline oxohydrocarbons. *Acta Crystallographica Section A*,37(3):293-301, May 1981.

[10] D. E. Williams and D. J. Houpt. Fluorine nonbonded potential parameters derived from crystalline perfluorocarbons. *Acta Crystallographica Section B*,42(3):286-295, Jun 1986.

[11] L.-Y. Hsu and D. E. Williams. Intermolecular potential-function models for crystalline perchlorohydrocarbons. *Acta Crystallographica Section A*,36(2):277-281, Mar 1980.

[12] C.C. Pantelides, C.S. Adjiman, and A.V. Kazantsev. General computational algorithms for ab initio crystal structure prediction for organic molecules. In Sule Atahan-Evrenk and Alan Aspuru-Guzik, editors, *Prediction and Calculation of Crystal Structures*, volume 345 of *Topics in Current Chemistry*, pages 25-58. Springer International Publishing, 2014.

383334 Basis Function Sampling for Material Property Computations

Abstract:

Wang--Landau sampling, and the associated class of flat histogram simulation methods, have been particularly successful for free energy calculations in a wide array of physical systems. Practically, the convergence of these calculations to a target free energy surface is hampered by reliance on biasing parameters which are unknown *a priori*, and which are often incommensurate with boundary conditions. We derive and implement a method based on orthogonal (basis) functions which is fast, parameter-free, and (importantly) geometrically robust. An important feature of this method is its ability to achieve arbitrary levels of description for the free energy. It is thus ideally suited to *in silico* measurement of elastic moduli and other quantities related to free energy perturbations. We demonstrate the utility of such applications by applying our method to calculation of the Frank elastic constants of the Lebwohl-Lasher model.

388336 Vapor Liquid Equilibria of Hydrofluorocarbons Via First Principles Monte Carlo Simulations

Abstract:

The Kohn-Sham density functional theory (DFT) is the most prominent way to compute condensed phase properties. In Kohn-Sham DFT, the local or semi local density functionals do not capture van der Waals interactions accurately. An accurate description of van der Waals interactions is essential in determing thermodyanmic properties of molecules. The development of fully non local van der Waals density functional adequatey describe dispersion interactions. In this work, we present first principles Monte Carlo simulations to obtain vapor liquid coexistence curves for hydrofluorocarbons by using Becke-Lee-Yang-Parr (BLYP) functional, dispersion corrected functionals, and with rVV10 nonlocal van der Waals density functional.

Development and Application of ab-initio based Force Fields for Noble Gas Adsorption in MOFs

Hakan Demir,[a] Jeffery A. Greathouse,[b] and David S. Sholl*[a]

[a]School of Chemical and Biomolecular Engineering, Georgia Institute of Technology, 311 Ferst Drive NW, Atlanta, Georgia 30332-0100

[b]Sandia National Laboratories, Albuquerque, New Mexico 87185

*Corresponding author: david.sholl@chbe.gatech.edu

Grand Canonical Monte Carlo (GCMC) simulation is a useful computational tool to predict adsorption in porous materials whose success heavily relies on the force field (FF) used. There are a variety of generic FFs in the literature that have showed some success for predicting adsorption, but the reliability of these FFs for MOFs is unclear. Ab-initio based FFs offer a potential path to overcoming these two issues. In this work, we derived ab-initio based FFs for noble gas (Ar, Kr, Xe) adsorption in M-MOF-74 (M=Co, Ni, Zn), ZIF-8 and CuBTC using periodic calculations with three DFT functionals (PBE-D2, vdW-DF, and vdW-DF2). Our simulation results show that there are cases where one DFT functional based FF can accurately predict experimental adsorption. However, none of the DFT functionals accurately describe interactions for all of the noble gas-adsorbent pair studied. It has been found that, on average, PBE-D2 and vdW-DF2 based FFs predict experiments at a similar quality of UFF at atmospheric and subatmospheric (0.01bar) pressure, while vdW-DF based FFs overpredict experiments significantly.

378187 Prediction of Adsorption and Diffusion Properties of CH4 in Zeolites Using Force Field Derived from DFT-CC Methods

Abstract:

Molecular simulations like Grand Canonical Monte Carlo (GCMC) and Molecular Dynamics (MD) are widely used to predict adsorption isotherms, heat of adsorption, and diffusion of adsorbate molecules in porous materials like zeolites. The accuracy of such predictions depends mainly on how accurately Force field represents adsorbate-zeolite interaction. For such simulations, either off-the-shelf force fields (FF) or adsorption fitted force fields are typically used. Developing force fields (FFs) from first principle methods is a powerful approach to developing transferrable FFs that improve upon these approaches. We have developed a transferrable force field for CH_4-zeolite interactions using the DFT-CC method. Here we first developed Coupled-Cluster (CC) corrections for CH_4-zeolite interactions using high level computational chemistry. The DFT-CC method adds corrections to DFT interaction energies to make them closer to CCSD(T) energies. We also tested other DFT methods, finding that PBE-D2, PBE-D3, and VDW-DF2 overestimate adsorption isotherms while the recently developed VDW-DF-CC method underestimates adsorption isotherms. We have applied our newly developed force field to obtain diffusion properties of CH_4 in silica-zeolites.

385574 Virial Coefficients of Flexible Molecules Using Path Integral Monte Carlo Methods to Capture Nuclear Quantum Effects

Abstract:

Virial coefficients and other cluster integrals can be used to compute various physical properties of gases, such as the pressure, critical properties, Joule-Thomson coefficient, and others. For small molecules, it becomes feasible to use first-principles methods, or potentials derived from first-principles considerations, to model the molecular interactions that are input to the cluster-integral calculation. Given this input, Mayer Sampling Monte Carlo (MSMC) can be used to evaluate virial coefficients efficiently. An important consideration when attempting to work with first-principles models is explicit treatment of nuclear quantum effects, which are significant for light atoms and/or low temperatures.

The Path-integral Monte Carlo (PIMC) methods provide a convenient way to incorporate these quantum effects, and interface well with the MSMC algorithm. PIMC operates by discretizing the positions of a molecule as "beads" that form a closed ring. Neighboring beads act as if they are connected by a harmonic spring with a spring constant that is proportional to temperature and mass. For multi-atomic molecules such as H_2 or H_2O, the PIMC is significantly more complicated to implement, because of the need for the path integrals to consider molecular orientation and internal degrees of freedom. These complexities directly affect the sampling efficiency involved in the MSMC method and capturing them accurately is a difficult task. In this presentation, we describe how we have accounted for both changing orientations and bond lengths in the PIMC technique, and used these developments to evaluate the virial coefficients using the MSMC method. We present results for second and third order virial coefficients of some multi-atomic molecules, examining both rigid and flexible potentials, and compare these results with experimental data as well as available values in the literature.

Ionic Liquids (ILs) have been widely studied as promising solvents for post-combustion carbon capture. ILs based on aprotic heterocyclic anions (AHAs) chemically complex with CO_2 at a 1:1 uptake ratio (moles CO_2:mole IL) and with a tunable strength of binding (heat of reaction) [1]. There is also current interest in the use of ILs in pre-combustion carbon capture. For example, Basha et al. [2, 3] have considered physically absorbing ILs for the pre-combustion case, using a pressure-swing absorption configuration.

In this study, we consider the use of chemically absorbing ILs in pre-combustion carbon capture, with the aim of determining chemical property (e.g., enthalpy of reaction) values that will lead to best performance. These values can then be used as targets in a computational molecular design (CMD) approach to determine candidate IL structures. In addition to considering ILs with 1:1 chemical uptake, we also consider ILs with a 2:1 chemical uptake ratio, and present experimental CO_2 uptake data for such a case. A two-site chemical uptake isotherm model based on structural cooperativity is proposed in order to explore the range of possible behavior for the higher uptake ILs. We use this isotherm model, together with a simple pre-combustion CO_2 capture process model, to determine target IL chemical property values that will minimize specified objective functions (e.g., parasitic power, processing cost, etc.) and to do related sensitivity studies. The process model is configured to include both temperature-swing absorption (TSA) and pressure-swing absorption (PSA), and includes energy integration.

References

[1] B. Gurkan, B. F. Goodrich, E. M. Mindrup, L. E. Ficke, M. Massel, S. Seo, T. P. Senftle, H.Wu, M. F. Glaser, J. K. Shah, E. J. Maginn, J. F. Brennecke, and W. F. Schneider, "Molecular Design of High Capacity, Low Viscosity, Chemically Tunable Ionic Liquids for CO_2 Capture," *J. Phys. Chem. Lett.*, 2010, 1, 3494–3499.

[2] O. M. Basha, M. J. Keller, D. R. Luebke, K. P. Resnik, and B. I. Morsi, "Development of a Conceptual Process for Selective CO2 Capture from Fuel Gas Streams Using [hmim][Tf2N] Ionic Liquid as Physical Solvents." *Energy Fuels* 2013, 27, 3905–3917.

[3] O. M. Basha, Y. J. Heintz, M. J. Keller, D. R. Luebke, K. P. Resnik, and B. I. Morsi, "Development of a Conceptual Process for Selective Capture of CO_2 from Fuel Gas Streams Using Two TEGO Ionic Liquids as Physical Solvents," *Ind. Eng. Chem. Res.* 2014, 53, 3184–3195.

Title:

Electrospray as a nanoscale gas-liquid contactor for carbon dioxide absorption in space habitats

Text:

Metabolic generation of carbon dioxide by astronauts in space habitats poses a health threat due to accumulation, and thus must be removed during missions. We have investigated the use of liquid amines for carbon dioxide removal by using electrospray as a novel gas-liquid contactor, which creates sub-micron droplets of liquid amine to improve diffusion-limited absorption processes. Through characterization of liquid properties, we can estimate the performance of liquid amine nanodroplets as a carbon dioxide scrubber. Further laboratory experiments of electrospraying mixtures of monoethanolamine into carbon dioxide atmospheres have demonstrated the effectiveness of this method.

Electrospray is an electrohydrodynamic phenomenon which permits sub-micron generation of droplets from liquids of varying properties by the application of a high voltage (1 kV/mm) electric field in the region of a liquid meniscus [1]. In doing so, the meniscus is deformed into a cone from which an electrically charged jet is emitted. In a short distance, the jet breaks up due to capillary instability (Plataeu-Rayleigh instability), forming tens of millions of droplets per second (see Figure 1). Most commonly used as an interface between liquid chromatography columns and mass spectrometers, electrospray has also been used or proposed for ionic propulsion of satellites [2,3,4], nanoparticle generation [5,6], nanofiber creation (electrospinning) [7], pharmaceutical encapsulation [8], nanocombustors [9], microelectronic cooling [10], dust mitigation technology [11,12,13], vacuum pumps [14], printers nozzles [15,16], as a chemical preparation method [17] and biological contaminant control [18].

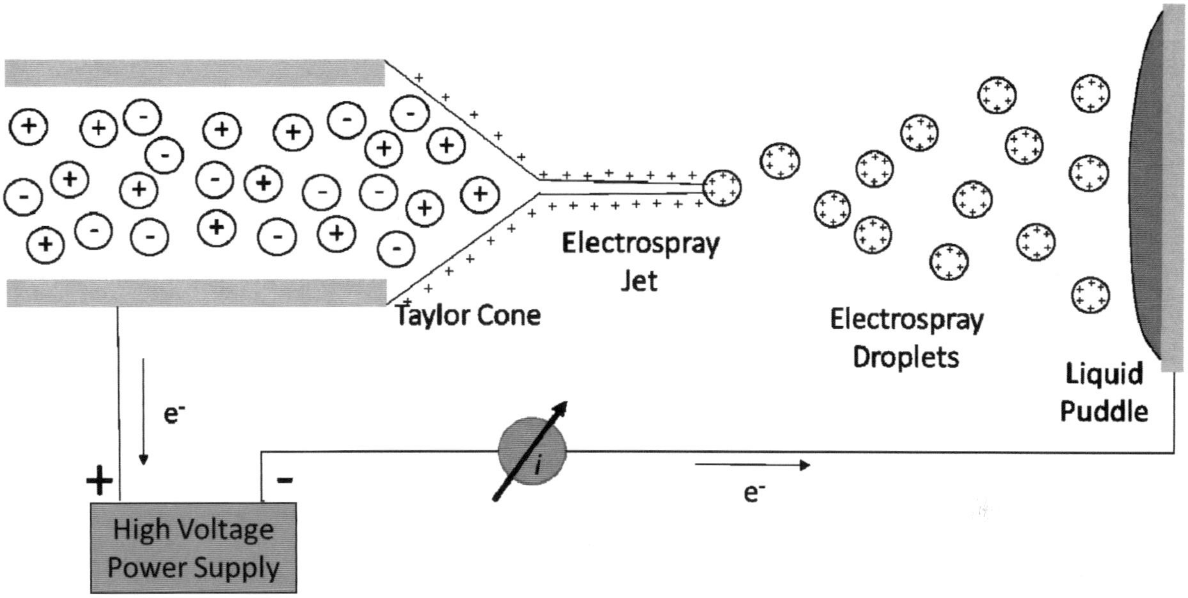

Figure 1: Schematic of the electrospray process

As a charged nanodroplet generator, electrospray is potentially ideal for gas-liquid contacting in space, since it overcomes the difficulty associated with controlling a liquid as the charged droplets can be manipulated by an electric field. As an atomization technique, electrospray requires approximately 1000x less energy to create a droplet of a given size than conventional atomization techniques [10] and permits generation of smaller diameter droplets than traditional techniques. Smaller diameter is important since increasing the surface area to volume ratio will help overcome the transport limitations associated with liquid absorption.

Mathematical descriptions of the electrospray [19] allow us to predict the droplet radius (R) and time of flight (τ) based on the following parameters: surface tension (gamma), density (rho), electrical conductivity (K), liquid flow rate (Q), gas viscosity (μ), droplet mobility (Z), distance to electrode (delta) and electric field (E).

$$R_{min} \cong \left[\frac{\gamma * \varepsilon_0^2}{\rho * K^2}\right]^{\frac{1}{3}}$$

$$Q_{min} = \frac{\gamma * \varepsilon_0}{\rho * K}$$

$$R \cong \left[\frac{Q * \varepsilon_0}{K}\right]^{\frac{1}{3}}$$

$$Z = \frac{2}{3} * \frac{(\gamma * \varepsilon_0 * R)^{\frac{1}{2}}}{\mu_{gas}}$$

$$\tau = \frac{\delta}{Z * E}$$

As we can see from the above equations, the radius of the droplet is dependent on Q and K, and these parameters can be varied to tune the performance of the electrospray gas-liquid contactor. The time of flight varies according to several parameters, but only the radius of the droplet can be easily adjusted over more than an order of magnitude. After the physical properties are known, we now have enough capability to predict the absorption of a single droplet. Indeed, absorption by electrospray droplets has been studied before for the purpose of understanding ionization modes [200], but not as a gas-liquid contactor.

Essential to predicting system performance and understanding gas absorption into nanodroplets, we must first understand the solutions that we are electrospraying.

Key properties include viscosity, density, CO2 equilibrium pressure and electrical conductivity against the loading of the solution (number of moles of CO2 absorbed per mole of amine). Figure 2 shows that for pure monoethanolamine, the viscosity increases by more than two orders of magnitude as monoethanolamine becomes loaded with CO_2.

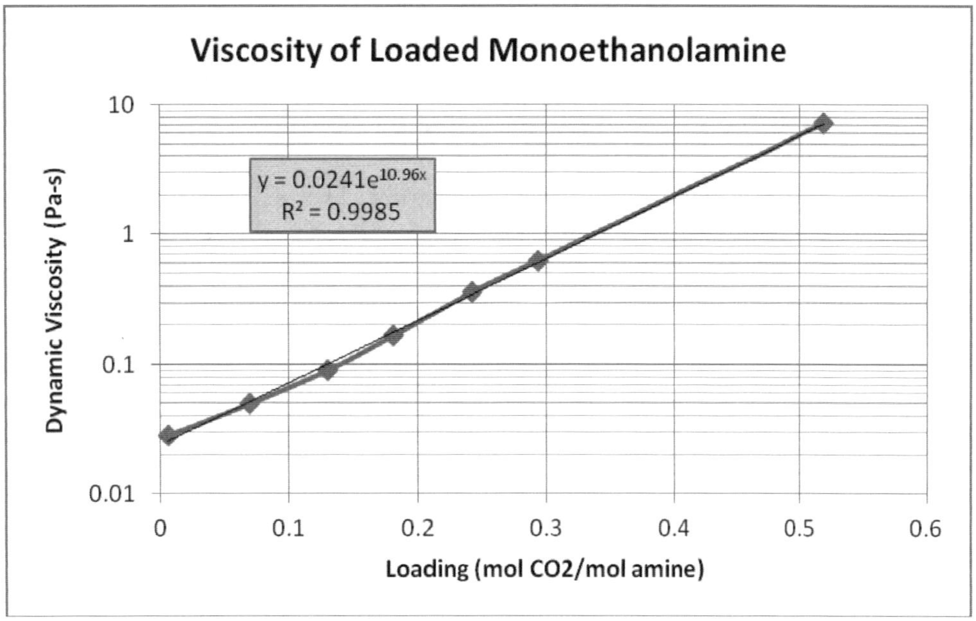

Figure 2: Viscosity of pure monoethanolamine vs. loading

By determining the fluid properties, one can estimate the diffusivity of CO_2 in MEA. The Stokes-Einstein relationship allows the estimation of diffusivity by assuming small spherical particles in a dilute solution. Even though loaded MEA is not dilute and highly ionic, the estimation of diffusivity may be confirmed by comparing measured electrical conductivity to the electrical conductivity predicted by the Nernst-Einstein equation. The diffusivity (D) may be used to describe transport within spherical nanodroplets, where q is charge, c is ion concentration, R is ionic radius, Z is the ion mobility, and mu is solvent viscosity [21].

$$D_{Stokes-Einstein} = Z * k_B * T = \frac{k_B * T}{6\pi * \mu_{solvent} * R_{ion}}$$

$$K = \frac{D}{k_B * T} q^2 * c_{solute}$$

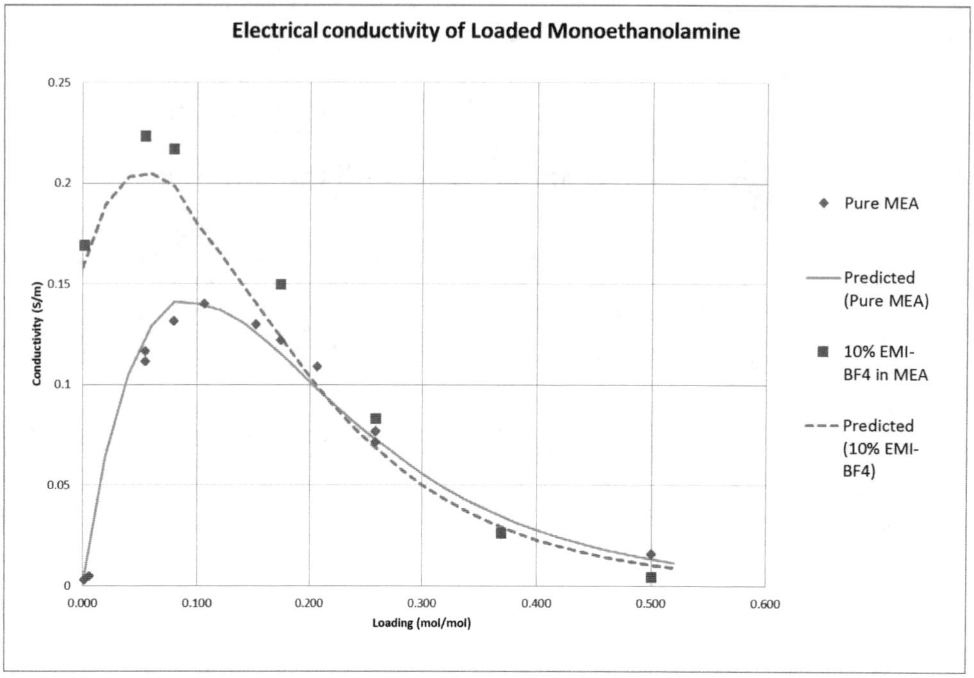

Figure 3: Electrical Conductivity of Loaded Monoethanolamine, data vs. predictions

As shown in Figure 3, the electrical conductivity is predicted with excellent correlation through use of the Stokes-Einstein and Nernst-Einstein equations for pure MEA, indicating a confidence in diffusivity better than 10%, substantially exciting given that loaded MEA is far from ideal and highly ionic. Slightly more error is noted with the more complex and phase-separating system of MEA/EMI-BF4 (1-ethyl-3-methylimidazolium tetrafluoroborate) partially due to a lack of viscosity data. Since diffusivity changes by two orders of magnitude over the entire loading range, a potential error of 10% in diffusivity should produce very small errors in mass transport calculations.

In order to experimentally measure the absorption by electrospray, a small (~10 mL) sealed vessel was constructed with a gauge pressure transducer and a septum for GC/MS measurements. The center of the vessel has a port for a silica capillary to be passed through the pressure vessel in order for an electrospray to be established. The chamber is initially filled with a gas mixture containing 1-10% CO_2 and sealed at ambient pressure. During operation of the electrospray source, the pressure will decrease as CO_2 is absorbed, becoming sub-ambient which eliminates false positives due to pressure loss by vessel leaks. By measuring the rate of decay (and comparing that to the feed rate of the solution), it is possible to determine a loading of each droplet. Pressure measurements can be compared with GC/MS measurements for confirmation of absorption of CO_2. An example of these data is found in Figure 4.

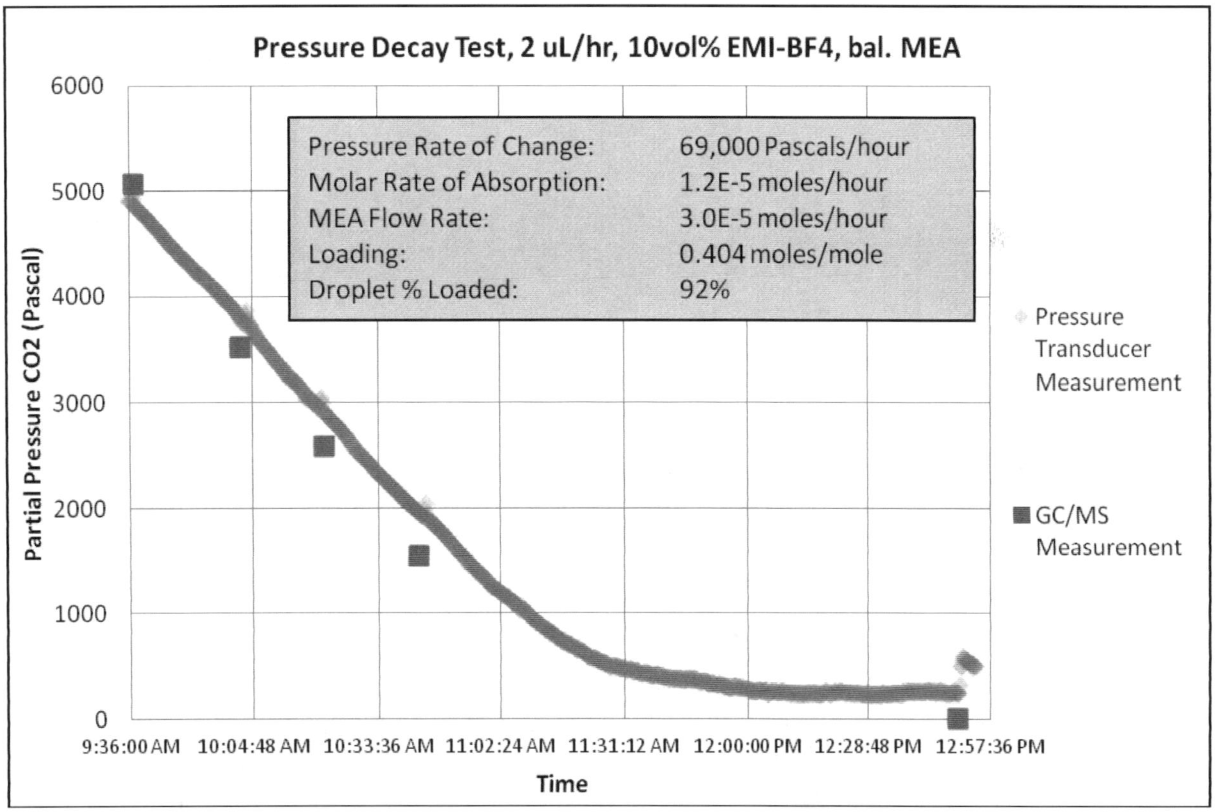

Figure 4: Pressure decay absorption measurements

A suite of experiments were conducted varying the liquid conductivity by adding an ionic liquid salt and by changing the liquid flow rate. The effect of modifying these parameters would be to modify nanodroplet radius according to $R \cong \left[\frac{Q * \varepsilon_0}{K}\right]^{\frac{1}{3}}$, and hence modifying the extent of reaction within the time of flight of the nanodroplet. When accounting for the small amount of absorption that also occurs within the electrospray cone (and modifies the electrical conductivity), the predicted absorption and measured extent of reaction line up well with each other across two orders of magnitude of flow rate and an order of magnitude of electrical conductivity (Figure 5).

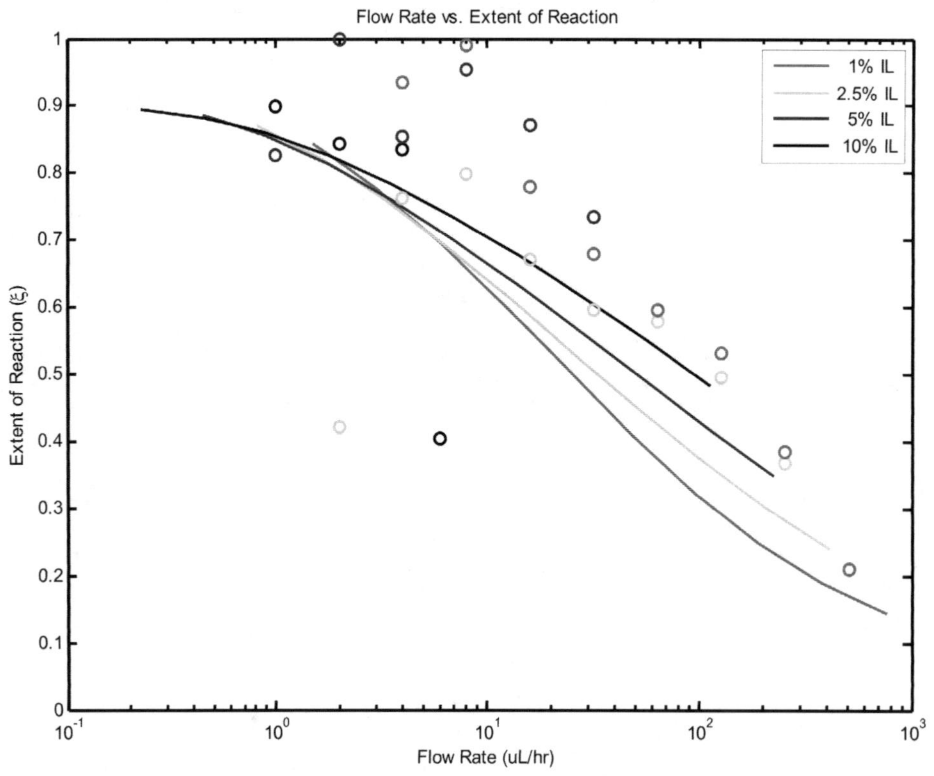

Figure 5: Absorption by nanodroplets as a function of flow rate and salt content

REFERENCES

1. J. Fernandez de la Mora, "Fluid Dynamics of Electrosprays," *Ann. Rev. Fluid Mech.*, 39, 217-243 (2007)
2. G. Lenguito, J. Fernandez de la Mora and A. Gomez, "Multiplexed Electrospray for Space Propulsion Applications," 46th AIAA Jt. Prop. Conf., AIAA-201-6521 (2010)
3. SBIR/STTR. "Variable Thrust/Specific Impulse Electrospray Propulsion." Award 90198, DOD F08B-T09-0141. http://www.sbir.gov/sbirsearch/detail/67451
4. G. Lenguito, "Multiplexed Electrosprays for Space Propulsion Applications." Ph.D. Thesis, Yale University, 2013.
5. A. Jaworek, Micro- and nanoparticle production by electrospraying, Powder Technology, Volume 176, Issue 1, 10 July 2007, Pages 18-35, ISSN 0032-5910, http://dx.doi.org/10.1016/j.powtec.2007.01.035.
6. A. Jaworek and A.T. Sobczyk, "Electrodraying route to nanotechnology: An overview." J. Electrostatics **66** (2008) 197-219
7. D. Li and Y. Xia, "Electrospinning of Nanofibers: Reinventing the Wheel?" Adv. Mat., Vol 16, 14, p. 1151-1170. DOI: 10.1002/adma.200400719
8. B. Almeria, Ph.D. Thesis, Yale University, 2012.
9. C.M. Waits et al, "Multiplexed electrospray scaling for liquid fuel injection," *J. Micromech. Microeng.* **20** (2010) 104010
10. W. Deng and A. Gomez, "Electrospray for Microelectronics Cooling." *Intl. J. of Heat and Mass Transfer*, 54 (2011) 2270-2275
11. Tepper et al. "An electrospray-based, ozone-free air purification technology." *J. App. Phys.* (2007) 113305
12. Kim et al. "Electrospray with electrostatic precipitator enhances fine particles collection efficiency," *J. Electrostatics* 68 (2010) 306-310
13. J. J. Bango et al, "Development of Electrospray Technology for the Removal of Lunar Dust from Habitable Space Atmospheres." 41st Intl. Conf. Env. Sys., 17-21 July 2011, Portland, Oregon.
14. P. Diaz Gómez Maqueo, "Electrospray emitters for diffusion vacuum pumps." Masters Thesis, Massachusetts Institute of Technology, 2011. http://hdl.handle.net/1721.1/67181

15. Wang K, Paine MD, Stark JPW (2010) . Novel electrohydrodynamic jet-printing of organic silver ink in unforced form. Optoelectronics and Advanced Materials, Rapid Communications vol. 4, (3) 365-368

16. M.D. Paine et al. “Controlled electrospray pulsation for deposition of femtoliter fluid droplets onto surfaces.” Aerosol Sci. 38 (2007) 315-324

17. T. Müller, A. Badu-Tawiah, and R.G. Cooks, “Accelerated Carbon-Carbon Bond-Forming Reactions in Preparative Electrospray.” Angew. Chem. Int. Ed 2012, 51, 11832-11835, DOI: 10.1002/anie.201206632

18. SBIR. “Continuous Ionization System for Electrostatic Collection of Bioaerosols in Building Protection Applications.” CBD12-107 (CBD). Accessed November 1, 2013. http://www.dodsbir.net/sitis/archives_display_topic.asp?Bookmark=42615

19. J. Fernández de la Mora, “The effect of charge emission from electrified liquid cones.” *J. Fluid Mech.* 243, 561-574, (1992)

20. M. Girod et al. “Basic Vapor Exposure for Tuning the Charge State Distribution of Proteins in Negative Electrospray Ionization: Elucidation of Mechanisms by Fluorescence Spectroscopy.” J. Am. Soc. Mass Spectrom. (2012) 23:1221-1231, DOI: 10.1007/s13361-012-0375-1

21. D. A. Saville, “ELECTROHYDRODYNAMICS: The Taylor-Melcher Leaky Dielectric Model,” *Annu. Rev. Fluid Mech.* 1997. 29:27–64

388246 Developing a Model for Carbon Dioxide Using the Modified Group Contribution-SAFT-VR with Dipole Theory

Abstract:

The statistical associating fluid theory (SAFT) [1] is a commonly used molecular-based equation of state that has been successfully applied to study a wide range of fluid systems. In recent work, the GC-SAFT-VR equation was developed, which combines the SAFT equation for potentials of variable range (VR) [2] with a group contribution (GC) approach [3] that allows for the description of hetero-segmented chain molecules. The GC-SAFT-VR approach has been shown to provide an excellent description of the phase behavior of pure associating and non-associating fluids and their mixtures, with minimal reliance on fitting the model parameters to experimental data [3-5].

Carbon dioxide (CO_2) is central to a wide range of environmental and energy applications resulting in the importance of understanding the phase behavior and thermophysical properties of carbon dioxide containing systems. Due to the symmetry of CO_2, it contains no dipole moment; however, a significant quadrupole moment is present. The presence of these multipoles can have an impact on the thermodynamics and phase equilibrium behavior and thus should be included in any molecular model of the molecule. In this work, a new model for CO_2 was tested in which the quadrupole in CO2 is represented by two point dipoles oriented away from each other at 180°and described using the GC-SAFT-VR approach, which has been extended to handle dipolar square-well fluids by implementing the dipolar term from the SAFT-VR+D equation [6] into the GC-SAFT-VR theory. Isothermal-isobaric (NPT) and Gibbs ensemble Monte Carlo (GEMC) simulations have been performed to test the ability of the theory to capture the thermodynamics of the proposed model. Predictions for the thermodynamic properties and phase equilibrium of dipolar square-well dimer fluids are considered and compared with simulation data at various dipole strengths, temperatures, and interaction strengths.

With the ability of the theory to accurately describe the proposed model verified, CO_2 parameters were obtained through fitting to experimental vapor pressure and saturated liquid density data. Results will be presented that show that the CO_2 model provides a good description of the phase behavior of pure CO_2 and its mixtures.

References:

[1] W.G. Chapman, K. E. Gubbins, G. Jackson, M. Radosz. Fluid Phase Equilibria, 52, 31-38 (1989).

[2] A. Gil-Villegas, A. Galindo, P.J. Whitehead, S.J. Mills, G. Jackson, and A.N. Burgess, Journal Of Chemical Physics, 106, 4168-4186. (1997).

[3] Y. Peng, K. D. Goff, M. C. dos Ramos, C. McCabe, Fluid Phase Equilibria, 277, 131- 144 (2009).

[4] Y. Peng, K. D. Goff, M. C. dos Ramos and C. McCabe, Industrial & Engineering Chemistry Research, 49, 1378-1394 (2010).

[5] M. C. dos Ramos, J. D. Haley, J. R. Westwood, and C. McCabe, Fluid Phase Equilibria, 306, 97-111 (2011).

[6] H. Zhao, Y. Ding, C. McCabe, Journal of Chemical Physics, 127, (2007).

388273 Development of a Group Contribution Methodology for Multi-Functional Amines in the Context of Carbon Capture Applications

Abstract:

Physical property models are developed within the framework of the statistical associating fluid theory (SAFT) to describe and predict the vapor-liquid equilibria of pure and mixed solvents of alkanolamines that are of relevance for carbon dioxide (CO_2) capture. Predictive approaches are essential when experimental data is limited, with the benefit that they become valuable approaches in the molecular and process design. The purpose of this work is the development of transferable intermolecular (square well SW) potential models that can be used within a group-contribution (GC) SAFT framework to identify mixtures that are good candidates for CO_2 absorption. The key advantage of the SAFT-γ GC-approach [1,2] is not only that the compounds of interest can be described in significant degree of molecular detail, but also the parameterization based on the contributions to the free energy of each like and unlike functional-group pair interactions can be transferred to a different compound that contains these common functional groups. The development of a molecular based model with transferable parameters is a key achievement compared to previous versions of the theory. In the proposed approach, we investigate the applicability of the SAFT-γ equation of state to study carbon dioxide, water and alkanolamine mixtures, and discuss how SAFT-γ opens avenues for thermodynamic property prediction, decreasing the dependence on experimental data. Systems that have been investigated include, but are not limited to primary and secondary amines and alcohols, and alkanolamines. Mixtures with amines are known to be very reactive and following earlier work [3], chemical equilibrium is treated here via a physical approach, so that reaction products are treated implicitly and their concentration in the mixtures is represented with appropriate Wertheim association schemes. We present calculations and predictions of the fluid phase behavior of these compounds and a number of their aqueous mixtures with and without CO_2.

References

[1] A. Lymperiadis, C. S. Adjiman, A. Galindo, and G. Jackson, *J. Chem. Phys.*, **2007**, 127, 234903.

[2] A. Lymperiadis, C. S. Adjiman, G. Jackson, and A. Galindo, *Fluid Phase Equilib.*, **2008**, 274, 85-104.

[3] J. Rodriguez, N. Mac Dowell, F. Llovell, C. S. Adjiman, G. Jackson, and A. Galindo, *Mol. Phys.*, **2012**, 110, 1325-1348.

Associated Lattice Fluid Equation of State for CO_2 – Polymer Systems

Mohammad Hossain, Yanhui Yuan, and Amyn S. Teja

School of Chemical and Biomolecular Engineering
Georgia Institute of Technology, Atlanta, GA
E-mail: mhossain32@gatech.edu, yanhui.yuan.algenol@gmail.com, and amyn.teja@chbe.gatech.edu

The phase behavior of CO_2 + polymer systems is of interest in polymer synthesis, flue and natural gas processing, polymer foam and nanoparticle processing, and drug delivery. Theoretical and experimental evidence suggests that CO_2 is able to interact with electron donating functional groups in polymers to form weak Lewis acid – base (EDA) complexes. These interactions significantly affect phase behavior in CO_2 + polymer systems. However, few thermodynamic models explicitly account for EDA complexes. This work presents a new Equation of State (EOS) for associating systems that incorporates complex formation in the Guggenheim-Huggins-Miller lattice fluid partition function. The new EOS is simple to use and most importantly one of the parameters of the EOS that describes specific interactions between the polymer and CO_2 is obtained from *in situ* Attenuated Total Reflection Fourier Transform Infrared (ATR – FTIR) spectroscopy measurement. Molecular level information is collected from the vibrational spectra of the polymer in contact with high pressure CO_2. These spectra are used to estimate specific interactions which are incorporated in the model. The significance of the new EOS is demonstrated by calculating the solubility and phase equilibria of CO_2 in associating polymers. Only one adjustable parameter is required to calculate both solubility and cloud point behavior in polymer + CO_2 systems.

368144 Kinetic Study of Carbon Dioxide Absorption By Aqueous Solutions of N-Methyldiethanolamine and Piperazine

Abstract:

Kinetic study of carbon dioxide absorption by aqueous solutions of N-methyldiethanolamine and piperazine

Alberto SERVIA[a], Nicolas LALOUE[a], Julien GRANDJEAN[a], Sabine RODE[b], Christine ROIZARD[b]

[a] *IFP Energies nouvelles, Rond-point de l'échangeur de Solaize BP3, 69360, Solaize, France*

[b] *LRGP-CNRS Université de Lorraine, 1 rue Grandville* - BP 20451 - 54001 Nancy, France

1. Introduction

The kinetics of CO_2 absorption with aqueous blends of N-MethylDiEthanolAmine (MDEA) and PiperaZine (PZ) has been widely studied (Zhang et al., 2001, Bishnoi and Rochelle, 2002, Edali et al., 2010, Samanta and Bandyopadhyay, 2011). Nevertheless, some discrepancies remain concerning the reaction mechanisms proposed in the literature. For instance, the synergy between both amines is still not completely understood and proposed kinetics models cannot fully explain the experimental data at very high loadings. Moreover, the models used in the literature to deduce the kinetics of CO_2 absorption by aqueous blends of PZ and MDEA from experiments present some limitations. They all assume a constant CO_2 partial pressure in the gas phase. This hypothesis can lead to noticeable deviations if high CO_2 quantities are absorbed.

The objective of this work is to investigate the kinetics of CO_2 absorption by aqueous blends of MDEA and PZ. Experiments are performed in a wetted wall column (WWC) under various operating conditions and simulated using a rigorous reactor model.

2. Experimental setup

A WWC was used to obtain experimental data on the kinetics of CO_2 absorption by aqueous blends of MDEA and PZ. Within the reactor, the gas phase flows counter-currently with the liquid that overflows from the inside of a cylinder to form a thin liquid film. The gas phase, composed of CO_2 and Nitrogen (N_2), is water-saturated before being in contact with the liquid in the reactor to prevent from water mass transfer in the reaction zone. The experimental flux is calculated using the variation of the CO_2 gas concentration between the inlet and the outlet of the reactor, measured by an in-line infra-red spectrometer.

The WWC validation and hydrodynamic characterization were presented in a previous work (Servia et al., 2013).

3. Model description

A 2D stationary model has been developed using COMSOL software to predict the absorption flux of CO_2 in aqueous blends of PZ and MDEA in the WWC. This model couples hydrodynamics, gas-liquid equilibrium, mass transfer and chemical reactions (Servia et al., 2013).

The liquid phase velocity profile is determined by the Navier-Stokes equation for an incompressible fluid associated with specific boundary conditions. Since CO_2 was transferred from the gas into the liquid phase, the gas velocity varied within the reactor. This evolution was modeled through a mass balance on the inert compounds (N_2 and water).

The concentration of each species at equilibrium conditions were provided by a thermodynamic model accounting for the deviations from the ideal behavior through an electrolyte NRTL approach provided by ASPEN. In this model, the CO_2 Henry constant is determined by the ratio between P_{CO2} and the molecular CO_2 concentration at equilibrium provided by the thermodynamic model, which assures the consistency between the mass transfer and the thermodynamic models.

Amines, water and bicarbonate dissociation were supposed to be instantaneously equilibrated, while CO_2 chemical reactions with amines and OH^- were assumed to be kinetically controlled and reversible.

Equation 1

$$CO_2 + OH^- \xleftrightarrow{k_1, k_{-1}} HCO_3^-$$

Equation 2

$$CO_2 + PZ + H_2O \xleftrightarrow{k_2, k_{-2}} PZCOO^- + H_3O^+$$

Equation 3

$$CO_2 + PZ + MDEA \xleftrightarrow{k_3, k_{-3}} PZCOO^- + MDEAH^+$$

Equation 4

$$CO_2 + PZCOO^- + H_2O \xleftrightarrow{k_4, k_{-4}} PZ(COO)_2^{2-} + H_3O^+$$

Equation 5

$$CO_2 + MDEA + H_2O \xleftrightarrow{k_5, k_{-5}} MDEAH^+ + HCO_3^-$$

As CO_2 is absorbed, P_{CO2} presents a decreasing profile within the reactor. Consequently, the proposed model takes into account the evolution of the CO_2 partial pressure through a plug-flow model combined with the film theory.

4. Results

This model has been validated by comparing its predictions with experimental data obtained on unloaded aqueous solutions of PZ ranging from 0,2 to 1 M and at temperatures between 24 and 60 °C (Servia et al., 2013). The reaction mechanism proposed by Bishnoi and Rochelle, 2002 and their kinetic constants associated to equations 2 and 4 were used to model the kinetics of the CO_2/PZ system.

4.1. CO_2 absorption on unloaded solutions

A first set of experiments were performed on unloaded solutions presenting MDEA concentrations of 2 and 3 M, PZ concentrations ranging from 0.2 to 1 M and temperatures from 297 to 328 K. The experimental results and model predictions were presented in Figure 1. The increase of MDEA concentrations has a negative effect on the overall CO_2 mass transfer, especially at low temperature. Indeed, the solution viscosity rises involving a decrease of the CO_2 diffusion coefficient within the liquid phase. The experimental results were used to regress the kinetic constant of the synergy chemical reaction (Equation 3), which plays a significant role at the tested conditions. Indeed, the model systematically underestimates CO_2 flux with an AAD of 10 %, if this chemical reaction is not considered in the reaction mechanism.

a)

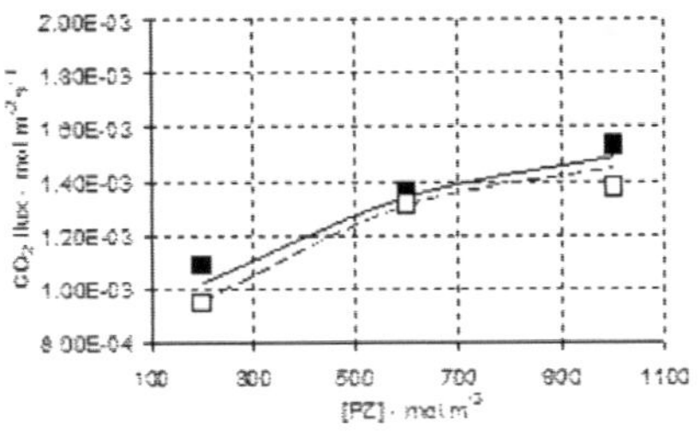

b)

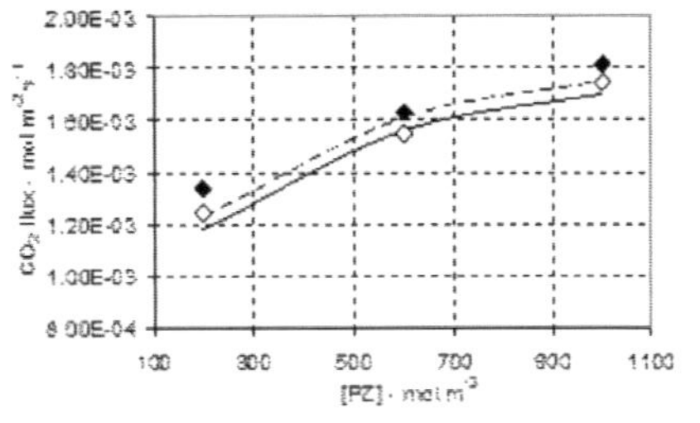

c)

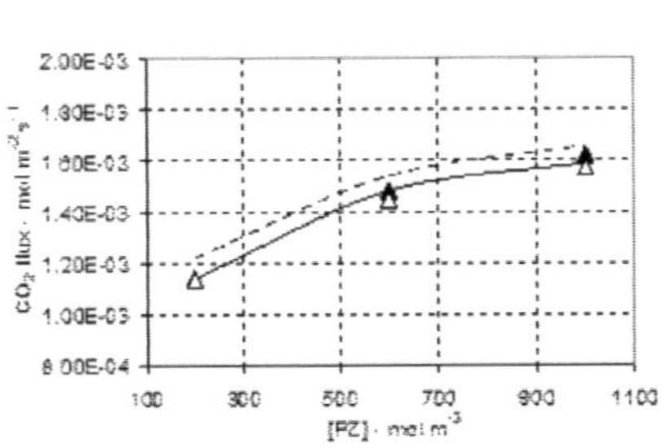

d)

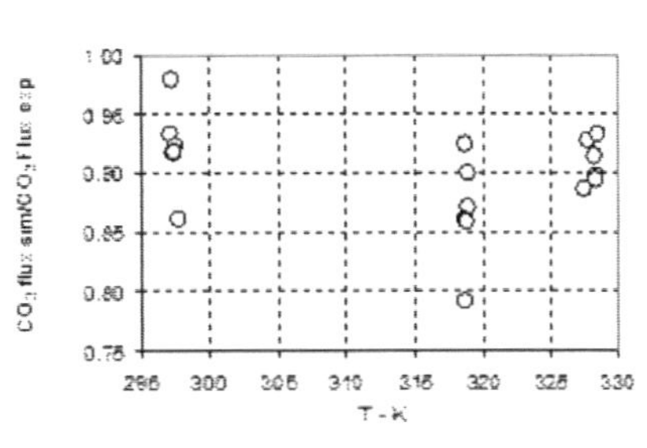

Figure 1 – CO_2 absorption flux evolution as a function of PZ concentration for different conditions of MDEA concentrations and temperatures. Filled symbols: [MDEA] = 2 M; empty symbols: [MDEA] = 3 M; continuous curves: [MDEA] = 2 M; discontinuous curves: [MDEA] = 3M. Curves represent model predictions whereas symbols represent experimental data. a – T = 297 K; b – T = 319 K; c – T = 328 K; d – Comparison between model predictions without considering the synergy chemical reaction between MDEA and PZ and experimental data as a function of temperature.

4.2. CO_2 absorption on loaded solutions

A second set of experiments were carried out on loaded solutions presenting MDEA concentrations of 2 and 3 M with a fixed PZ concentration of 1 M, temperatures ranging from 297 to 328 K and CO_2 loadings up to 0.25 mol_{CO2}/mol_{amine}, at two different P_{CO2}. The experimental results and model predictions were presented in Figure 2 .The model systematically overestimates CO_2 flux at low P_{CO2}, even if the deviation between experimental and model data is lower than the determined experimental error. A methodology based on the experimental data acquisition at operating conditions allowing the experimental error and the model limitations impact to be reduced was then proposed and validated. This methodology is based on the experimental data acquisition at higher P_{CO2}. Model predictions are in good agreement with the experimental data at high solutions loadings indicating that the reaction between MDEA and PZCOO⁻ can be neglected at these conditions. Moreover, the model accurately reproduces the overall CO_2 flux, even in presence of mass transfer limitations.

a)

b)

THE GRAPHIC ON THIS PAGE WAS NOT AVAILABLE ON THE E-MEDIA VERSION OF THIS PAPER.

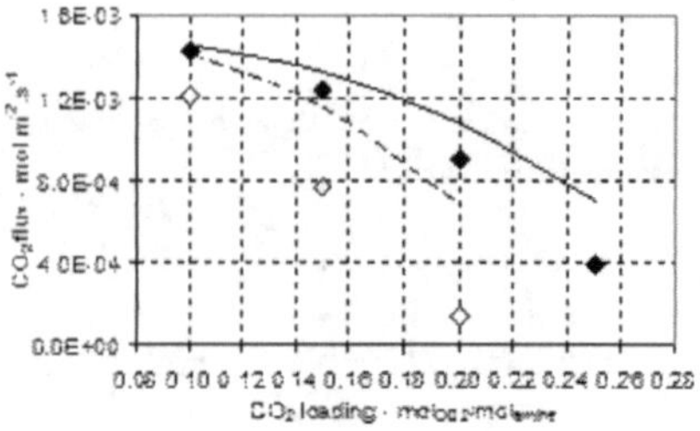

c)

d)

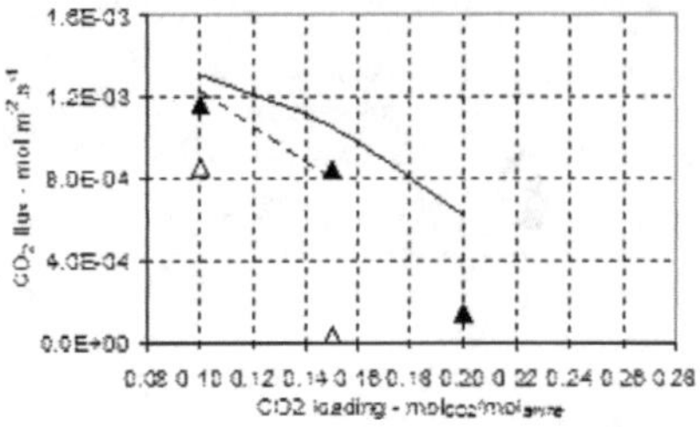

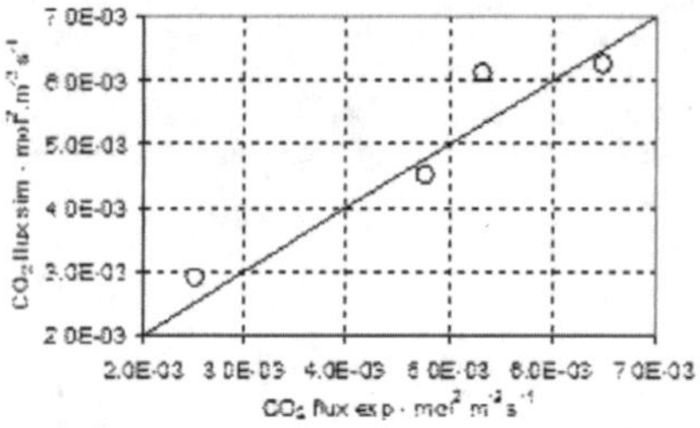

Figure 2 – CO_2 absorption flux evolution as a function of solution loading for different conditions of MDEA concentrations and temperatures. Filled symbols: [MDEA] = 2 M; empty symbols: [MDEA] = 3 M; continuous curves: [MDEA] = 2 M; discontinuous curves: [MDEA] = 3M. Curves represent model predictions whereas symbols represent experimental data. a – T = 297 K, y_{CO2}^{inlet} = 7000 ppm; b – T = 319 K, y_{CO2}^{inlet} = 7000 ppm; c – T = 328 K, y_{CO2}^{inlet} = 7000 ppm; d – Parity chart at y_{CO2}^{inlet} = 50000 ppm.

4.3. Sensitivity analysis

A sensitivity analysis (see Figure 3) was carried out involving all kinetics and diffusion parameters. At low loadings, the kinetic constant associated to the interaction between PZ, CO_2 and water as a significant impact. An accurate prediction of the reaction between $PZCOO^-$, CO_2 and water, as well as the liquid phase species diffusion are required at higher CO_2 loadings. The enhancement factor is sensitive to the synergy reaction (Equation 3) at low loadings, high temperatures, and high MDEA concentrations.

a)

b)

THE GRAPHIC ON THIS PAGE WAS NOT AVAILABLE ON THE E-MEDIA VERSION OF THIS PAPER.

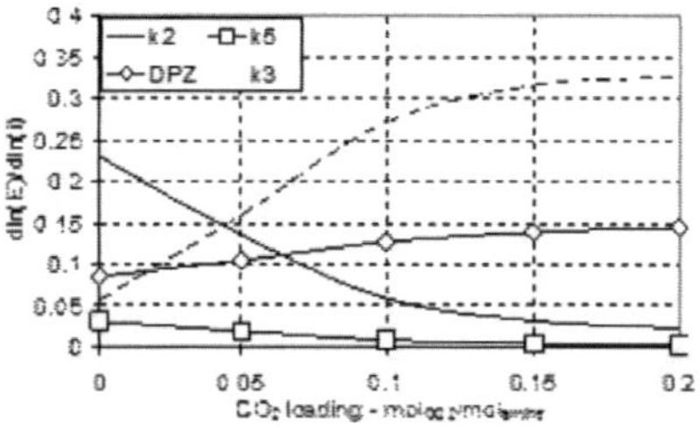

c)

d)

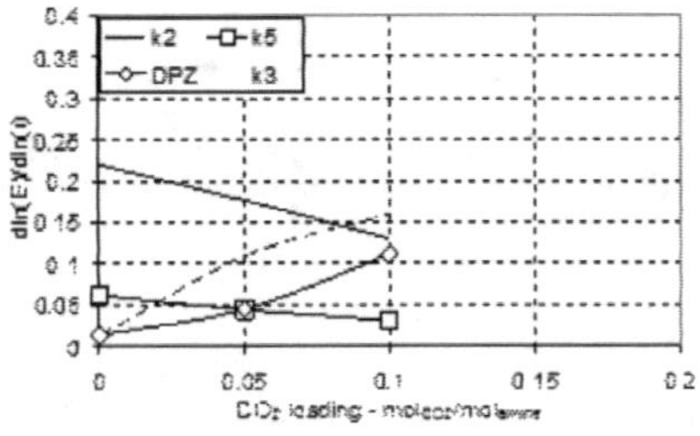

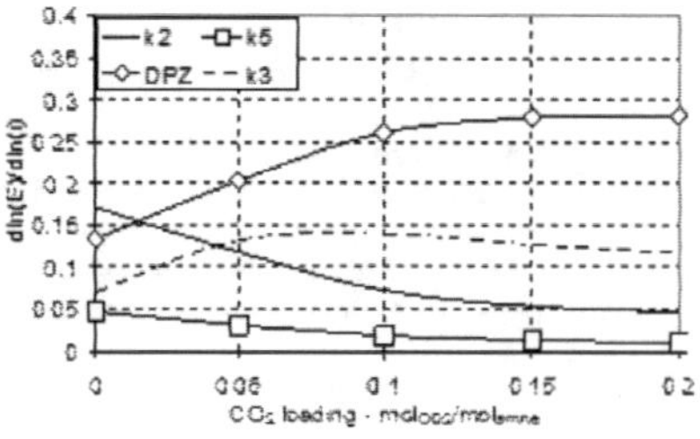

Figure 3 – Sensitivity analysis of the kinetics parameters and liquid phase species diffusion coefficient. a – T = 298 K, y_{CO2}^{inlet} = 7000 ppm; b – T = 298 K, y_{CO2}^{inlet} = 50000 ppm; c – T = 330 K, y_{CO2}^{inlet} = 7000 ppm; d – T = 330 K, y_{CO2}^{inlet} = 50000.

5. Conclusion

This paper presents an experimental and numerical investigation on the CO_2 absorption by aqueous blends of MDEA and PZ. A synergy exists between both amines to enhance CO_2 capture, whereas the chemical reaction between the $PZCOO^-$ and MDEA to accelerate CO_2 capture can be neglected at high loadings.

A sensitivity analysis on the kinetic and diffusion parameters indicates that an accurate description of the PZ interaction with CO_2 is required at low loadings, whereas diffusion phenomenon and $PZCOO^-$ interactions with CO_2 become essential at higher loadings.

References

Bishnoi, S., Rochelle, G.T., (2002). Absorption of carbon dioxide in aqueous piperazine/methyldiethanolamine. Aiche Journal, 48, 2788-2799.

Edali, M. et al. (2010). 1D and 2D absorption-rate/kinetic modeling and simulation of carbon dioxide absorption into mixed aqueous solutions of MDEA and PZ in a laminar jet apparatus. International Journal of Greenhouse Gas Control, 4, 143-151. Samanta, A., Bandyopadhyay, S. S., (2011). Absorption of carbon dioxide into piperazine activated aqueous N-methyldiethanolamine . Chemical Engineering Journal, 171, 3, 734-741. Servia, A., Laloue, N., Grandjean, J., Rode, S. and Roizard, C. (2013). Modeling of the CO_2 absorption in a wetted wall column by piperazine solutions. Oil & Gas Science and Technology. DOI: 10.2516/ogst/2013136 Zhang, X. et al. (2001). A kinetics study on the absorption of carbon dioxide into a mixed aqueous solution of methyldiethanolamine and piperazine. Industrial & Engineering Chemistry Research, 40, 3785-3791.

CO_2 absorption into solutions of amino-2-methyl-1-propanol (AMP) mixed with piperazine and glycine

Abdelbaki Benamor and Mohammed Jaber Al-Marri

Gas processing Center, Faculty of Engineering, Qatar University, P. O. Box 2713
Doha, Qatar
e-mail: benamor.abdelbaki@qu.edu.qa

A blended solution of sterically hindered amine 2-amino-2-methyl-1-propanol (AMP) mixed with piperazine and glycine was studied in this work to absorb Carbon dioxide from natural and flue gases. The obtained results show that the new solution represents a promising absorbent for CO_2 capture. Up to now, there is no reliable information on the reaction between CO2 and AMP-PZ-G in the literature. In this work, the reaction mechanism and kinetics were investigated using a stirred-cell reactor at a temperature range varying between 298 and 313 K. The reaction pathway was described using both zwitterion and termolecular mechanisms. The kinetic data obtained from the absorption process occurring in the fast reaction regime for amine concentration in the range of 0.5 - 2M were analyzed and the kinetic parameters associated to the reaction were obtained and presented here.

385767 Restraint-Free Coarse-Grain Modeling of a Crystalline Cellulose Fibril Based on Force Matching

Abstract:

As a major component of ligno-cellulosic biomass, crystalline cellulose fibrils are resistant to hydrolysis (biomass recalcitrance) and this is a bottle neck for any process that hopes to decompose cellulosic feedstocks into glucose that can later be converted to bio-fuels. To design efficient processes to degrade the biomass, it is important to understand the molecular forces that bind the cellulosic units. Coarse-grained simulations are useful to explore the residue scale motions in large molecules for long periods of time. The multi-scale coarse-graining (MS-CG) method based on force matching is used to derive effective coarse-grained forces from all-atom molecular dynamics trajectories to model a crystalline cellulose fibril. Unlike for many coarse-graining methods, the force matching method systematically optimizes coarse-grained forces from all-atom simulation trajectories without fitting them to a predetermined analytical function. Two restraint-free coarse-grained mapping schemes with two and eight unique coarse-grain bead types are evaluated and tested against the same all-atom reference system. For the performance analysis of coarse-grained models such properties as crystalline lattice parameters, radial distribution functions, and root mean square fluctuations and displacements are compared. The choice of coarse-grained mapping scheme significantly influences the performance of the coarse-grained molecular dynamics simulation in predicting the properties computed for the reference all-atom crystalline cellulose. A rationale strategy for designing two-type and eight-type coarse-grained mapping schemes in a crystalline cellulose fibril is developed and presented. Using this approach, both CG mapping schemes predicted the sheet structure of a cellulose fibril well, while only eight-type CG model is capable to accurately represent the intersheet ordering and therefore a complete crystalline structure for a crystalline cellulose fibril.

TITLE:

Ligand Binding Dynamics and Directionality in *Cellulomonas fimi* Family 4 Carbohydrate Binding Modules

AUTHORS: Abhishek A. Kognole[1] and Christina M. Payne[1,2*]

1. Department of Chemical and Materials Engineering, University of Kentucky, Lexington, KY 40506, USA
2. Center for Computational Sciences, University of Kentucky, Lexington, KY 40506, USA

Abstract

Enzymatic hydrolysis of biomass is often accomplished with multi-modular cellulase enzymes consisting of a catalytic domain appended to one or more carbohydrate binding modules (CBMs) by peptide linkers. CBMs play significant roles in modulating the function of cellulases, and understanding the protein-carbohydrate recognition mechanisms by which the CBMs selectively bind their substrate is critical to further development of enhanced biomass conversion technology. CBMs frequently exhibit a limited range of specificity and appear to bind polysaccharide substrates in a directional fashion dictated by the position of the ring oxygen relative to the protein fold. The family 4 CBMs of *Cellulomonas fimi* CenC (*Cf*CBM4) are reported to preferentially bind both amorphous cellulose and soluble cellooligosaccharides. However, experimental evidence suggests these CBMs may not exhibit a thermodynamic preference for a particular orientation, having significant implications for chimeric cellulase applications. The apparent lack of a thermodynamic preference is counterintuitive and has been the source of some debate in the literature. Here, we use molecular modeling and free energy calculations to investigate protein-carbohydrate recognition mechanisms in *Cf*CBM4-1 and *Cf*CBM4-2 and to elucidate the preferential ligand binding orientation. We evaluate four different cellopentaose orientations including that of the crystal structure and three others suggested by NMR. These four orientations differ from each other based on position of reducing end of ligand and pyranose ring orientations relative to the core of the protein. Using molecular dynamics, we find that the plausible orientations reduce to two cases. Ligand binding free energies calculated through free energy perturbation with Hamiltonian replica-exchange molecular dynamics indicates the two orientations are equally favorable. The calculated free energies are in excellent agreement with isothermal titration calorimetry measurements of binding free energy. Comparison of dynamics from MD simulations further suggests the approximate structural symmetry of the oligosaccharides relative to the amino acids along the binding cleft plays a role in the promiscuity of ligand binding. Evaluation of additional ligand-bound structures insinuates this phenomenon may be generic to proteins of the β-sandwich fold.

387168 Understanding the Interactions of N-Methylmorpholine N-Oxide with Cellulose

Abstract:

The abundance and energy content of native cellulose makes it an extremely important raw material and source for sustainable biofuels. A major challenge in processing natural cellulose lies in its strong resistance to dissolution. Only a limited number of conventional solvents have been found to effectively dissolve cellulose and the mechanisms of action still remains unclear. We study the dissolution of cellulose in N-methylmorpholine N-oxide (NMMO) hydrate, the key solvent of one of the most successful commercial processes for dissolving cellulose, via molecular simulation. The solubility of cellulose is highly sensitive to the concentration of water and we examine the behavior of these solutions over the full concentration range. We calculate the thermodynamic driving forces leading to dissolution, the disruption of the hydrogen bonding network, the role that water plays in this process and the underlying molecular interactions that make these solvents effective in dissolving cellulose. Finally, we discuss what can be learned from understanding of NMMO that can be applied to the search for more better, cheaper and more environmentally-friendly alternatives.

Transport of Pyrolysis Fragments in Cellulose – A Molecular Dynamics Study

Abdul Salam Mohammad[1], Jessica Murillo[1], Joseph J. Biernacki[1], S. Northrup[2]

Department of Chemical Engineering, Tennessee Tech University, Cookeville, TN 38505
Department of Chemistry, Tennessee Tech University, Cookeville, TN 38505

Pyrolysis of lingocellulosic biomass could become a significant source of renewable chemicals and fuels in the near future. In addition to bio-oil production, fast pyrolysis of biomass produces significant quantities of permanent gases, including H_2, CO, CO_2, and CH_4, also known as synthesis gases. Heating of wood, grass and other cellulose rich plants matter in the absence of air also produces chemical fragments that resemble the lingocellulosic starting materials. The processes that control these events are complex and not yet well understood. Among the many questions is the rate at which the bio-pyrolysis fragments are transported through the lingocellulosic frame work. The purpose of this study was to use molecular modeling to simulate the transport of pyrolysis product fragments through a model cellulose structure. Materials Studio was used to create and optimize a model for crystalline cellulose. To simulate the diffusion of product gases, a molecule of CO_2 was placed within the cellulose matrix as a simple test bed. Using the FORCITE module and the molecular mechanics force field, pcff, molecular dynamics was performed on the CO_2-cellulose system at 598 K with an NPT ensemble. The mean-square displacement at each time step was measured and the diffusivity of CO_2 calculated. At 598 K, the diffusivity of CO_2 through cellulose was calculated to be 0.4403×10^{-12} cm^2 s^{-1}. It appears that the effect of temperature on the diffusivity of CO_2 is small. At 298K, the diffusivity of CO_2 through crystalline cellulose was found to be 0.1720×10^{-12} cm^2 s^{-1}.

383203 A Comparison of Coarse-Graining Approaches for Modeling Morphology in Push-Pull Polymers

Abstract:

Predicting the morphology of semiconducting polymers used in organic photovoltaics is difficult because of their long relaxation times. Coarse-grained models permit access to the time-scales required to observe equilibration, but can introduce structural artifacts. Here we present and compare the computational costs and structural differences of three models of benzodithiophene and cyclopenta[c] thiophene-4,6,-dione based copolymers, including the degree to which they reproduce pi-stacked and lamellar structural features observed in experiments. We find that the use of rigid bodies to represent the aromatic groups of each copolymer subunit can decrease computational cost by a factor of two without introducing packing artifacts that arise from the representation of planar groups with spherical simulation beads.

384863 Molecular Properties for Renewable Energy Materials from Big Data

Abstract:

Computational quantum chemistry is a useful tool to assess the utility of novel compounds for use in renewable energy materials or processes. While these characterizations are often valuable for the specific systems under investigation, their broader impact tends to be more limited.

Our work focuses on merging computational chemistry and molecular modeling with Big Data ideas. The virtual high-throughput screening of compound libraries allows us to generate huge quantum chemical data sets, which can be employed for data-driven discovery. To extract an understanding of the underlying structure-property relationships from these data sets we develop chemical data mining tools. These have their roots in machine learning, statistical learning, and information science. We will discuss the utility of these tools and the resulting models for the prediction of molecular properties without the need for expensive quantum chemical calculations.

AUTHOR INDEX

AUTHOR INDEX

AUTHOR INDEX

AUTHOR INDEX